INVESTIGATION OF
Aeronautical and Engineering Component Failures

INVESTIGATION OF
Aeronautical and Engineering Component Failures

A. Venugopal Reddy

CRC PRESS

Boca Raton   London   New York   Washington, D.C.

Library of Congress Cataloging-in-Publication Data

Reddy, A. Venugopal.
    Investigation of aeronautical and engineering component failures / A. Venugopal Reddy.
       p. cm.
    ISBN 0-8493-2314-2 (alk. paper)
       1. Airplanes--Parts--Defects. 2. Aerospace engineering--Materials--Defects. 3. System failures (Engineering) 4. Metals--Fracture. 5. Materials--Defects. I. Title

TL672.R44 2004
629.134'3--dc22

                                                             2004045163

This book contains information obtained from authentic and highly regarded sources. Reprinted material is quoted with permission, and sources are indicated. A wide variety of references are listed. Reasonable efforts have been made to publish reliable data and information, but the author and the publisher cannot assume responsibility for the validity of all materials or for the consequences of their use.

Neither this book nor any part may be reproduced or transmitted in any form or by any means, electronic or mechanical, including photocopying, microfilming, and recording, or by any information storage or retrieval system, without prior permission in writing from the publisher.

The consent of CRC Press LLC does not extend to copying for general distribution, for promotion, for creating new works, or for resale. Specific permission must be obtained in writing from CRC Press LLC for such copying.

Direct all inquiries to CRC Press LLC, 2000 N.W. Corporate Blvd., Boca Raton, Florida 33431.

Trademark Notice: Product or corporate names may be trademarks or registered trademarks, and are used only for identification and explanation, without intent to infringe.

Visit the CRC Press Web site at www.crcpress.com

© 2004 by CRC Press LLC

No claim to original U.S. Government works
International Standard Book Number 0-8493-2314-2
Library of Congress Card Number 2004045163
Printed in the United States of America 1 2 3 4 5 6 7 8 9 0
Printed on acid-free paper

Dedication

To my wife Aruna, my daughters Dr. Haritha and Dr. Neelima, and my grandson Rohit

Foreword

Analysis of failures of components and systems in service can be a daunting task. The Columbia shuttle disaster is currently under investigation, and we do not need another example to educate us on the complexity of such exercises. A major challenge is to reconstruct the events preceding the failure from whatever can be gathered from the widely scattered debris. Once this is reasonably accomplished, further analysis requires a whole range of skills and multidisciplinary knowledge associated with the system *per se* as well as the components that go into the making of it. Sometimes, the investigation can lead to a new finding that may spur a significant advance in the field. The embrittling role of undesirable traces of the V-group elements in iron was unraveled (in 1969) when the rotor of the turbine disintegrated in the Hinkley Point Power Station turbine generator. This important finding came from a remarkable application of Auger electron spectroscopy (AES), then quite a novel experimental tool. AES, in turn, came to be regarded as a valuable technique to study segregation phenomena.

Investigating a component failure is therefore an absorbing exercise; however, it demands considerable breadth of knowledge of the science of materials as well as their engineering. It is not merely knowledge of the principles that underlie these disciplines that one has to bring to bear to solve a problem, but also a sense of the history of past failures.

It should be clear from the foregoing that a failure analyst must possess an excellent academic background and substantial relevant experience. Such is the case with the author, Dr. A. Venugopal Reddy, whom I had valued greatly as a superb professional during my years at the Defence Metallurgical Research Laboratory (DMRL) in Hyderabad. The compilation painstakingly put together by him is based on his intimate knowledge of the subject and nearly three decades of valuable investigative experience.

The reader is provided herein a set of chapters introducing the attributes expected of an expert and credible analyst, the experimental tools and techniques available, and a database of characteristic failure features related to the material properties and the operating conditions of stress, temperature and the environment. Gathered here are also the commonly encountered manufacturing and maintenance defects. Given this preparation, the reader is taken through a number of failure cases, primarily of aerospace components, arranged in a way that allows illustration of the introductory material.

The author's treatment of the complex subject is systematic and scientific. The progression from the basics of failure analysis to the tools and techniques available to the failure analyst and their effective utilization in identifying the root cause of failure is smooth and supported with examples of exceptional quality. The author's attempt to illustrate fault tree analysis in investigating major system failures using the complex example of jet engine flame-out is commendable. The fracture feature and causative factor analysis concept presented in this book is useful to enumerate the various possibilities associated with an observed fracture feature and provides a systematic method of identifying the correct cause by elimination of other possibilities based on supportive experimental evidence. The frequently encountered design deficiencies and defects that trigger premature service failures are well described with appropriate examples.

The case studies presented in this book, though focused on aeronautical components, cover a wide spectrum of engineering components and the commonly observed failure processes in metallic materials. The case studies reflect the complex correlation that exists between composition, processing, microstructure, mechanical properties and fracture behavior and how the interplay of these variables influences the performance of metallic components.

This handy volume of service failures is a readable document that can serve as a useful tutor to acolytes in the field of service failure analysis and as a helpful reference for the experienced materials engineer.

Prof. P. Rama Rao
Former Secretary, Department of Science and Technology
Government of India

Preface

During the early days of its development as a discipline, failure analysis suffered from two major setbacks arising out of false conceptions: (i) it was looked at by the industry as an academic exercise, and (ii) it was considered by most of the metallurgists and material scientists that practising failure analysis was less rewarding. However, there has been a gradual change in the attitude of the industry and the material scientists toward failure analysis, and today, failure analysis is regarded as one of the highly specialized fields in materials science. Over the years, failure analysis has grown enormously in its scope and utility. Developments in materials characterization techniques have made the job of a failure analyst a lot easier and more precise. With the concept of "Preventive Failure Analysis" gaining popularity, the involvement of a failure analyst in the design and development of a component is accepted as a beneficial practice.

As a failure analyst for more than three decades, I feel that failure analysis is extremely challenging, immensely interesting and satisfying, and as rewarding as any other field in materials science. I have had the opportunity to deal with more than 500 failure cases, covering components of military aircraft and defense hardware ranging from simple bolts to highly critical turbine aerofoil castings. Considering the complexity of the aeroengine and the possibility of failures due to a wide variety of failure processes, which often operate simultaneously, aeronautical component failures present a special challenge to failure analysts requiring special care and precise interpretation during investigation. Although a lot of literature is available on the topic of failure analysis in general, information available on the analysis of aeronautical component failures is very scarce. It is this fact which has motivated me to make an attempt to write this book based on my own experience for the benefit of young practicing failure analysts.

This book was written assuming certain level of prior knowledge in metallurgy and materials science among the readers. The book was divided into two major parts. In the first part, which was further divided into six chapters, various aspects of failure analysis were discussed in some detail. In Chapter 1, the consequences of premature failures, the importance and objectives of failure analysis and the qualities required in a failure analyst were discussed. In Chapter 2, various critical steps involved in failure analysis and the usefulness of various tools and experimental techniques available to the failure analyst were discussed. Chapter 3 was dedicated to the examination and interpretation of fracture surfaces for different types of failure processes. A method, Fracture Feature and Causative Factor Analysis, for identifying the most probable cause of failure based on the observed fracture features was also presented. Finally, the role played by the deficiencies in component design and material selection, manufacturing defects, and operational and maintenance deficiencies in causing premature service failures was discussed in Chapters 4, 5, and 6, respectively. Illustrations and examples were included throughout the text, wherever it felt necessary.

In the second part of this book, a number of case studies were presented covering the most frequently encountered failures in the aeronautical industry along with a few universal failures. The case studies were categorized under different headings based on the principal cause of failure. The major emphasis was placed on the methodology, analysis of experimental results and underlying logic in identifying the root cause of failure. Remedial measures in most of the case studies were deliberately not suggested keeping in view the danger of inexperienced failure analysts picking up the suggestions without comprehensive study on the pros and cons of implementation.

I hope that this book will help the young failure analysts in developing the right kind of attitude toward the subject and in developing a methodical and systematic approach in analyzing service failures. Although this book is not intended to be a textbook for materials science students, I hope that the case studies presented will serve as useful classroom examples for teachers and as an interesting study material for postgraduate students undergoing a course in failure analysis. Any suggestions and comments on the contents of the book are welcome.

A. Venugopal Reddy
Hyderabad, India
August, 2003

Acknowledgments

I wish to express my gratitude for the assistance of the late Dr. D. P. Lahiri, who guided me through my formative years as a practicing metallurgist. I had the privilege of working with the eminent metallurgists Dr. V. S. Arunachalam, former Scientific Adviser to the Defence Minister, and Prof. P. Rama Rao, former Secretary of the Department of Science and Technology, Government of India. I am grateful to them for encouraging me to practice failure analysis as a profession, which is generally considered not to be materialistically rewarding but technologically important and scientifically satisfying. I am thankful to Dr. D. Banerjee, Director, Defence Metallurgical Research Laboratory, Hyderabad, India, who encouraged me to persevere in the field of failure analysis, participated in the analysis of critical failures and accorded me permission to include relevant case histories in this book.

This book is an attempt to consolidate my experience of over three decades in conducting metallurgical analysis of failed components. I am indeed indebted to many of my colleagues at DMRL, Hyderabad, without whose support I would not have been a successful failure analyst. I am especially grateful to Dr. K. P. Balan, who worked with me on failure analysis for nearly two decades. Though it is difficult to name all those with whom I worked in failure analysis over such a long period, I acknowledge with gratitude the support extended by Vydehi Joshi, B. Venkateshwara Rao, N. B. Jagtap, G. Sundar Sharma, Dr. Trilok Singh, S. R. Sahay, S. V. Athavale and V. V. Rama Rao.

I express my profound thanks to Dr. A. C. Raghuram, former head of the Material Science Division, National Aerospace Laboratory, Bangalore, and a peer in the field of failure analysis, for encouraging me to pen my experiences as a failure analyst and also for reviewing the manuscript during the course of its development. I am grateful to C. Jagannathan, Regional Director, RCMA (GTRE), Bangalore, for help and guidance in formulating FTA and FMAAM illustrations on aeroengine flame-out.

I am thankful to Durga Bhavani, C. Santhi, Dr. V. Gopala Krishna and G. D. Janaki Ram, my colleagues at the Regional Centre for Military Airworthiness (Materials), Hyderabad, for their painstaking efforts during the preparation of the manuscript. I am thankful to J. K. Sharma, Chief Executive (Airworthiness), Centre for Military Airworthiness and Certification, Bangalore, for according permission to publish the book.

A. Venugopal Reddy

Contents

Art and Science of Failure Analysis

1. **Failure Analysis** 3
   1.1 Introduction 3
   1.2 Qualities of a Failure Analyst 4
   1.3 Ethics in Failure Analysis 5

2. **Tools and Techniques** 7
   2.1 Introduction 7
   2.2 Tools in Failure Analysis 7
   2.3 Techniques in Failure Analysis 8
       2.3.1 Visual Examination 8
       2.3.2 Nondestructive Testing 16
       2.3.3 Fractography 16
       2.3.4 Microanalysis 16
       2.3.5 Chemical Analysis 17
       2.3.6 Microstructural Examination 17
       2.3.7 Mechanical Testing 18
       2.3.8 Analysis and Interpretation
             of Experimental Data 18

3. **Fracture Feature Analysis** 19
   3.1 Introduction 19
   3.2 Intergranular Fracture 19
       3.2.1 Creep and Stress Rupture 21
       3.2.2 Liquid Embrittlement 23
       3.2.3 Heat Cracks 23
       3.2.4 Embrittlement 24
       3.2.5 Intergranular Corrosion 25
       3.2.6 Stress Corrosion Cracking and
             Hydrogen-Induced Failures 25
   3.3 Transgranular Fracture 28
       3.3.1 Ductile Fracture 28
       3.3.2 Brittle Fracture: Quasi-Cleavage 28
       3.3.3 Brittle Fracture: Cleavage 28
       3.3.4 Fatigue 31
       3.3.5 Transgranular Stress Corrosion Cracking 34
   3.4 Corrosion-Induced Failures 34
   3.5 Wear-Related Failures 40

4. Deficiencies in Design and Material Selection 47
    4.1   Introduction 47
    4.2   Design Concepts and Concerns 49
            4.2.1   Stress Raisers 50
            4.2.2   Residual Stresses 54
    4.3   Material Selection 60

5. Manufacturing Defects 71
    5.1   Introduction 71
    5.2   Melting and Teeming Defects 71
    5.3   Casting Defects... 74
    5.4   Metal Working Defects 77
            5.4.1   Inherited Defects.. 77
            5.4.2   Generated Defects.. 78
    5.5   Heat Treatment Defects... 84
            5.5.1   Internal Stresses and Quench Cracks....................... 94
            5.5.2   Tempered Martensite Embrittlement and
                     Temper Embrittlement 95
    5.6   Defects Generated in Finishing Operations 98

6. Operational and Maintenance Defects 105
    6.1   Introduction 105
    6.2   Maintenance Defects 110
    Summary.. 115

Case Studies

Part 1. Failures Due to Improper Material Selection and Heat Treatment 123
    1.1   Failure of Emitting Electrodes of an Electrostatic
           Precipitator.. 123
            Introduction 123
            Experimental Results 123
            Discussion 127
            Conclusion 128
            Remedial Actions.. 128
    1.2   Failure of Impellers of a High-Pressure Water Pump......... 128
            Introduction 128
            Experimental Results 128
            Discussion 134
            Conclusion 134
    1.3   Failure of a Track Shoe of an Infantry Combat Vehicle 134
            Introduction 135
            Experimental Results 135
            Discussion 139
            Conclusion 139

	1.4	Failure of a Draw Hook of a Railway Car	140
		Introduction	140
		Experimental Results	140
		Discussion	143
		Conclusion	144
	1.5	Failure of a 30-mm Armor-Piercing Shell	144
		Introduction	144
		Experimental Results	144
		Discussion	147
		Conclusion	147
		Recommendations	147
	1.6	Failure of the Balancing Gear Rod of a Gun Carriage	147
		Introduction	147
		Experimental Results	148
		Discussion	151
		Conclusion	151
	1.7	Failure of a Gigli Saw	152
		Introduction	152
		Experimental Results	152
		Discussion	157
		Conclusion	157
	1.8	Failure of the Camshaft of a Combat Vehicle Engine	157
		Introduction	157
		Experimental Results	158
		Discussion	163
		Conclusion	164

Part 2. Fatigue Failures 165

	2.1	Failure of First-Stage Compressor Blades of an Aeroengine	165
		Introduction	165
		Category 1	165
		Discussion	170
		Conclusion	171
		Category 2	171
		Discussion	178
		Conclusion	178
	2.2	Failure of Third-Stage Compressor Blades of an Aeroengine	178
		Introduction	178
		Experimental Results	178
		Discussion	184
		Conclusion	184
	2.3	Failure of the High-Pressure Turbine Blades of an Aeroengine	184
		Introduction	184

		Case 1	185
		Discussion	191
		Conclusion	192
		Case 2	192
		Discussion	196
		Conclusion	196
	2.4	Failure of the Fuel Pipes of a Jet Engine	196
		Introduction	196
		Experimental Results	197
		Discussion	202
		Conclusion	203
	2.5	Failure of the Impeller of a Turbo Starter	203
		Introduction	203
		Experimental Results	203
		Discussion	207
		Conclusion	208
	2.6	Failure of the Landing Gear Component of an Aircraft	208
		Introduction	208
		Experimental Results	208
		Discussion	213
		Conclusion	213
	2.7	Spalling of the Work Roll of a Cold Rolling Mill	214
		Introduction	214
		Experimental Results	214
		Discussion	219
		Conclusion	219
	2.8	Failure of the Main Gear Box Pinion of a Helicopter	220
		Introduction	220
		Experimental Results	220
		Discussion	225
		Conclusion	225
	2.9	Failure of the Bevel Gear of an Aeroengine Gear Box	225
		Introduction	225
		Experimental Results	225
		Discussion	231
		Conclusion	231
Part 3. Failures Due to Embrittlement			**233**
	3.1	Delayed Cracking of Maraging Steel Billets	233
		Introduction	233
		Experimental Results	233
		Discussion	238
		Conclusion	238
	3.2	Failure of Large-Caliber Gun Barrels	238
		Introduction	238
		Experimental Results	239

		Discussion	244
		Conclusion	244
		Remedial Measures	244
	3.3	Failure of the Nose Fairing of an Aeroengine	245
		Introduction	245
		Experimental Results	245
		Discussion	250
		Conclusion	250

Part 4. Failures Due to Overheating ... 253

	4.1	Failure of the HP Turbine Blade of an Aeroengine	253
		Introduction	253
		Experimental Results	253
		Discussion	258
		Conclusion	258
	4.2	Failure of a Mounting Bolt of the Second-Stage NGV of an Aeroengine	258
		Introduction	258
		Experimental Results	259
		Discussion	263
		Conclusion	263
	4.3	Failure of the Nozzle Guide Vane of an Aeroengine	263
		Introduction	263
		Experimental Results	263
		Discussion	269
		Conclusion	269
	4.4	Failure of an Aeroengine Center Support Bearing	270
		Introduction	270
		Experimental Results	270
		Discussion	273
		Conclusion	274
	4.5	Failure of a Drive Shaft	274
		Introduction	274
		Experimental Results	274
		Discussion	278
		Conclusion	279
	4.6	Failure of the Center Main Bearing of an Aeroengine	279
		Introduction	279
		Experimental Details	279
		Discussion	284
		Conclusion	284

Part 5. Failures Induced by Corrosion ... 263

	5.1	Failure of the Blow-off Vanes of an Aeroengine	285
		Introduction	285
		Experimental Results	285

	Discussion	289
	Conclusion	289
5.2	Failure of the Undercarriage Cylinder of an Aircraft	289
	Introduction	289
	Experimental Results	289
	Discussion	294
	Conclusion	294
5.3	Failure of the Flame Tube Retainer Bolts of an Aeroengine	295
	Introduction	295
	Experimental Results	295
	Discussion	302
	Conclusion	302
5.4	Failure of the Side Strut of a Helicopter	302
	Introduction	302
	Experimental Results	303
	Discussion	306
	Conclusion	306
5.5	Failure of Fourth-Stage Stator Casing Bolts	306
	Introduction	306
	Experimental Results	306
	Discussion	311
	Conclusion	311
5.6	Failure of the NGV Bolts of an Aircraft	311
	Introduction	311
	Experimental Results	312
	Discussion	315
	Conclusion	315
5.7	Exfoliation Corrosion in an Aircraft Structural Member	315
	Introduction	315
	Experimental Results	316
	Discussion	319
	Conclusion	319
	Remedial Measures	319
5.8	Failure of a High-Pressure Oxygen Cylinder	320
	Introduction	320
	Experimental Results	320
	Discussion	322
	Conclusion	323
5.9	Failure of the Boiler Tube of a Thermal Power Station	323
	Introduction	323
	Experimental Results	323
	Discussion	326
	Conclusion	327

Part 6. Failures Initiated by Wear 329
    6.1   Failure of a High-Pressure Turbine Rotor Blade 329
           Introduction 329
           Experimental Results 329
           Discussion 333
           Conclusion 333
    6.2   Failure of the Reheater Tube of a 220-MW
          Coal-Fired Boiler 333
           Introduction 333
           Experimental Results 333
           Discussion 338
           Conclusion 338

References 339

Index 345

Art and Science of Failure Analysis

1
Failure Analysis

1.1 Introduction

The abundant availability of raw materials in nature is the gift of God to humanity. The invention of technologies to convert raw materials into useful products is the contribution of science to modern civilization. Scientific principles and laws of nature broadly dictate the conversion of materials into components and, thus, the performance of the components in an engineered product. Disregard for these laws and principles leads to deficiencies in design, manufacture and maintenance. The deficiencies may be flaws or defects; these two faults are often incorrectly considered to be synonymous. While a flaw is a deviation from the perfect, a defect is a deviation from the acceptable. All the defective parts are flawed, but very few flawed parts are defective. Similarly, all defects need not result in failure, but all failures originate at defects. It is, therefore, necessary to identify the nature and source of the defects, which ultimately cause failures in service.

An air crash in the Himalayas; a train accident down south; an unscheduled shutdown of a power generation plant; immobilization of a tank in the battlefield; a gun incapable of firing ammunition. There is something common to these seemingly unconnected events: failure of an engineering component to function as its designer predicted. The term *failure* has negative connotations and signifies malfunctioning, leading to unsatisfactory performance of an engineering system. Even in modern times, in spite of all our technological advances, failures do occur, leading to loss of expensive equipment and invaluable human lives.

The consequences of premature failures are quite damaging to society and result in loss of lives and materials. There are myriad reports quantifying the losses caused by material degradation processes such as corrosion and wear. The losses due to corrosion alone are estimated to be at 5% of the gross domestic product (GDP) of any nation, and the loss is about $170 billion per year in the U.S. The loss due to wear in the U.S. is estimated to be approximately 1% of its GDP. If any such estimates of the direct and indirect losses due to unpredicted failures were made much before achieving the design lives of systems, the damages to the economy of any country would be mind-boggling.

For example, the seemingly trivial failure of a boiler tube leads to the shutdown of an entire power generation unit. The direct consequence of the event is the loss of power generation. The indirect effects of the unavailability of power are loss of production in the manufacturing sector, stoppage of railway traction, immense inconvenience due to nonfunctionality of all electrical gadgetry, etc. The total loss can be significant if one considers all the consequential damages of the power failure.

It is, therefore, imperative and customary to minimize losses by careful failure analysis and correct implementation of remedial measures. Though finding the cause of failure has been a fascination of humans from the beginning of history, the necessity of learning from mistakes continues to be the driving force for failure analysis. Avoidance of failures or their recurrence through failure analysis is a sure way to minimize economic losses. Thus, a failure analyst has a significant role to play in society.

1.2 Qualities of a Failure Analyst

Failure analysis is a daunting and demanding task. The analyst has to take a comprehensive approach, covering a large spectrum of disciplines, to arrive at the correct cause of failure. He or she should also be endowed with certain qualities to be successful: comprehensive knowledge of the field; adequate knowledge of related fields; knack for gathering information; ability to filter disinformation; an eye for minor details; absolute honesty and integrity; and lots of common sense.

The failure analyst must possess adequate professional qualifications and hands-on experience in the relevant field and should strive to keep abreast of latest developments. Inadequacy in either the qualifications or the experience of the failure analyst invariably leads to disastrous consequences. It is not uncommon to hear of *sulphate* inclusions in steel purely because a qualified chemist does the metallurgist's job!

The failure analyst should possess broad understanding of the operation of the system to which the failed component belongs. Though it is not possible for any failure analyst to know about every system, adequate knowledge can be assimilated by discussing the functional aspects of the system with the users. It is absolutely necessary that the failure analyst be familiar with the functional aspects of the system being analyzed, such as loads, operating conditions, etc., by thoroughly interacting with the shop floor engineers. Such an interaction with shop floor engineers not only helps in developing better understanding of the functional aspects of the system, but also helps the failure analyst to win their confidence. While dealing with the failure of complex systems, it is desirable to create a failure analysis team, drawing experts from design, production, operation, maintenance and materials science.

The failure analyst should be able to seek complete information about the design, engineering and operational aspects of the system in which the failed component was a subsystem. Ideally, it is best to collect information from different interest groups, though there is an inherent danger of the failure analyst becoming biased for or against the theories proposed. The information channels should be kept open until completion of the investigation. Experience suggests that after preliminary tests, more specific queries can be put forth and correct information can be obtained in support of the line of investigation.

Like a forensic expert, the failure analyst must critically examine all the information. The failure analyst's ability to filter out distorted information furnished by different agencies is of paramount importance. It is necessary to remember that the culprits intentionally leave clues at the site of the crime to mislead the sleuths. Though it is impossible to identify useful information from the total collected *a priori*, it is relatively easy to differentiate between a lie and the truth after the completion of a preliminary investigation. The expertise to look for seemingly insignificant scientific detail is the quality that differentiates successful failure analysts from the rest. When a research metallurgist examines the microstructure of a quenched and tempered low-alloy steel, he sees how nice the laths are, how fine the carbides are, and how good the overall microstructure is. In contrast, when a forensic metallurgist examines the same sample, he not only observes the overall microstructure but also looks for evidence of decarburization, grain size, inclusion morphology and population, the presence of cracks and the path they take in relation to the microstructural features, and so on. This in effect means that failure analysts look at all the good and bad features of a component.

The process of finding out the correct cause of failure through a systematic and scientific investigation centers around the scientific capability, honesty and integrity of the investigators. If the failure analyst lacks integrity, the scientific data generated on the failed component can intentionally be misinterpreted by putting forth wrong reasoning. Once a wrong cause is reported, it is difficult to correct the damage even if the correct diagnosis is made subsequently by more competent analysts, because of the usual tendency of interested parties to use the findings to suit their convenience. All successful people are gifted with extraordinary common sense, and a successful failure analyst is no exception. The need for practical wisdom will be felt by the failure analysts and the assisting team at every stage of the failure analysis, be it in filtering out disinformation or in disregarding misleading experimental evidence.

1.3 Ethics in Failure Analysis

The efficiency of the solutions derived from the failure analysis to avoid recurrence of failures depends on the ethics followed by the involved institutions and honesty of the individuals associated with the investigation. To ensure that the investigation is undertaken without bias or prejudice, it is necessary to

1. entrust the responsibility of failure analysis to a team whose members are endowed with sound technical judgment and irrefutable integrity,
2. ensure independence of the investigation team,
3. avoid the presence of pressure groups in the investigation team, and
4. empower the investigation team to implement the remedial measures and reassess the performance.

If the investigation team is not competent and honest, the findings tend to be ambiguous and do not address the critical issues, leading to no real benefit from the whole exercise. Intrainstitutional investigation teams are generally constrained and seldom exhibit the independence needed to record the facts. Failing to find relevant information and distorting the facts related to the failure is too common a practice to be ignored. It is, therefore, desirable to consider all of these factors while creating an investigation team.

2

Tools and Techniques

2.1 Introduction

Failure analysis is a complicated activity, demanding systematic and careful probing into every minor detail to identify the root cause of failure. Investigation into failure of major systems, such as an aircraft accident, requires multidisciplinary knowledge and thus can best be handled by a multidisciplinary expert team. The team should consist of a designer to reconstruct the failure scene by analysis of stress pattern, service loads, environmental influences, etc.; a manufacturer to analyze the probability of the incidence of manufacturing defects and lapses in quality control; a field service engineer to investigate operational and maintenance abuses; and a materials engineer to identify failure mode and material abnormalities. While investigating the failures of major systems, it is important first to correctly identify the subsystem that is responsible for the failure event based on careful cause and effect analysis. Further investigation focuses on that particular subsystem to identify the root cause of failure.

2.2 Tools in Failure Analysis

Most failure analysts intuitively develop a mental picture of the cause and effect scenario. Such methodologies, though adequate in the analysis of simple failures, are grossly inefficient while dealing with major system failures. Accordingly, it is always beneficial to establish all possible causes by a fault tree analysis (FTA) and evaluate the probability of each event of failure by developing a failure mode assessment and assignment matrix (FMAAM). The FMAAM identifies the major event and gives a description, probability assessment of the occurrence, and procedure to be adapted to evaluate the event. A number of similar tools suited to specific failure situations are available with different names and acronyms that have been derived from FTA and FMAAM principles. The fundamental purpose of these tools is to list all possible causes, assign probability to each of them based on investigation, and identify the most probable cause of failure.

The usefulness of FTA can be gauged by taking the example of flame-out of a jet engine, probably the most complicated engineering system. The flame-out can result either from a combination of rich fuel and lean air or lean fuel and rich air. The possible causes for jet engine flame-out are given in Figure 2.1 (for rich fuel and lean air combination) and Figure 2.2 (for lean fuel and rich air combination). Once all the causes are listed in the fault tree, the FMAAM has to be prepared. If such an analysis shows that a low engine rpm caused by the seizure of the center support bearing is the most probable cause for flame-out, then the cause for bearing seizure must be identified. The possible causes for center support bearing seizure are depicted in Figure 2.3. If further analysis using FMAAM shows that the bearing seizure was caused by blade failure leading to engine vibrations, then the causes for blade failure must be evaluated. The possible causes for turbine blade failure are shown in Figure 2.4. If subsequent analysis shows that the turbine blade failure was due to overheating caused by blocked cooling holes, then through backward integration the jet engine flame-out can be attributed to blade failure due to overheating with reliability beyond reasonable doubt.

In practice, turbine blade failures result from overheating, overload, fatigue and degradation caused by corrosion and erosion. In some instances, the primary damage may be due to spalling of coating, corrosion or wear, etc., leading to fatigue crack initiation. Extraneous metal deposition on turbine blades due to overheating and melting of upstream components is yet another common occurrence. A typical example is shown in Figure 2.5.

2.3 Techniques in Failure Analysis

Various experimental techniques are available to help failure analysts in identifying the cause of failure. It is possible to accurately define the failure mode, the origin of the failure and the material abnormalities and operational damages at the fracture origin by combining experimental techniques with the expertise of the failure analyst. In order to avoid disputes, it is often necessary to utilize modern experimental tools to generate further evidence in support of the findings obtained from the commonly used experimental techniques. The ingenuity of the failure analyst lies in selecting the right type of tests and examinations. The sequence of conducting these tests and examinations is also important. A flow chart for carrying out failure analysis of metallic components is given in Figure 2.6.

2.3.1 Visual Examination

Visual examination is the first and foremost step in failure analysis. An experienced person with an eye for defects can derive a lot of information

Tools and Techniques

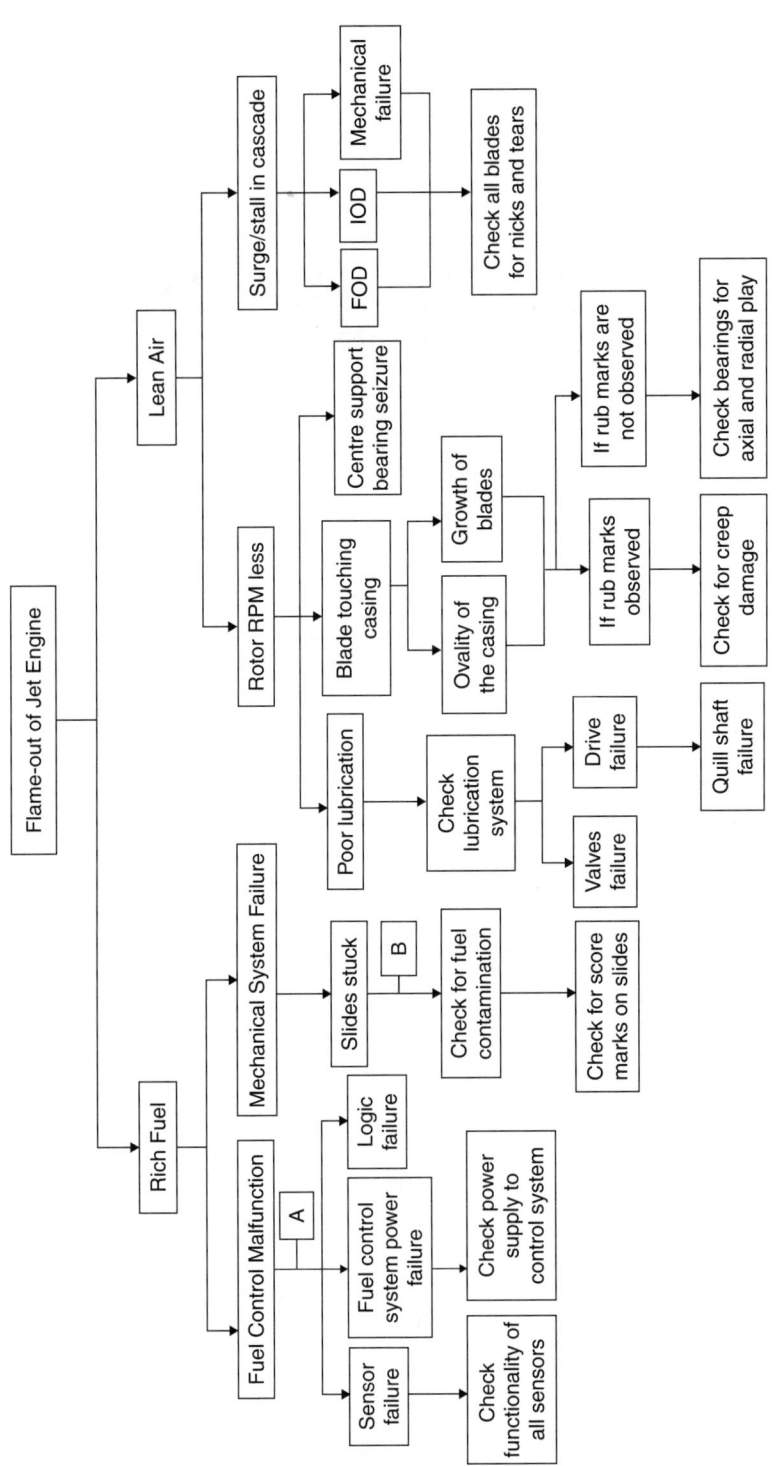

FIGURE 2.1
Possible causes for jet engine flame-out (rich fuel – lean air combination).

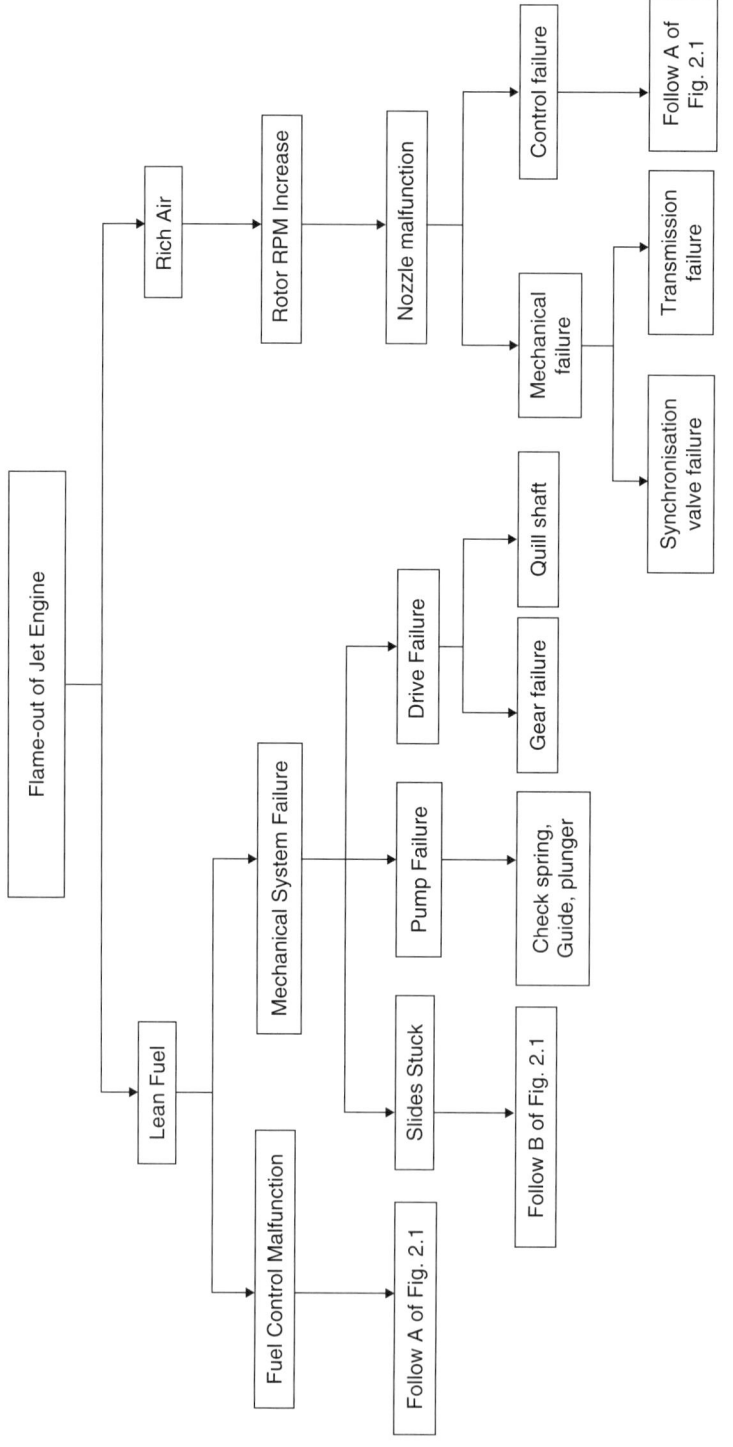

FIGURE 2.2
Possible causes for jet engine flame-out (lean fuel – rich air combination).

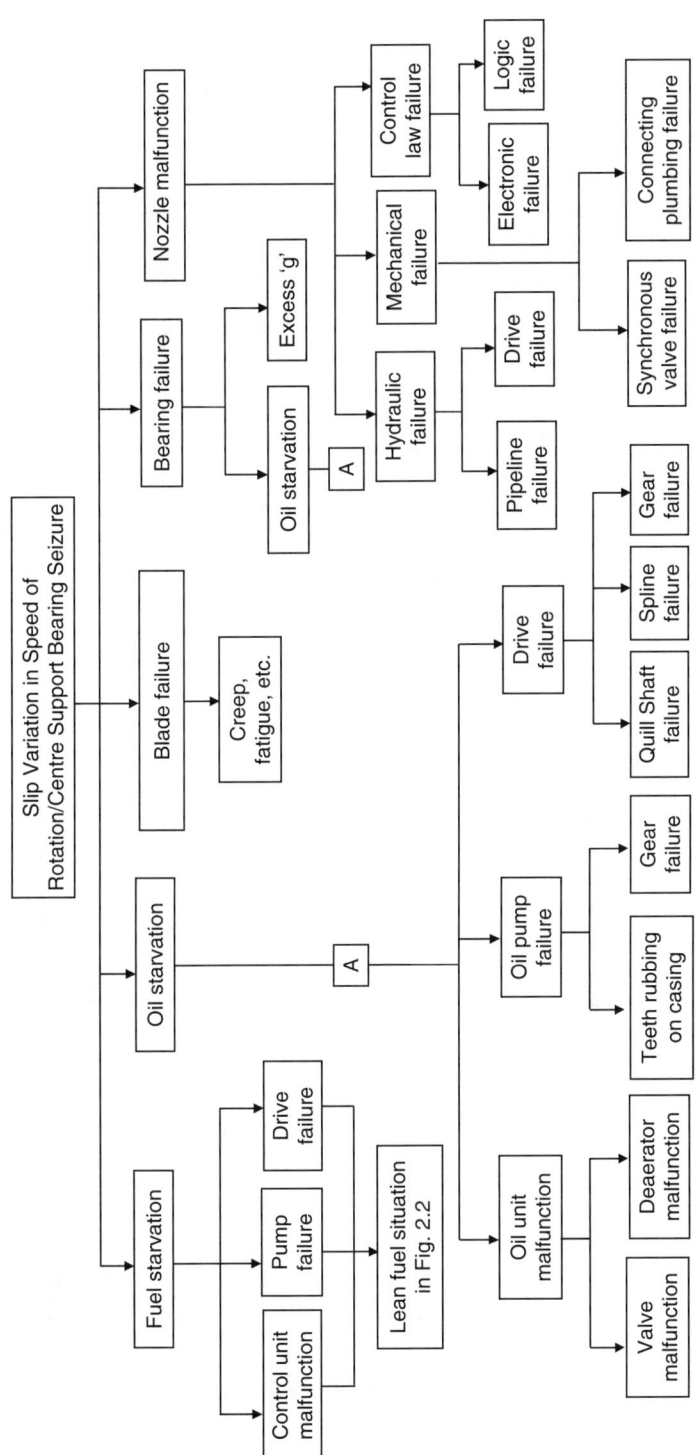

FIGURE 2.3
Possible causes for center support bearing seizure in a jet engine.

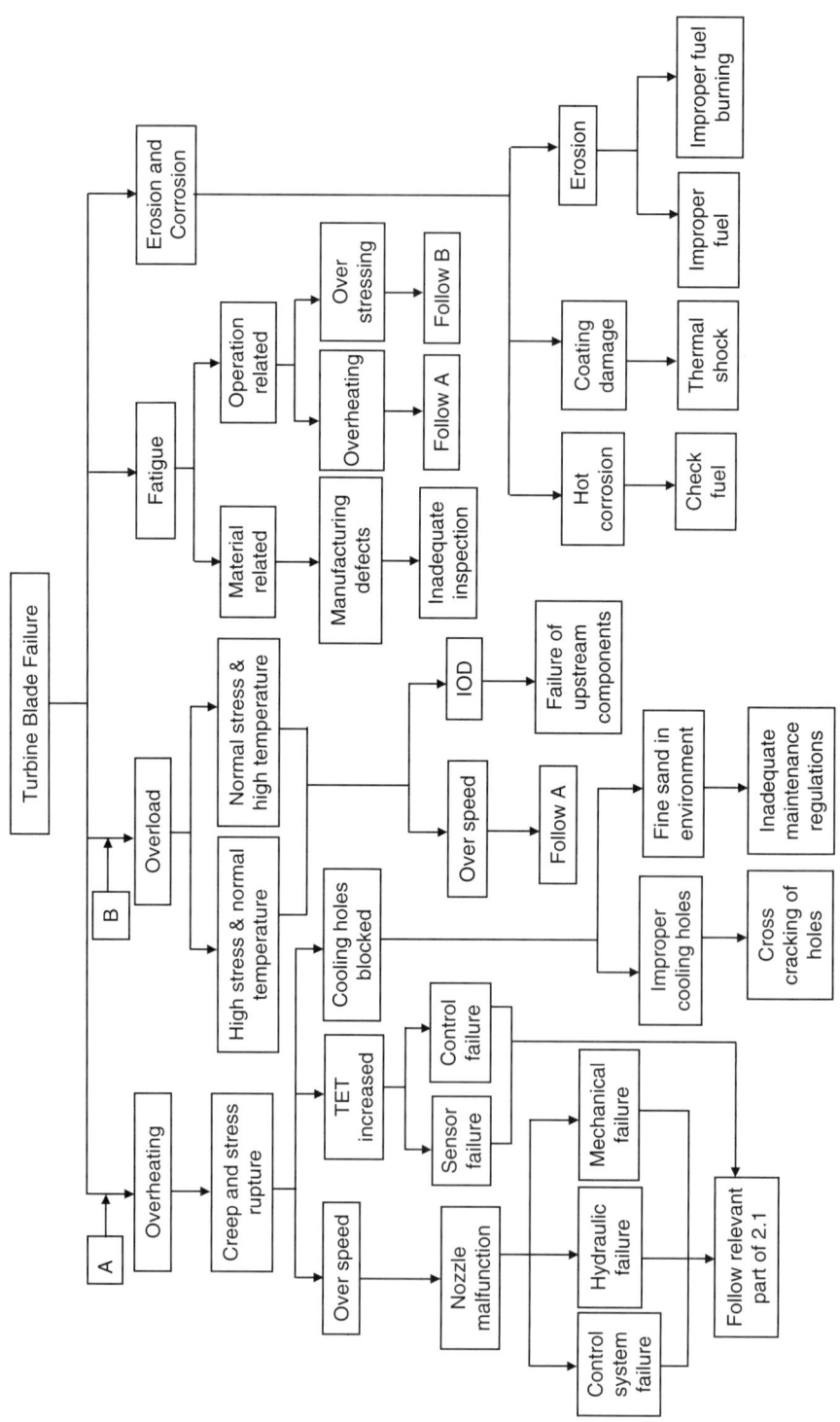

FIGURE 2.4
Possible causes for turbine blade failure.

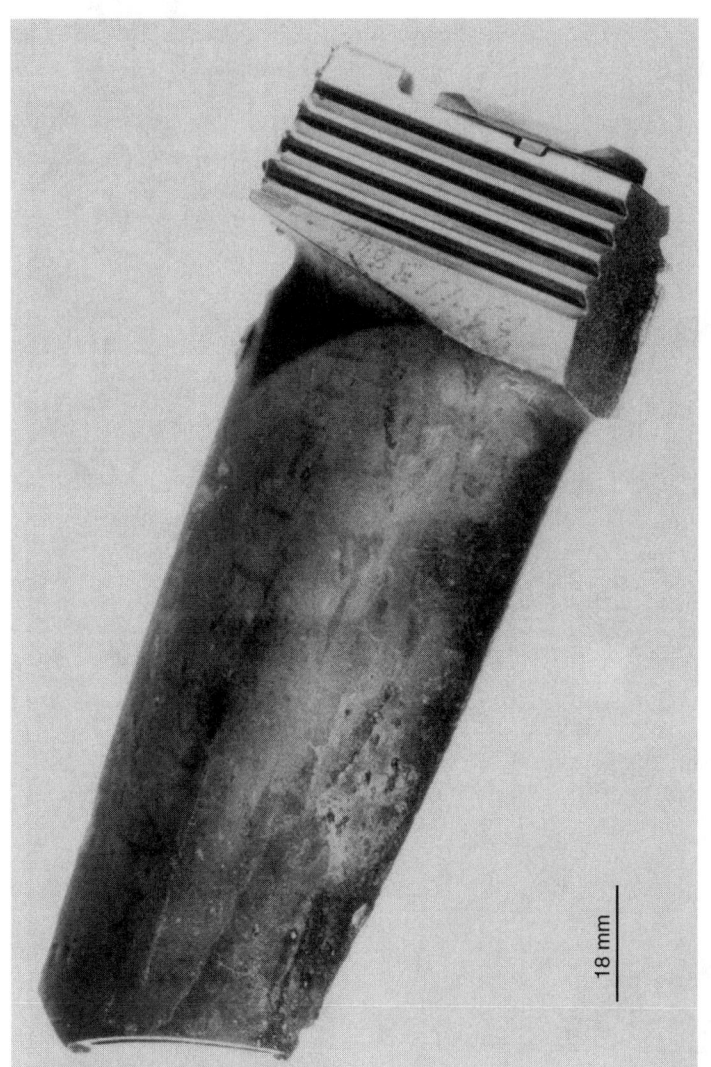

FIGURE 2.5
(a) Failed turbine blade. (b) Section of the failed turbine blade showing extraneous metal deposition on the blade surface due to melting of upstream components.

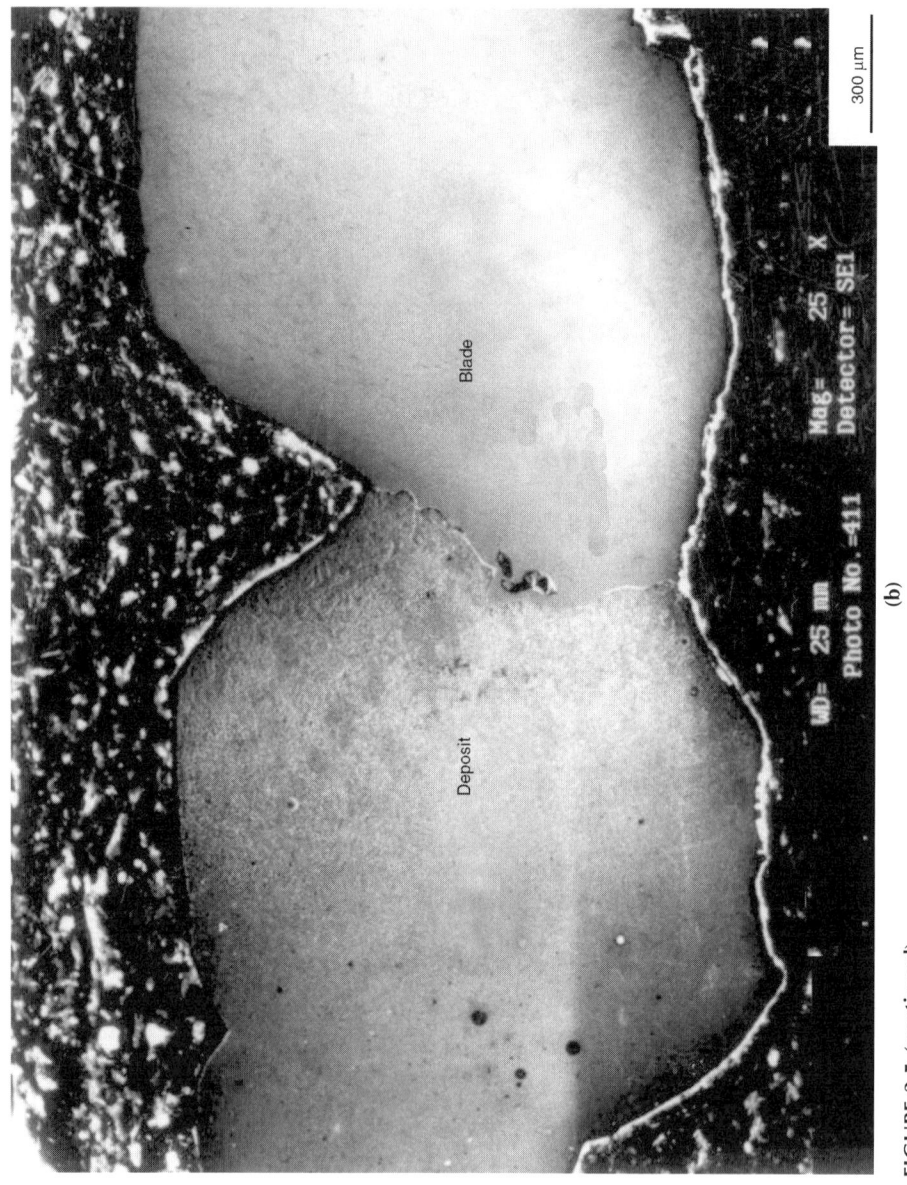

FIGURE 2.5 (continued)

Tools and Techniques

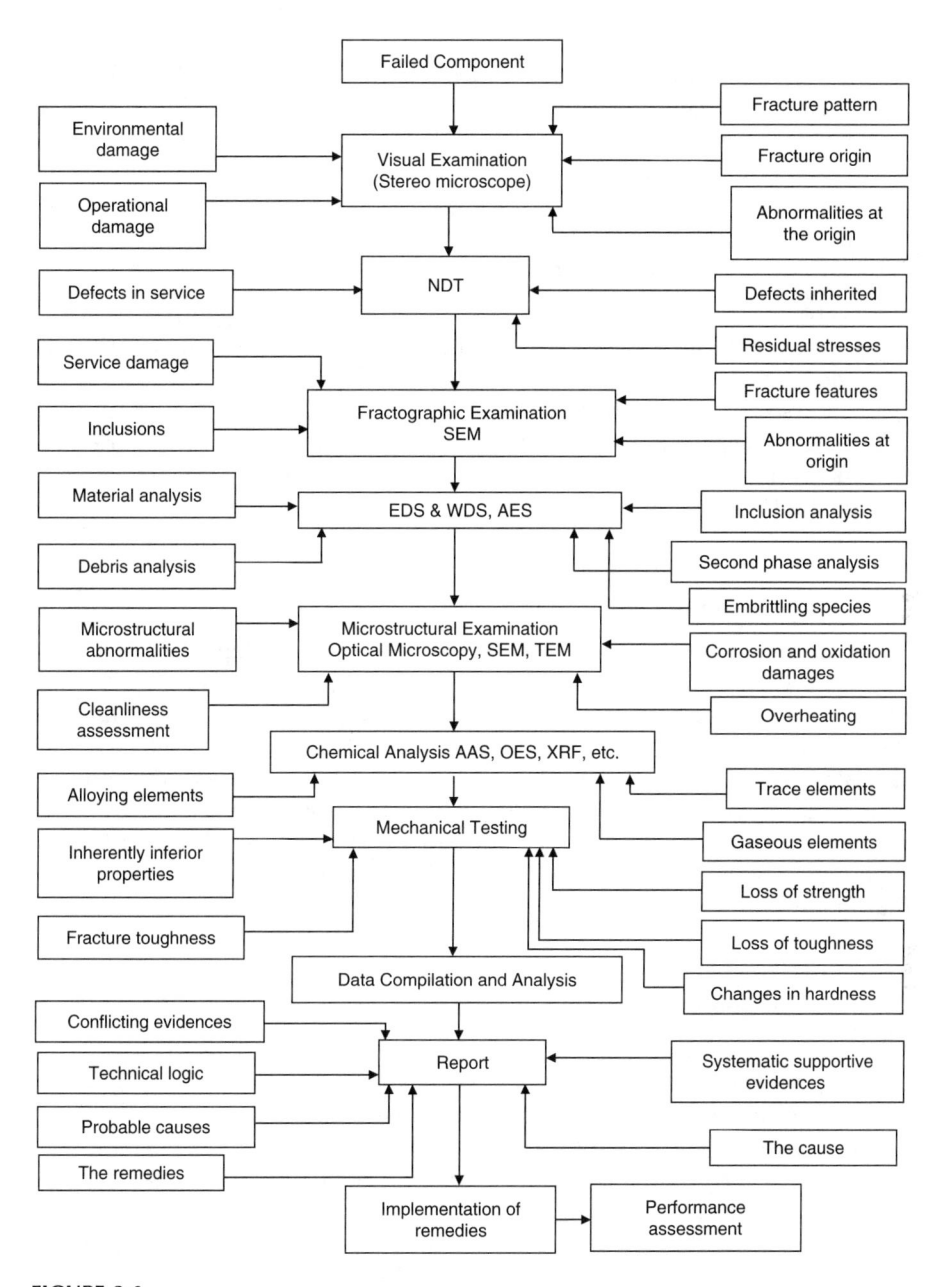

FIGURE 2.6
Flow chart for carrying out failure analysis of a typical metallic component.

by critically examining the failed component. By studying the fracture surface, it is possible, as a first approximation, to identify the type of fracture (ductile, brittle, fatigue, etc.). It is also possible to locate the fracture origin by studying the fracture pattern.

Examination of the fracture origin and the surface perpendicular to the fracture surface at the origin gives clues regarding abnormalities and damages that could have triggered the initiation of the crack. The use of a stereomicroscope and magnifying lens to aid the eye in resolving finer details is extensively practiced.

2.3.2 Nondestructive Testing

Nondestructive examination of the failed component along with an unused component provides information on the type of defects, their inheritance from the production stage and their generation during service. Dye penetrant tests, radiography and ultrasonic tests are useful techniques to provide this information. It is desirable to analyze the indications and isolate the primary defects from the secondary damages. Residual stress measurement, as needed, yields valuable information.

2.3.3 Fractography

The scanning electron microscope (SEM), with its high depth of focus and resolution capabilities, is an important tool and is regarded as an eye for the failure analyst. By fractographic examination using SEM, the mode of failure, the fracture origin, the abnormalities that initiated the failure, etc., can be precisely identified. In failures originating at the free surface of the component, because of the high depth of focus achievable with a SEM the free surface where the crack originated and the fracture features at the origin can be simultaneously examined to identify the type of damage or abnormality that initiated the crack.

2.3.4 Microanalysis

The energy dispersive spectroscopic (EDS) analysis facility, an accessory available with all modern SEMs, can be used to analyze the component for material composition and to identify the constituent elements in inclusions, slag pockets, corrosion debris, extraneous deposits, etc., that may be located at the origin. The analytical data generated on rough surfaces need to be used with caution. Depending on the criticality of the compositional information generated by EDS, complementary techniques such as wavelength dispersive spectroscopy (WDS) can be used to isolate elements with overlapping energy levels in EDS, such as sulfur in alloys containing molybdenum. Electron probe microanalysis (EPMA) is an extremely useful technique for quantitative analysis of microstructural features. X-ray images generated with EPMA on features of interest such as slag pockets, corrosion debris, oxidation, etc., provide useful information to identify the source and mechanism of the formation of the feature of interest. Auger electron spectroscopy

(AES) is an excellent technique to identify embrittling species on fractured samples *in situ*. Temper embrittlement caused by segregation of phosphorus, tin, arsenic, and antimony to prior-austenite grain boundaries and sulphide embrittlement caused by precipitation of sulphides on prior-austenite grain boundaries are two of the many examples that can be unambiguously identified by AES.

2.3.5 Chemical Analysis

In situations where nonconformance of the material composition to the specified grade is identified as the primary cause of failure, it is necessary to accurately determine the composition of the failed component. A plethora of analytical techniques based on atomic absorption and emission principles is available for estimation of elemental concentrations at levels of a few tens of a percentage to a few parts per billion.

X-ray fluorescence (XRF) spectrometry is used for shop floor analysis for the control of melt composition and raw material analysis because of the ease with which a large number of elements can be analyzed simultaneously on a solid sample. Atomic absorption spectrometry and its modern variants are extensively used as confirmatory tests, especially for the analysis of trace elements. Hydrogen, oxygen and nitrogen are analyzed by vacuum and inert gas fusion techniques and carbon and sulfur by the combustion method.

2.3.6 Microstructural Examination

Microstructural examination of a failed component gives valuable information. It is well known that microstructure dictates the mechanical properties and the fracture behavior of a metallic material, which is in turn dependent on composition, processing and heat treatment. By careful analysis of the microstructure, it is possible to assess deficiencies in composition, processing, heat treatment, etc. Microstructural damages are not obvious in many situations, and a failure analyst must be trained in identifying them. Grain boundary films and cavities, improper second-phase distribution, existence of brittle phases, surface damages (caused by oxidation, corrosion, wear and erosion), nonmetallic inclusions, shrinkage porosity, etc., are some of the defects that can easily be identified by metallographic examination. Sometimes, it may be necessary to generate supportive or confirmatory evidence for the observed microstructural abnormality by subjecting the material to specific tests. A few situations arise during the course of failure analysis where the resolutions and magnifications available in optical microscopy are not adequate for examining very fine microstructural details. For example, if transformation of retained austenite films at lath boundaries to carbides, which causes tempered martensite embrittlement or dissolution of γ' precipitates in nickel-based superalloy turbine blades exposed to high operating temperatures, are to be studied it is necessary to use high-resolution techniques such as transmission electron microscopy (TEM). SEM also can be used to study fine microstructural features if the desired contrast can be obtained by back-scattered electron imaging or the deep etching technique.

2.3.7 Mechanical Testing

Although extensive mechanical testing is seldom conducted as a routine requirement in failure analysis, specific tests may be necessary to generate data in support of the logic built in the analysis of the case. Hardness measurements, being simple and the least demanding in terms of sample requirements for test specimen preparation, give adequate information on the property variations due to microstructural changes. Microhardness measurement on microstructural features of interest is extremely useful in failure analysis.

2.3.8 Analysis and Interpretation of Experimental Data

The most critical step in failure analysis is the analysis and interpretation of the data generated using varieties of experimental techniques. It is necessary to (a) list all the data generated, (b) analyze the data based on scientific principles, (c) eliminate contradictions based on evidence and by generating supportive and confirmative experimental evidence, (d) consider all possible causes for the observed fracture mode and (e) arrive at the most probable cause of failure. Once the cause is identified, certain remedial measures become obvious and the implementation methodology for the most appropriate remedy has to be worked out in consultation with the designer, manufacturer, and user.

3
Fracture Feature Analysis

3.1 Introduction

Any tool used for identifying the primary cause of failure should be applied judiciously, taking into account the possible lapses during manufacture, operation and maintenance of the suspect subsystem. Due attention should also be paid to the uncertainties associated with human involvement at all stages.

In the analysis of metallurgical failures, it would be extremely useful to practice fracture feature and causative factors analysis (FFCFA), which involves listing all possible reasons and systematically identifying the most probable cause of the observed fracture features. Most of the metallurgical failures can broadly be classified into intergranular failures and transgranular failures. While mixed transgranular and intergranular fracture features are often observed in many components failed in service, the fracture origin generally possesses distinct features that can easily be labeled considering the predominance of either of the features.

3.2 Intergranular Fracture

Intergranular fracture (IGF) at high operating temperatures is caused either by time-dependent phenomena such as creep and stress rupture or by time-independent effects such as grain boundary attack by liquid metals and molten salts. IGF also occurs at room temperature due to (a) grain boundary embrittlement, (b) stress corrosion cracking and corrosion fatigue in certain material–environment combinations, and (c) development of high residual tensile stresses as a result of phase transformations and thermal effects. The probable reasons for the occurrence of IGF are shown in Figure 3.1. While attempting to identify the cause for IGF, it is essential to consider the operating stresses, environment and temperature in addition to the metallurgical aspects.

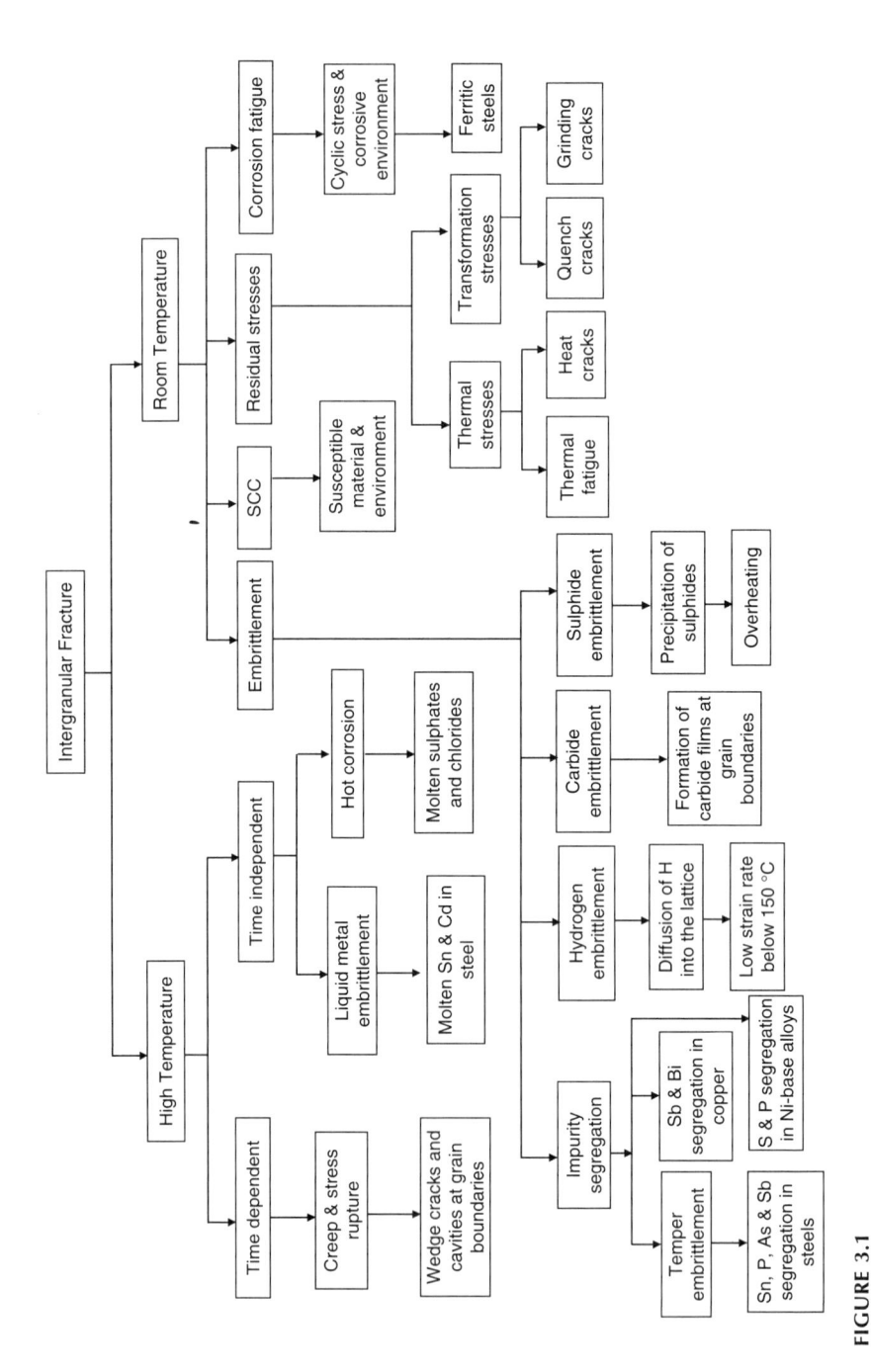

FIGURE 3.1
Fracture feature and causative factor matrix – intergranular fracture.

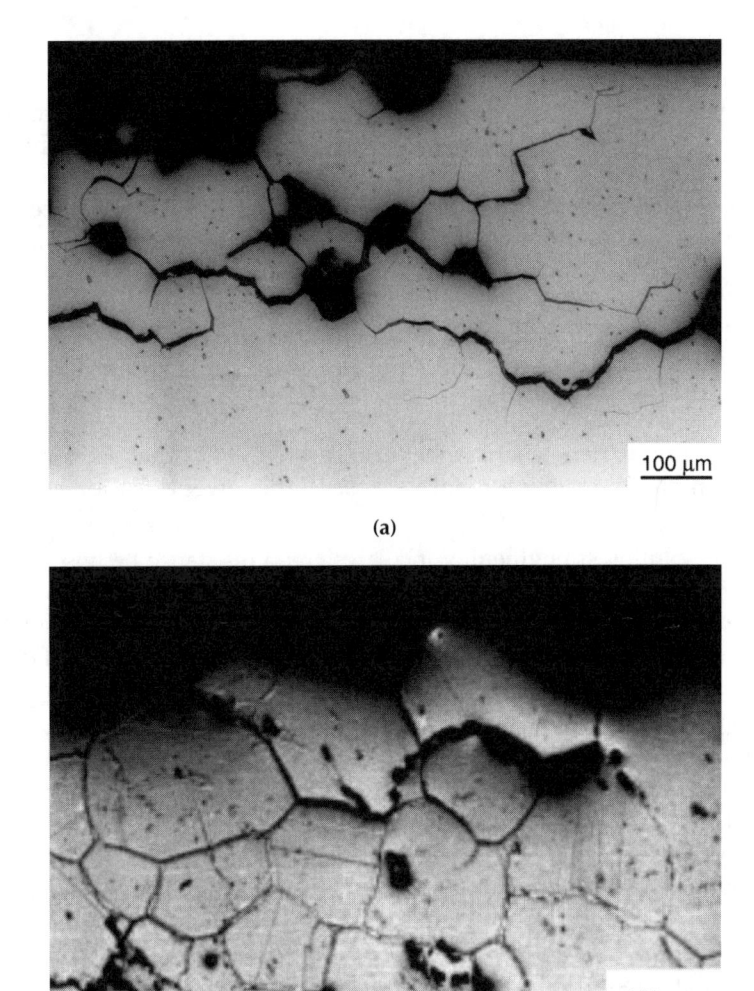

FIGURE 3.2
(a) Wedge-type grain boundary damage. (b) Round-type grain boundary damage.

3.2.1 Creep and Stress Rupture

Metallic materials undergo time-dependent plastic deformation when subjected to temperatures above 0.4 T_m, even when the stress levels are below their yield stress — a phenomenon known as *creep*. Creep deformation in polycrystalline materials occurs by dislocation climb and glide and by mutually accommodating grain boundary sliding and transport of matter. Failure under creep conditions is caused by the formation of (a) wedge-type cracks at the triple points under conditions of high stress and short rupture times

(Figure 3.2a) and (b) round/elliptical cavities at grain boundaries under conditions of low stress, high temperatures and long rupture times (Figure. 3.2b). When the temperature is high and the stress is low, failure occurs by nucleation and growth of creep voids at the grain boundaries. If the stress and temperature conditions are such that no steady-state creep occurs, the failure is termed *stress rupture*. Both wedge-type and round-type creep damages at the grain boundaries, oriented perpendicular to the stress axis, are reliable secondary evidence of intergranular failures due to creep and stress rupture. These features can easily be observed by microstructural examination of the section perpendicular to the fracture surface.

Many metallurgical factors influence the creep resistance of the metallic materials:

- FCC and HCP metals have better creep resistance than BCC metals of comparable melting point because of the lower diffusion rates in densely packed structures.
- Solid solution strengthening increases creep resistance because of the resistance offered by strain fields around solute atoms to the movement of dislocations. If solute additions result in reduction of stacking fault energy, creep resistance is further enhanced because of the difficulty associated with the movement of widely spaced dislocation partials.
- Increasing the grain size increases the creep resistance by lowering the grain boundary area since the strength of the grain boundary decreases faster than that of the grain interior with increasing temperature. Creep rates are inversely proportional to grain diameter. Directionally solidified structures with dramatically reduced grain boundary areas and single crystals with no high-angle grain boundaries are obviously well suited for creep-resisting applications.
- Enhancing the strength of grain boundaries increases the creep strength. Addition of Zr and B to superalloys is known to increase creep resistance.
- Precipitation hardening improves creep resistance as the fine, closely spaced, hard and coherent precipitates retard motion of dislocations. Precipitation of γ' ($Ni_3(Al, Ti)$) in Ni-base superalloys and various carbides in creep-resistant steels are typical examples.
- Dispersion hardening involving uniform dispersion of fine and inert particles increases creep resistance. ThO_2 dispersed Ni-Cr alloys and Y_2O_3 dispersed Fe-Cr-Al alloys are examples of dispersion-hardened, high-temperature alloys.

A good creep-resistant material can fail prematurely in creep or stress rupture mode as a result of microstructural degradation during service. For example, hot end components of gas turbines made of creep-resistant, precipitation-strengthened Ni-base superalloys fail in creep or stress rupture mode as a result of dissolution of γ' precipitates due to overheating.

Fracture Feature Analysis

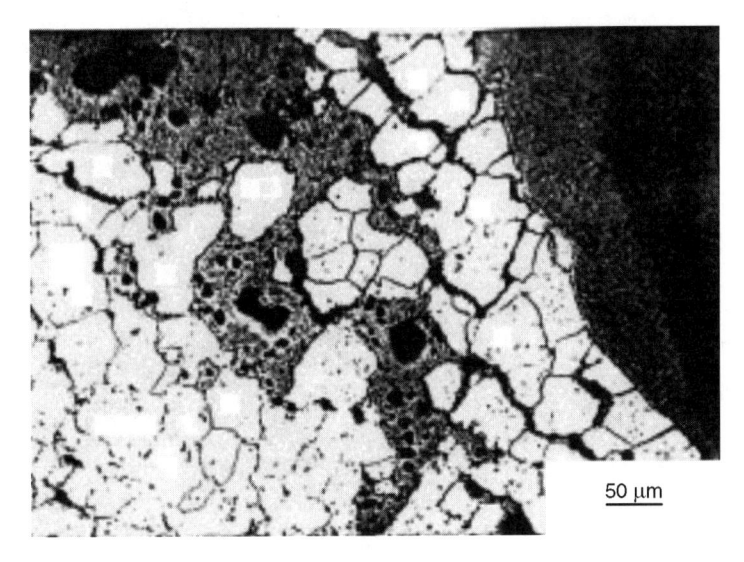

FIGURE 3.3
Intergranular attack by molten oxides in a boiler tube.

3.2.2 Liquid Embrittlement

When liquid metal or molten corrosive salt is in intimate contact with a component's surface, grain boundary damage occurs, leading to a reduction in cohesive strength and consequent intergranular fracture. Grain boundaries can be attacked by molten salts; typical examples include the damage caused to nickel-based superalloy turbine blades by the molten sulphates and chlorides generated from the combustion of the aviation fuel and operation in a marine environment. Damage to boiler and super heater tubes by molten oxides of vanadium and sulphates that form as a result of combustion of fossil fuels is another example. The fracture path in equiaxed polycrystalline materials under these conditions is generally intergranular (Figure 3.3).

Liquid metal embrittlement is encountered in systems such as brazed steel joints; hot dip (tin and cadmium) coated steels; brass components in contact with liquid mercury; gun barrel bores due to frictional melting of the copper driving band; and friction-induced melting in bearings. In some cases, melting occurs at the interfaces of moving components due to frictional heat and the molten metal of the low melting component spreads along the grain boundaries of the mating part, initiating intergranular fracture. The penetration of liquid inner race material (1C-17Cr-0.5Mo steel) along the prior-austenite grain boundaries of the shaft (0.35C-4Ni-1Cr-0.5Mo steel) due to bearing seizure is shown in Figure 3.4.

3.2.3 Heat Cracks

Yet another frequent cause of intergranular fracture is thermal shock, resulting from thermal stresses generated due to sudden heating and cooling,

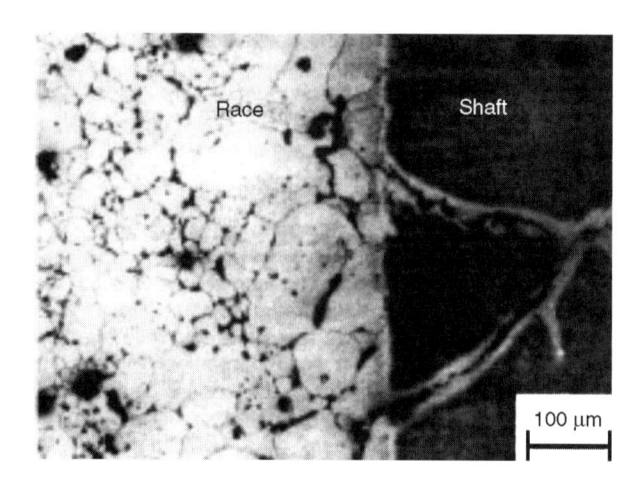

FIGURE 3.4
Molten race material at the prior-austenite grain boundaries of the shaft. Cracks are also evident.

typically noticed in the bores of gun barrels. If the heating and cooling are cyclic in nature, as in gas turbines, thermal fatigue can occur.

Intergranular cracks can also develop due to a combination of thermal stresses and transformation stresses. Quench cracks and grinding cracks fall under this category.

3.2.4 Embrittlement

Intergranular failure can also occur under ambient conditions. Because of the local disturbances in atomic arrangement, the stored energy at the grain boundaries of a polycrystalline material is higher than in the grain interior at ambient temperature. This makes the grain boundaries chemically more reactive and promotes segregation of impurity elements to grain boundaries. Segregation of (a) P, Sn, As and Sb at prior-austenite grain boundaries in quenched and tempered low-alloy steels during tempering; (b) Sb and Bi in copper and its alloys; and (c) Bi, Te and Pb in nickel-based alloys, etc., lead to grain boundary embrittlement and predominantly intergranular fracture. Intergranular fracture in overheated steels due to either grain boundary burning or sulphide embrittlement (caused by the dissolution and reprecipitation of sulphides at prior-austenite grain boundaries) are often encountered. Formation of carbide films at grain boundaries, as a result of improper processing and heat treatment, can lead to carbide embrittlement and intercrystalline fracture in many steels and cobalt-based alloys. The formation of grain boundary α films in α-β titanium alloys is known to promote intergranular fracture.

3.2.5 Intergranular Corrosion

Intergranular fracture is promoted by corrosion in susceptible alloys. Selective depletion of alloying elements at grain boundaries makes a corrosion-resistant alloy prone to grain boundary attack. For example, depletion of chromium due to formation of chromium carbide at the grain boundaries of austenitic stainless steels either due to welding or improper annealing promotes grain boundary attack, a phenomenon recognized as *sensitization*. Depletion of beneficial alloying elements close to the grain boundaries due to the formation of grain boundary intermetallic compounds results in intergranular corrosion in many other alloy systems. Intergranular corrosion can also occur as a result of segregation of impurity elements to grain boundaries, as in many copper, zinc, and aluminum alloys.

Intergranular failure under the combined influence of corrosion and cyclic loading, known as corrosion fatigue, is often encountered. Failures due to corrosion fatigue are common in ferritic steels operating under the conditions of low alternating stress levels and frequencies.

3.2.6 Stress Corrosion Cracking and Hydrogen-Induced Failures

Stress corrosion cracking (SCC) and hydrogen-induced failures can be intergranular or transgranular, depending on the material and environment. SCC occurs under the combined influence of tensile stress and a corrosive environment. SCC failures are characterized by (a) lack of any plastic deformation prior to failure, (b) presence of extensive secondary cracks, and (c) presence of corrosion debris. While austenitic stainless steels are prone to SCC in chloride-containing environments and are immune to SCC in ammonia, Cu-Zn alloys suffer SCC in ammonia but not in a chloride environment. The mode of failure depends on the metallurgical characteristics of the material and the chemistry of the environment. In nonsensitized, annealed, austenitic steels and cold-worked, austenitic stainless steels, SCC is in transgranular mode in a chloride environment and is in intergranular mode in acidified copper sulphate. Most of the high-strength aluminum alloys fail by intergranular SCC in most corrosive environments. The susceptibility of a material to SCC depends on the corroding species. The fracture modes in SCC are listed in Table 3.1 for different combinations of materials and corroding media.

Hydrogen-induced failures can be intergranular or transgranular, depending on the material and strength level. The observed fracture modes in various materials due to hydrogen embrittlement are listed in Table 3.2. Hydrogen embrittlement (HE) is a serious problem in steels, which pick up hydrogen during finishing operations. The higher the strength of the steel, the lower the critical hydrogen level necessary to induce embrittlement. HE failure can be transgranular with quasi-cleavage features or intergranular

TABLE 3.1

Material – Environment – Fracture Mode in SCC

Material	Environment	Mode[a] TG	IG
Plain carbon steels	Sodium hydroxide, ammonia, ammonium nitrate		✔
AISI 4340 steel	Salt water		✔
AISI 4140 steel	Hydrogen sulphide		✔
Ni-Mo steel	Iodides	✔	
Hy 180 steel	3.5% Sodium chloride		✔
AISI H11 tool steel	Sodium chloride		✔
Low-strength steel	Hydrogen sulphide	✔	
High-strength steel	Hydrogen sulphide		✔
Ferritic stainless steel	Chloride	✔	
AISI 304 stainless steel	Copper sulphate + sulfuric acid		✔
	Sea water (at low temperature)	✔	
	Boiling magnesium chloride	✔	
AISI 303, 304, 316 and 321 austenitic stainless steels	Chloride + Water	✔	
AISI 316, 25% cold-worked, sensitized austenitic steels	Boiling magnesium chloride, Marine environment		✔
High-Ni alloys, IN 738	Boiling dilute chlorides, Sulphates		✔
	Chloride	✔	
Ni-Cu alloys	Fluorides (at high strains)	✔	
	Fluorides (at low strains)		✔
Precipitation-hardenable Ni and Co alloys	Chlorides + Sulphides	✔	
Solid solution-strengthened Ni alloys	Sodium hydroxide		✔
Inconel 600	Hydrogen sulphide		✔
All Ti alloys	Red fuming nitric acid		✔
α-β Ti alloys	Aqueous chloride at RT, Aqueous brine, Halides >220°C	✔	
α Ti alloys	Salt water	✔	
	Red N_2O_2		✔
β Ti alloys	Alcohol		✔
High-strength Al alloys	Marine environment, Aqueous halides		✔
Cast 308 Al-4.5Cu-5.5Si	Chloride		✔
All Cu alloys	Concentrated ammonia	✔	
Brass	Marine environment		✔
	Ammonia	✔	
Al-brass, As presence	Acidified citrate chloride	✔	
Al-brass, P presence	Acidified citrate chloride		✔

[a] TG: transgranular; IG: intergranular.

with ridges resembling crows' feet and micropores on grain facets. Predominantly intergranular fracture caused by HE in a cadmium-plated AISI 4340 steel component heat-treated to a strength level of 1320 MPa is shown in Figure 3.5.

TABLE 3.2

Fracture modes in hydrogen embrittlement

Material	Condition	Mode	
		TG	IG
AISI 1018	Cold-worked	✓	✓
AISI 1074	Quenched and tempered		✓
AISI 4340	Quenched and tempered (low-strength)		✓
	Quenched and tempered (high-strength)	✓	
AISI 4130	Quenched and tempered		✓
AISI 4135	Quenched and tempered	✓	
AISI 15B22	Quenched and tempered		✓
AISI 8740	Quenched and tempered		✓
API 5L X 52	Annealed		✓
ASTM A 228 wire	Hardened and tempered		✓
13-8 Precipitation-hardenable steel	Hardening, sub-zero treatment and tempering		✓
AISI H 11 tool steel	Quenched and tempered	✓	
18% Ni maraging steel	Solution-treated	✓	
Sb-doped, Ni-Cr low-alloy steel	Austenitization, tempering and aging		✓

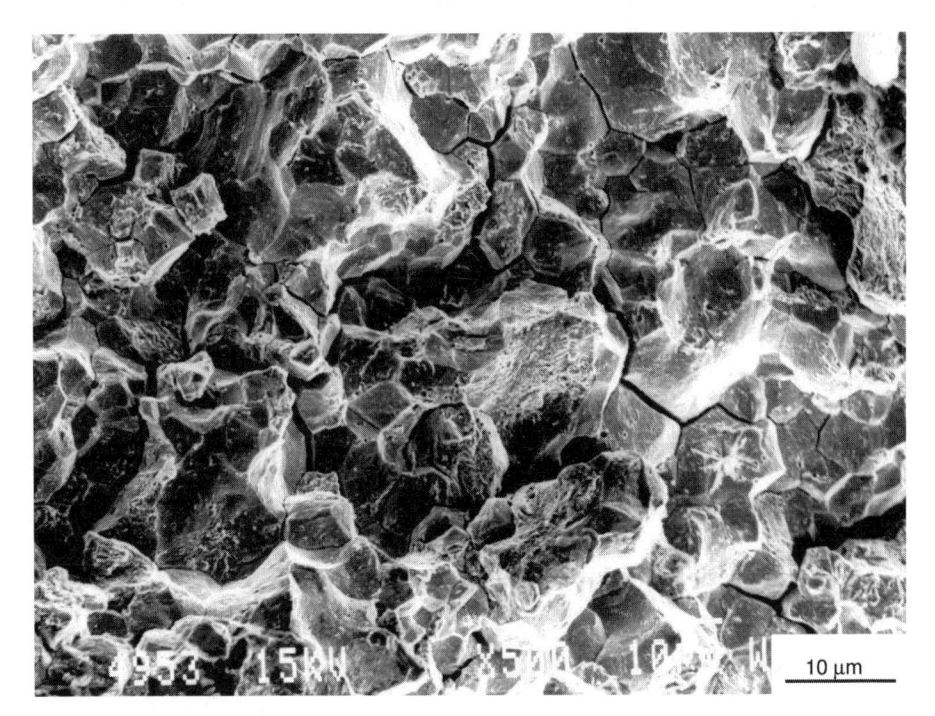

FIGURE 3.5
Intergranular HE failure in AISI 4340 steel.

3.3 Transgranular Fracture

The fracture features in transgranular failure vary depending on the loading conditions, material characteristics and environment. Transgranular fracture can be broadly divided into four categories: ductile failure, brittle failure, fatigue failure and stress corrosion cracking. The fracture features in each type and causative factors are shown in the FMCFA chart shown in Figure 3.6.

3.3.1 Ductile Fracture

Ductile failures are caused when the component experiences stresses exceeding the design limits. The fracture occurs by nucleation, growth and coalescence of microvoids.

The voids nucleate at inclusions or second-phase particles and grow in a direction perpendicular to the major stress axis. Such an overload ductile fracture is characterized by the presence of equiaxed dimples under plane strain conditions and elongated dimples under plane stress conditions (Figure 3.7).

3.3.2 Brittle Fracture: Quasi-Cleavage

Transgranular quasi-cleavage fracture results in a ductile material under conditions that favor brittle fracture. Quasi-cleavage fracture occurs at moderate strain rates and low temperatures in high-strength steel by the formation of a large number of microcracks and their coalescence and is characterized by the presence of rosette-type concave facets with tear ridges (Figure 3.8). In certain materials, HE also leads to failure in quasi-cleavage mode.

3.3.3 Brittle Fracture: Cleavage

Transgranular brittle fractures are spontaneous and are associated with little or no plastic deformation. Macroscopically brittle fractures are crystalline and flat. These fracture surfaces often bear a chevron pattern pointing to the fracture origin. The fracture occurs along certain crystallographic planes under conditions in which cleavage stress is lower than the flow stress. While BCC and HCP metals readily cleave, cleavage fracture seldom occurs in FCC metals, although cleavelike fracture features are often observed in service-failed Ni-based superalloy components. Fracture surfaces formed by the propagation of cleavage cracks show distinct cleave steps with a fanlike pattern in the grain and a river pattern in polycrystalline materials (Figure 3.9). A triaxial stress state created by cracks and notches, high strain rates

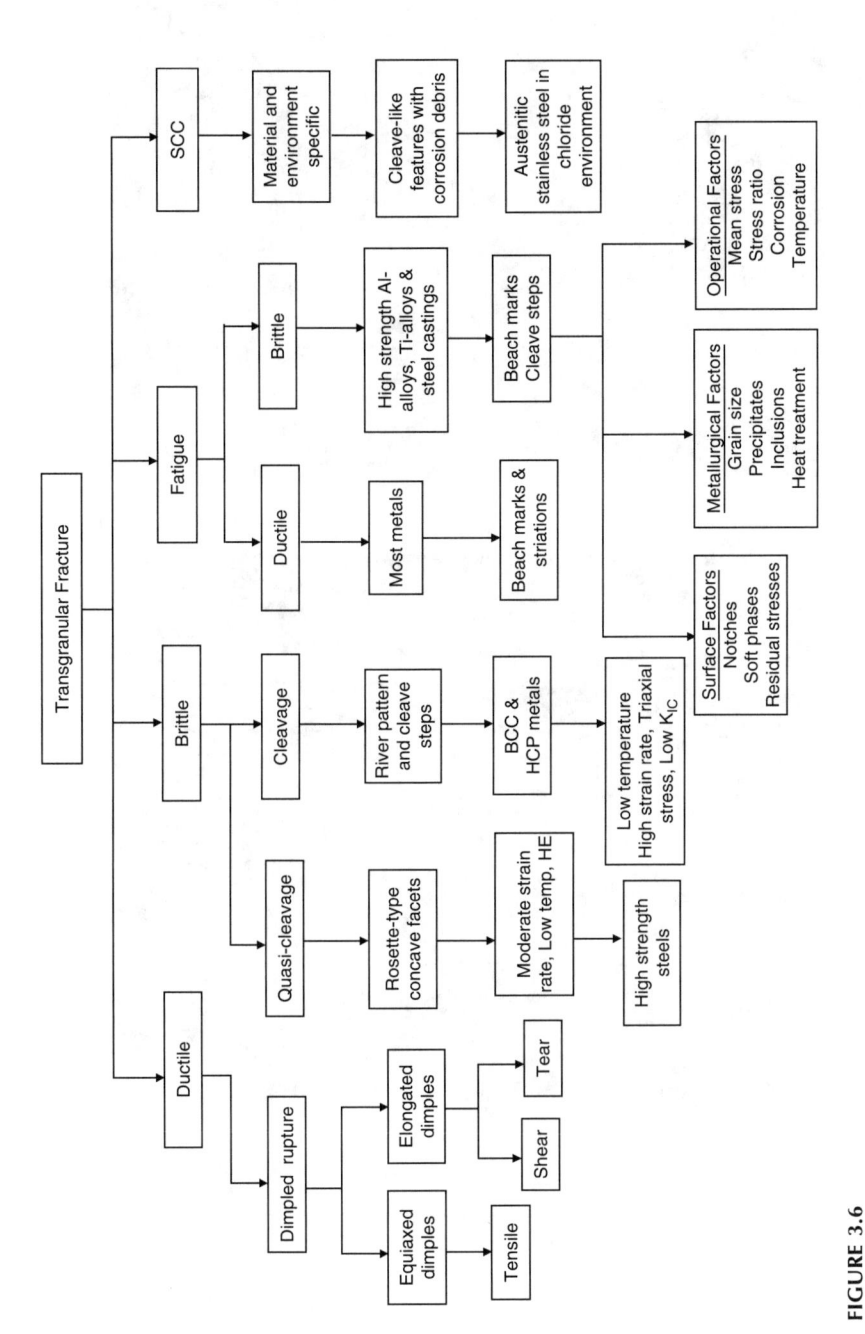

FIGURE 3.6
Fracture feature and cause analysis – transgranular fracture.

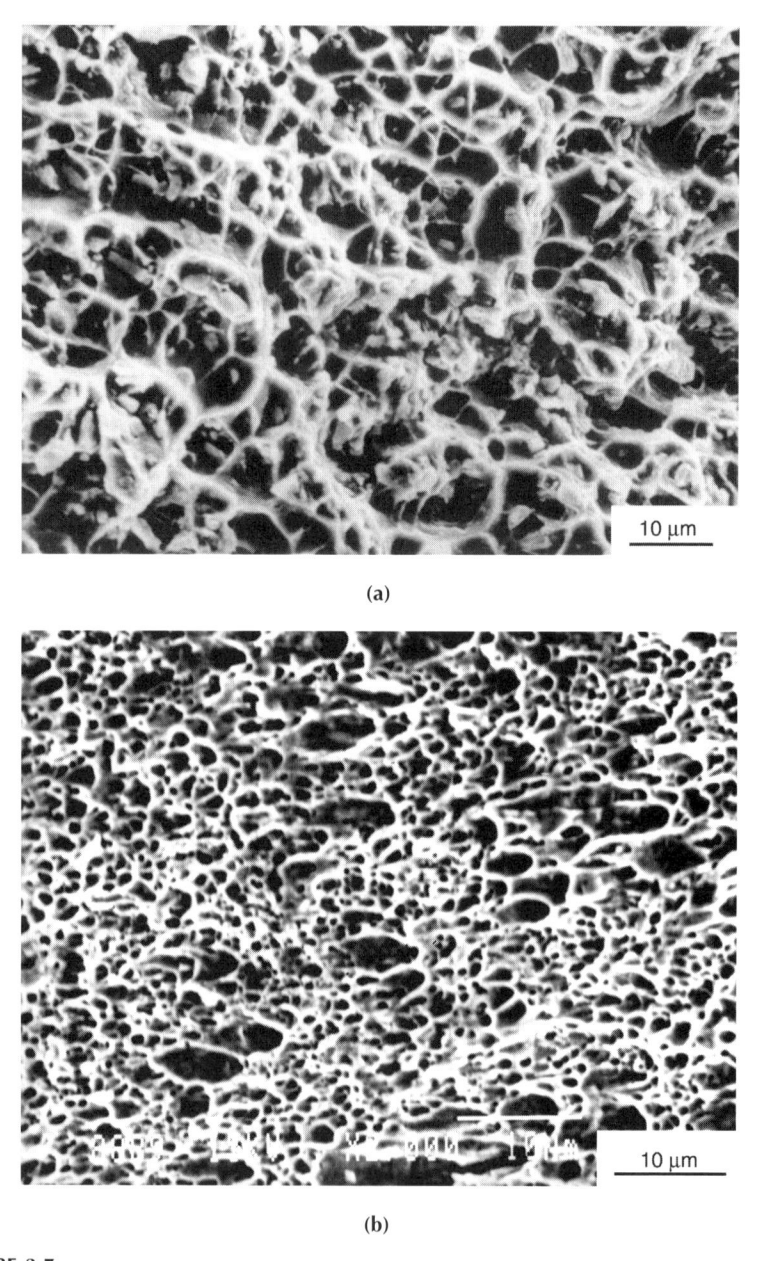

FIGURE 3.7
(a) Equiaxed dimples in Al-2.1Cu-1.5Mg-1.1Ni alloy in solution treated and aged condition. (b) Elongated dimples in a failed low alloy quenched and tempered steel tubular structure.

and low temperatures at which mobility of dislocations is low and all metallurgical parameters that reduce fracture toughness of the material promote brittle fracture.

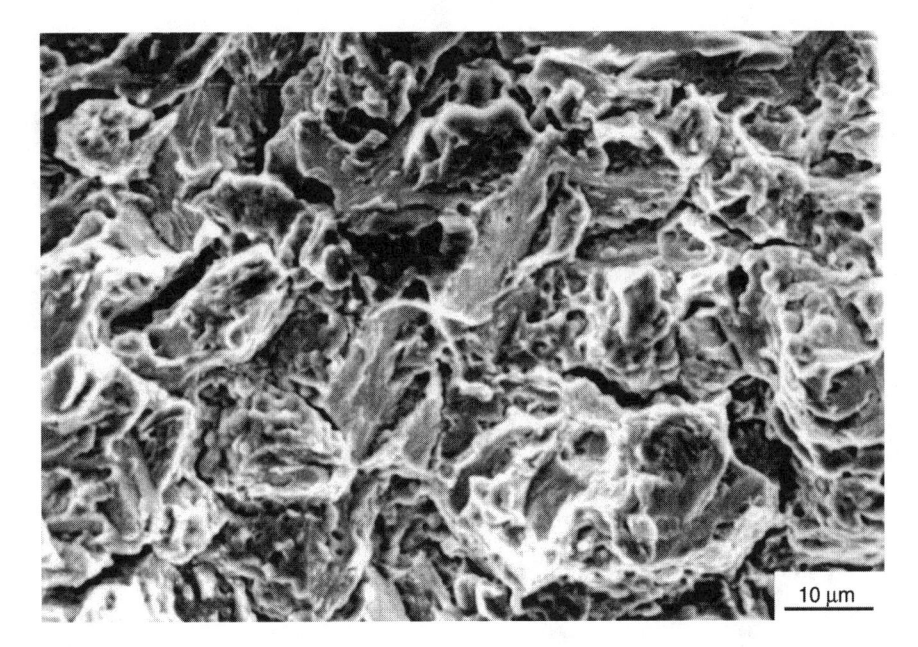

FIGURE 3.8
Quasi-cleavage fracture caused by HE in Maraging steel 250 (solution treated and aged).

3.3.4 Fatigue

Fatigue failure is by far the most important of the transgranular fractures because of the fact that a majority of the service failures are due to fatigue. Fatigue failures occur as a result of cumulative damage caused by cyclic stresses, generally below the yield stress of the material. If the cyclic stresses generate elastic strain cycles, the fatigue life is long and the process is called high cycle fatigue (HCF). If the load cycles lead to the development of plastic strain cycles, the fatigue lives are short and this is called low cycle fatigue (LCF). Fatigue cracks originate at free surfaces of the component, although subsurface crack initiation at gross metallurgical defects is not uncommon.

A major part of the total fatigue life of a component is utilized for fatigue crack initiation and only a minor portion is expended in its propagation. All the variables that promote fatigue crack initiation, therefore, bring down the total fatigue life dramatically. The factors that influence fatigue life are listed below.

Surface Effects

Since most fatigue failures originate at free surfaces, the surface condition plays a significant role in dictating the fatigue life:

- The smoother the surface, the longer the fatigue life.
- Notches, sharp fillets, machine marks and inclusions lower fatigue life because of the stress concentration effect.

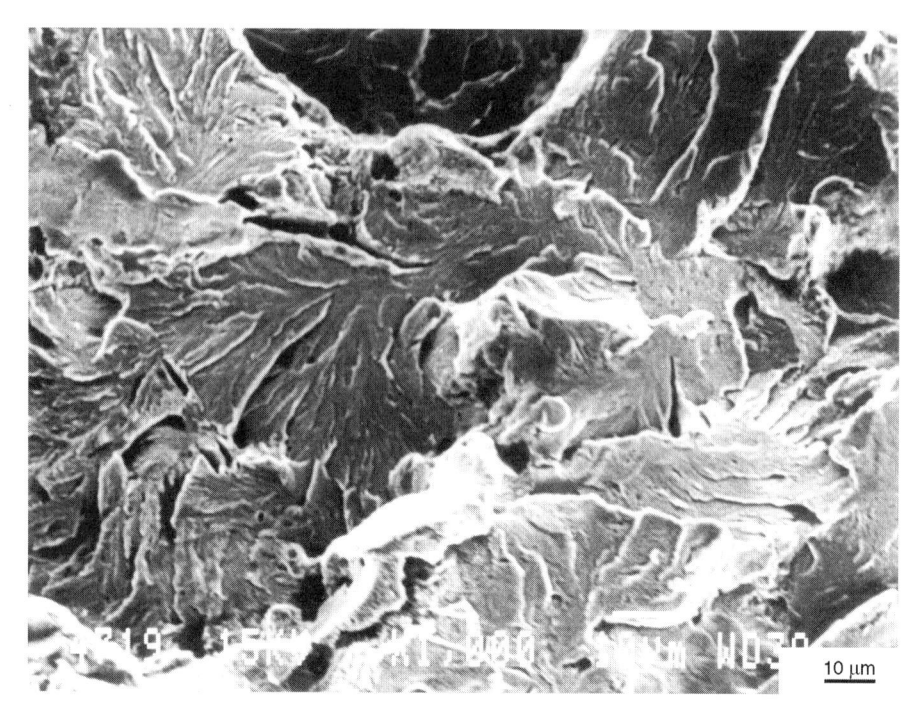

FIGURE 3.9
Cleavage fracture in 1C-1Cr steel in quenched and low temperature tempered condition.

- Coatings that induce residual tensile stresses at the surface, such as chromium on steel, reduce fatigue life.
- Surface treatments that result in compressive residual stresses at the surface, such as shot peening, surface alloying, laser surface modification, etc., increase fatigue life.
- The presence of soft surface layers reduces fatigue life (e.g., decarburized layer in steels and Al clad on high-strength Al-alloy components).
- Environmental damages such as grain boundary oxidation, corrosion, etc., decrease fatigue life.

Microstructural Effects

The fatigue behavior of metallic materials critically depends on the metallurgical characteristics:

- Decreasing the grain size, because of the beneficial effect of delaying fatigue crack initiation, increases fatigue life.
- Precipitation hardening decreases fatigue crack growth rate and increases fatigue life.

- Banding and grain flow in forged and rolled products result in lower fatigue life if the loading direction is across the grain flow.
- Coarse and nondeformable inclusions such as calcium aluminosilicates are more detrimental than fine and deformable inclusions such as manganese sulphide.
- In steels, bainitic structures have higher fatigue life than tempered martensitic structures.
- Welding generally decreases fatigue life as a result of weld defects and formation of undesirable phases in the weld metal and HAZ.

Operational Factors

Many operational factors such as cyclic stress pattern, environmental conditions, etc., influence the fatigue life:

- Increasing the mean stress from compressive to tensile reduces fatigue life.
- Increasing the stress ratio from negative to positive increases fatigue life.
- Increasing the thermal gradient increases the thermal stress and reduces fatigue life.
- An aggressive environment damages the surface and decreases fatigue life.

While investigating a fatigue failure, the failure analyst has to consider not only the influence of the individual factors, but also their interactive effects on the fatigue life. Such an analysis would be beneficial to identify clearly the primary cause for fatigue crack initiation and to suggest appropriate corrective action in the form of design modification, altering the operating conditions and improvement in the quality of material.

Fatigue failures are extremely easy to identify. They are associated with no macroscopic plastic deformation and bear characteristic features. The facture surface of a typical fatigue-failed component consists of two distinct regions: a smooth zone containing beach marks representing the fatigue cracked area (Figure 3.10) and a rough zone, usually granular or fibrous in appearance, representing the final catastrophic overload failure.

The beach marks radiating from the origin represent the boundaries between the period of crack growth and the period of small or no crack growth. In between beach marks, striations of varying spacing appear, which represent fatigue crack growth in each stress cycle, when observed at higher magnifications in SEM. While well recognizable striations are developed during fatigue crack growth in ductile materials (Figure 3.11a), ill-developed striations associated with cleave facets are observed in materials in which fatigue crack propagates in brittle mode (Figure 3.11b).

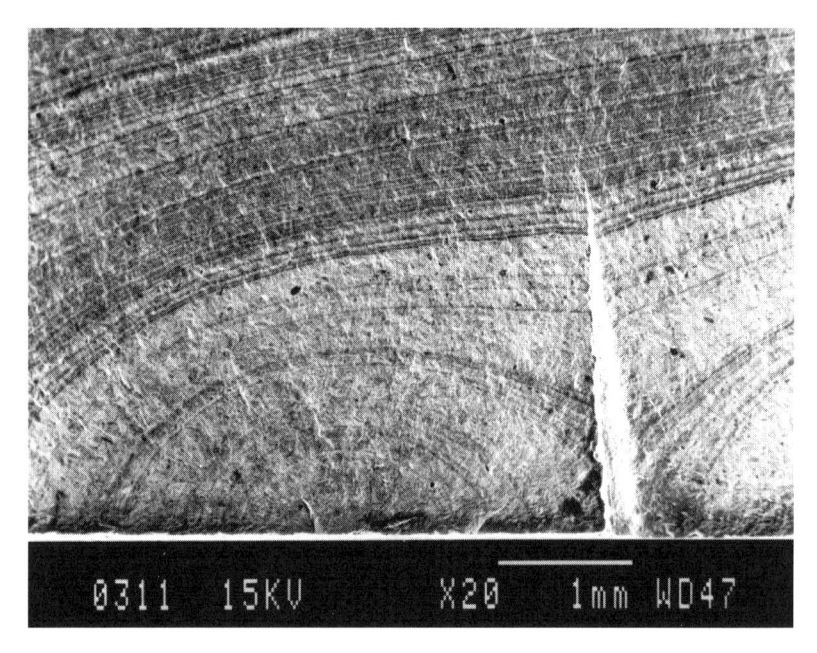

FIGURE 3.10
Beach marks radiating from the fatigue origins.

3.3.5 Transgranular Stress Corrosion Cracking

Transgranular cracking can also occur due to stress corrosion in certain materials under the combined action of stress (applied and residual) and corrosion. Certain specific stress–material–environment combinations promote failure by transgranular mode in SCC. The fracture surface of transgranular SCC exhibits cleavelike features with corrosion debris (Figure 3.12). Microstructural examination reveals a large number of transgranular branching cracks. Austenitic stainless steels and Inconels are prone to transgranular SCC in aqueous media containing chlorides and sulphides. Hydrogen-induced transgranular fractures generally occur in martensitic steels and are characterized by quasi-cleavage features.

3.4 Corrosion-Induced Failures

Various technological developments necessitate usage of materials in aggressive environments. While designing components for ensured performance over a specified period, one of the following three options is generally considered: (a) select materials with inherently high resistance to corrosion damage in the operating environment, (b) protect the components by suitably

Fracture Feature Analysis

(a)

(b)

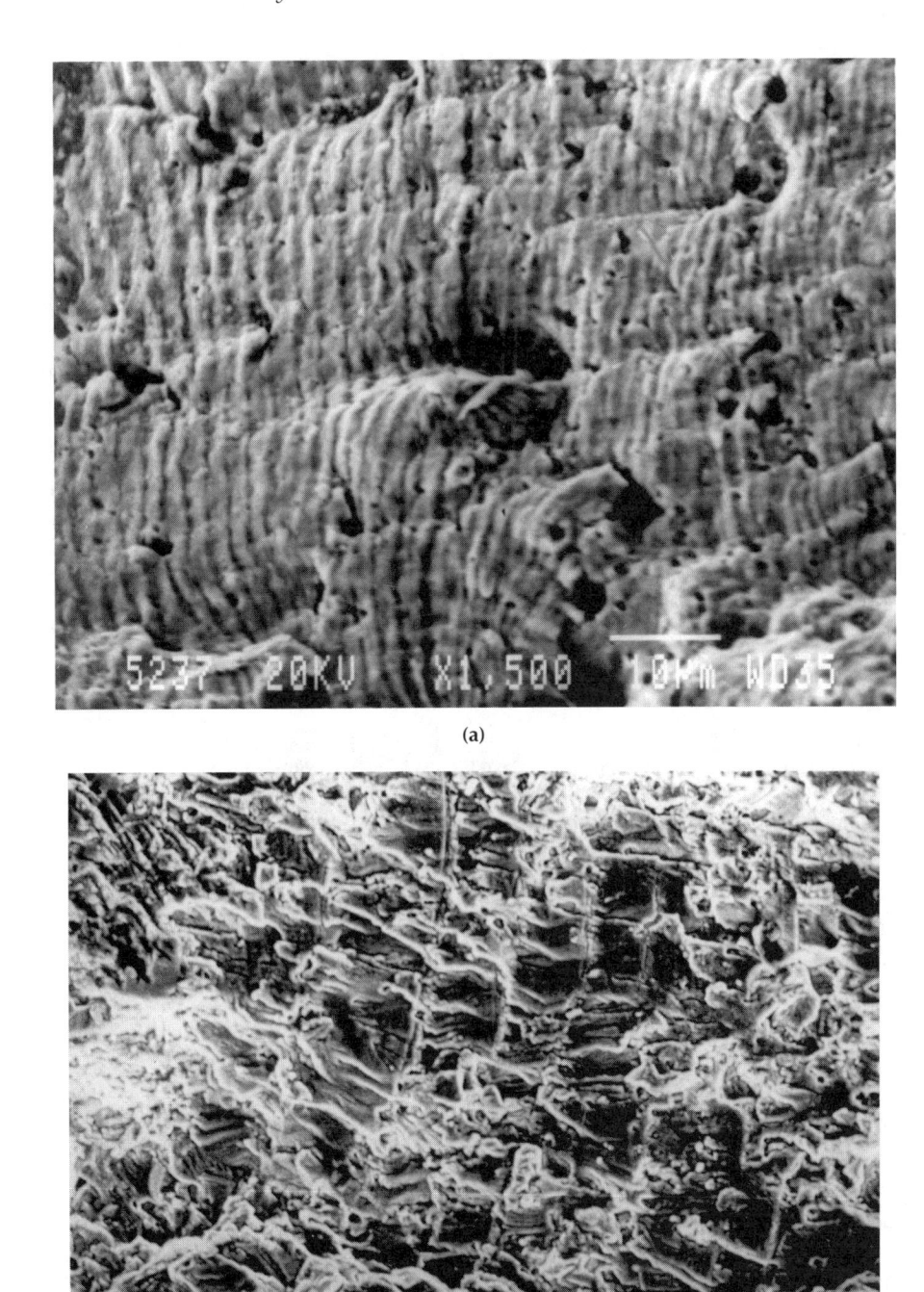

FIGURE 3.11
Fatigue striations in (a) ductile and (b) brittle modes of fatigue fractures.

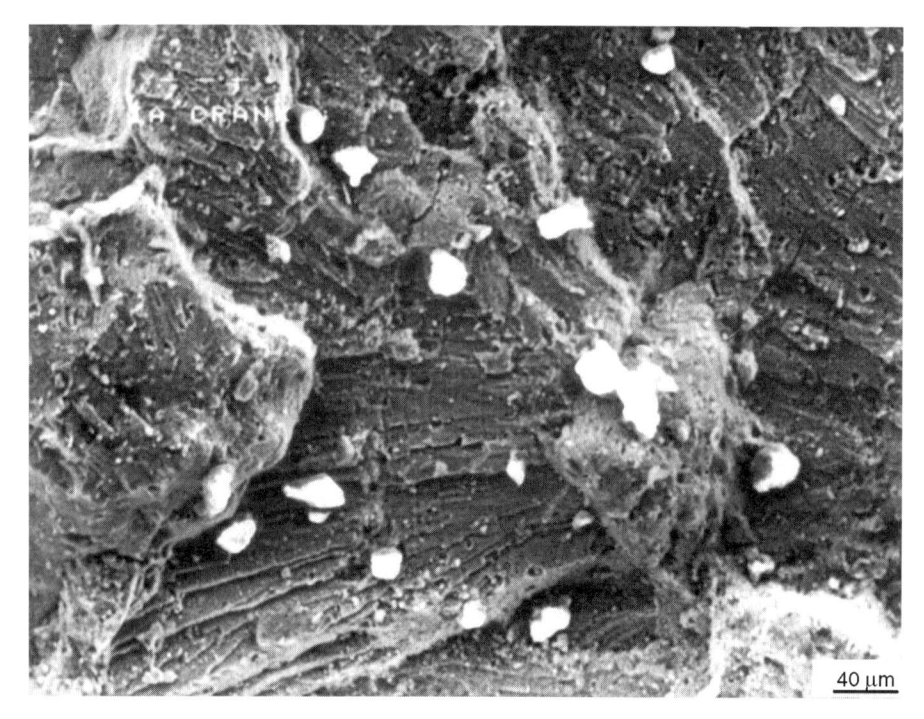

FIGURE 3.12
Typical transgranular SCC fracture surface exhibiting cleavelike features with corrosion debris.

modifying the surfaces to improve corrosion resistance, and (c) permit corrosion to occur at a predetermined rate and replace the component at preset time intervals. The choice essentially depends on the criticality of application, cost and desired service life.

There are many forms of corrosion and a variety of mechanisms operate during degradation of metals due to corrosion. Various forms of corrosion in metallic materials are shown in Figure 3.13. General corrosion occurs either because of chemical attack of the medium resulting in uniform loss of material or due to galvanic corrosion as a result of electrochemical reaction between two dissimilar metals in an electrolyte. Galvanic corrosion can be quite severe if one metal in the couple is highly anodic to the other and if the surface area of the anode is significantly smaller than the surface area of the cathode. The most economical method for protecting underground and underwater steel structures, in addition to employing suitable protective coatings, is cathodic protection achieved by employing either a sacrificial anode system or impressed current system.

Localized corrosion is relatively more dangerous than general corrosion as it leads to the formation of pits or holes as a consequence of highly localized attack. While crevice corrosion and pitting corrosion leave macroscopic evidences of corrosion damage, indications of intergranular corrosion and selective leaching of alloying elements can only be observed with the

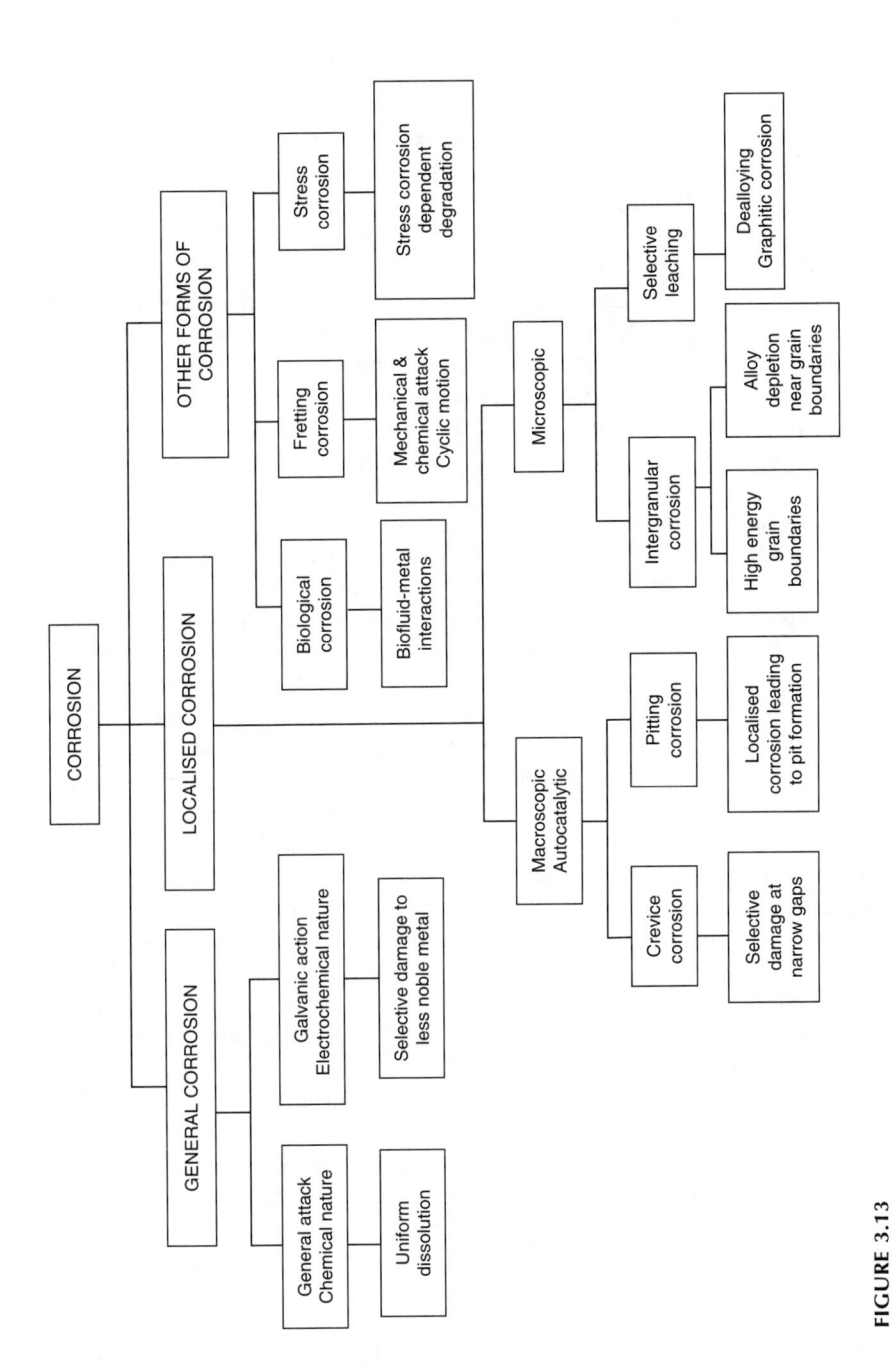

FIGURE 3.13
Various forms of corrosion in metallic materials.

aid of a microscope. Localized corrosion is initiated by localized changes in the fluid composition and/or by damage to the protective layer of the component due to tiny defects and mechanical damages. Once localized corrosion sets in, the degradation occurs at an accelerated rate as a result of sustained stagnant conditions developed within the damaged regions. The incidence of pitting corrosion, and failures due to it, are high in the aeronautical and petrochemical industries. An example of failure caused by pitting corrosion is illustrated below.

Balancing weights placed at the VI stage compressor of an aeroengine were made from a martensitic stainless steel (0.12C-11.8Cr-1.8Ni-1.82W-0.21Nb) and were protected with 20 μm thick silver coating (construction details shown in Figure 3.14a). Extensive pitting and dislodgement of some of the balancing weights were observed during overhaul. Analysis of the failure revealed that it had occurred due to extensive pitting in a chloride-bearing environment as a result of local damages to the silver coating. The damage to the balancing weights due to pitting corrosion is shown in Figure 3.14b.

Intergranular corrosion essentially occurs in certain materials under specific environmental conditions. As the stored energy of the grain boundaries is higher than the grains, they tend to be chemically more reactive than the grains. Segregation of impurity elements at the grain boundaries and development of alloy-depleted zones close to the grain boundaries under certain metallurgical processing conditions make a corrosion-resistant material prone to grain boundary attack. In some materials, alloying elements are selectively leached out due to corrosion, leaving a spongy substrate. Dezincification of brasses, denickelification in many stainless steels and nickel-based superalloys, and selective dissolution of matrix leaving a skeleton of graphite flakes in corroded gray irons are typical examples.

Interest in the corrosion of materials caused by body fluids is growing with the ever-increasing application of implantable devices. While general corrosion is seldom a serious problem in the case of body implants, as the material selection is based on extensive *in vitro* tests, failures occur due to crevice corrosion and stress corrosion cracking. Fretting corrosion is the damage caused to the contact surfaces by combined mechanical and chemical action due to relative cyclic motion. Fretting damage is encountered in many engineering systems and a few body implants such as total hip prosthesis.

The failure rate as a result of corrosion alone is not very high compared to the number of failures encountered due to other failure modes initiated by corrosion. Fatigue failures initiated by pitting and intergranular corrosion, failures due to stress corrosion cracking, perforation of high-pressure pipelines due to pitting, etc., account for a large percentage of service failures. Hence, it is necessary to consider not only corrosion but also its influence in triggering other failure modes during design and material selection.

Fracture Feature Analysis

(a)

(b)

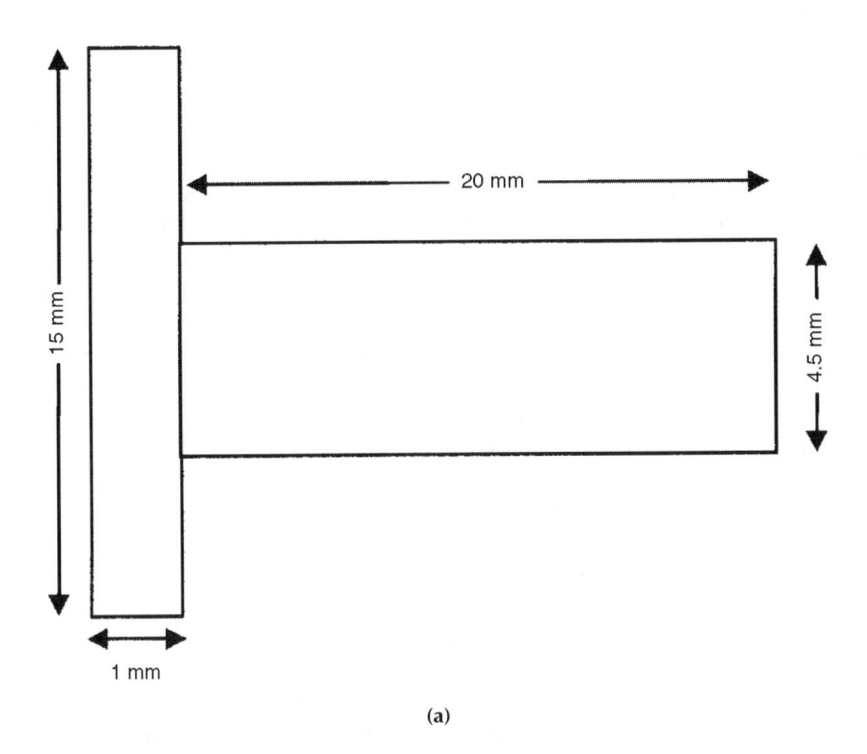

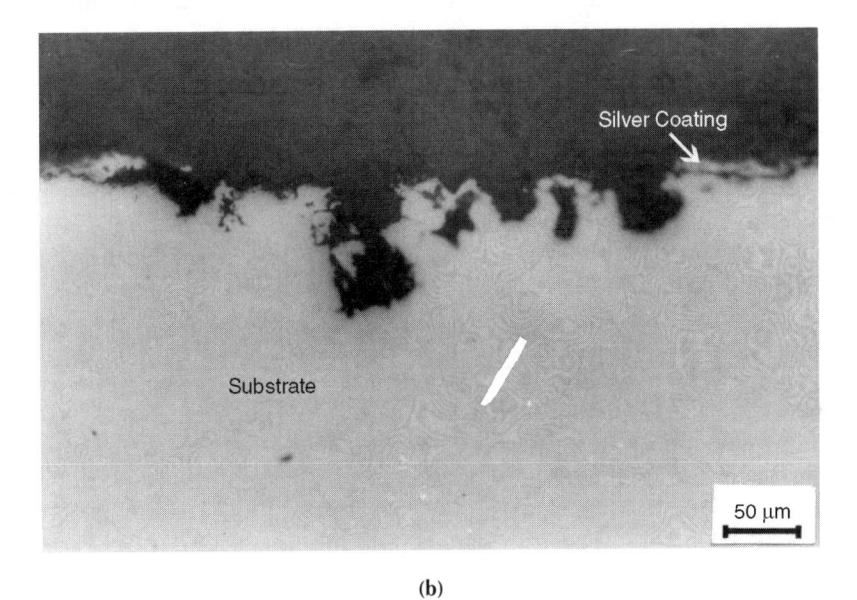

FIGURE 3.14
(a) The construction details of the balancing weight. (b) Pitting corrosion on the component; local absence of Ag coating is evident.

3.5 Wear-Related Failures

Material degradation due to mechanical damage is known as wear. Wear is a progressive phenomenon and failures due to wear are generally encountered in the last quarter of the component life. As the material removal under conditions of wear is uniform and damage accumulation is progressive, wear failures are generally predictable. However, if wear damage triggers other modes of failures, catastrophic failures are encountered.

There are many forms of wear, each being mechanistically different from the rest. The most important forms of wear and the operative mechanisms responsible for material removal are shown in Figure 3.15. Adhesive wear occurs when two moving mechanical elements are in contact under pressure. Material removal in adhesive wear occurs due to welding of surface asperities, deformation and rupture of the welded zones under the influence of contact pressure and the stresses generated by relative motion. In abrasive wear, material removal occurs in moving members under contact pressure due to cutting action of abrasive particles present in the operating environment or generated *in situ*. The presence of characteristic tracks in the direction of motion developed by the ploughing action of the abrasive particles is the evidence of damage due to abrasive wear. Typical surface features observed in adhesive wear and abrasive wear are shown in Figure 3.16.

A relatively less recognized form of wear is erosion damage. Innumerable failures occur due to erosion, particularly in power generation units run with fossil fuels and aeroengines operated in a dusty environment. There are two types of erosion: cavitational erosion and solid particle erosion. Cavitational erosion is caused by implosion of bubbles in flowing fluids on component surfaces causing plastic deformation and material removal. Material removal in solid particle erosion occurs because of repeated impacts of high-velocity particles in gas streams causing formation of craters and lips on the component surface as a result of plastic deformation and strain localization. A typical solid particle erosion damaged surface is shown in Figure 3.17.

Fretting wear causes removal of material from contact surfaces subjected to normal pressure and oscillating motion and mechanisms similar to adhesive or abrasive wear operate. The debris generated due to fretting gets oxidized due to frictional heat, causing extensive damage by abrasive action. As damage due to fretting is generally invisible until strip examination of the system is carried out, failures appear to be catastrophic, even though the damage was progressive. Corrosion wear and oxidation wear are caused by the combined action of chemical attack and any one of the many forms of wear. Galling and spalling result in removal of chunks of material (Figure 3.18). While galling is an intense adhesive wear form, spalling occurs in rolling contact elements under the influence of cyclic loads.

The incidence of wear failures much before the life stipulated by the designer is quite common. The reasons for the unpredicted behavior are (a) overheating caused by frictional heat due to lubricant failure, (b) usage of

Fracture Feature Analysis   41

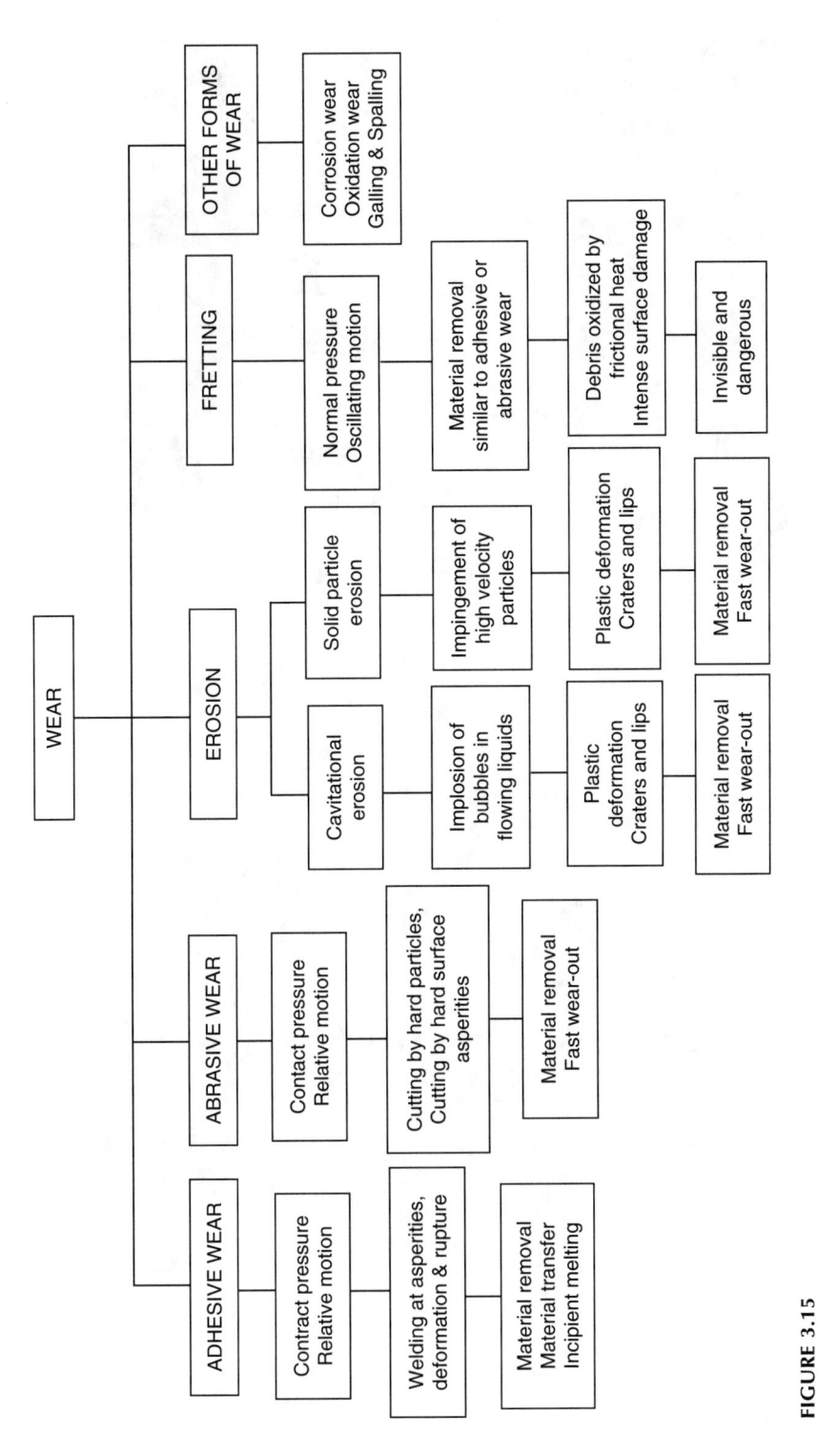

FIGURE 3.15
Various forms of wear of metallic materials.

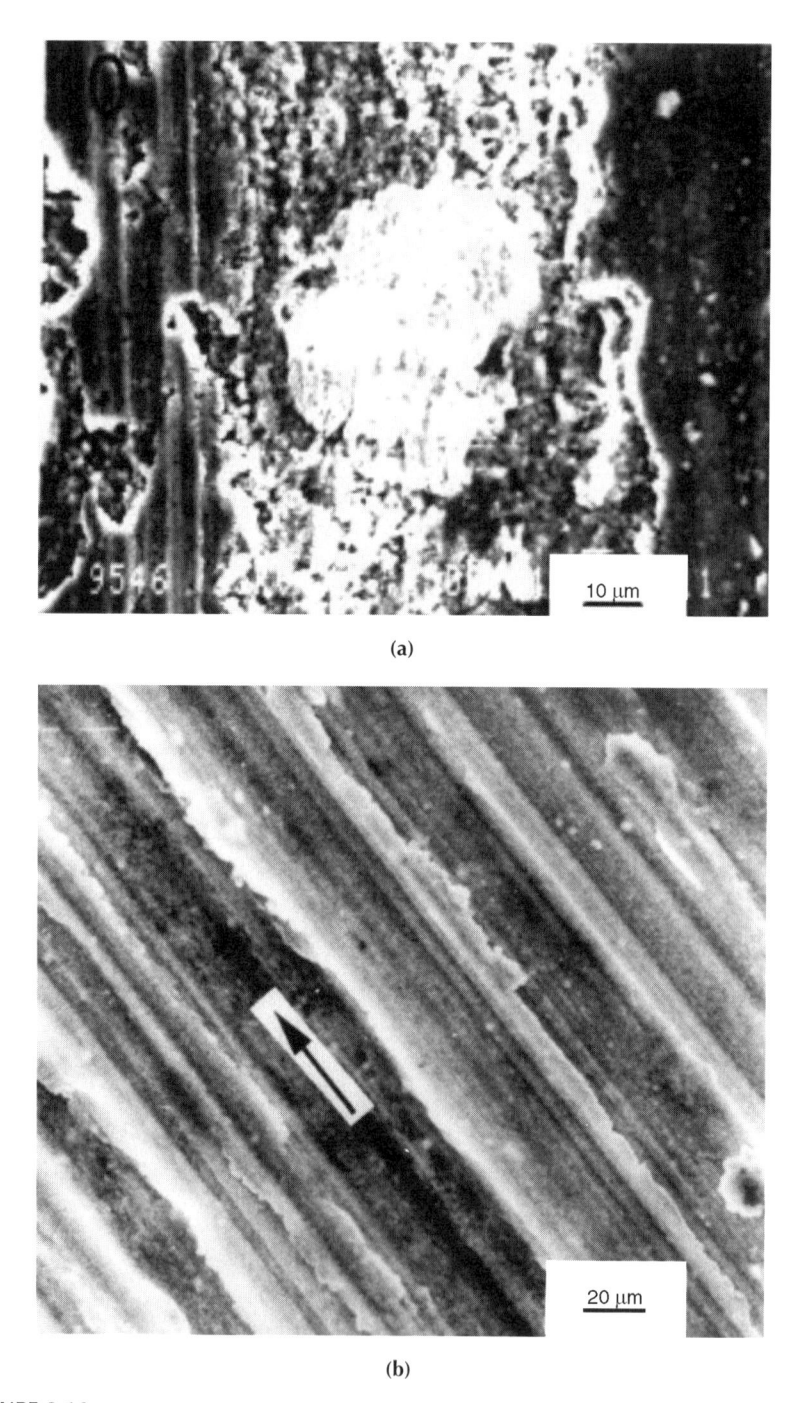

FIGURE 3.16
(a) Adhesive wear observed in an aluminum-steel wear couple. (b) Abrasive wear caused by oxides of aluminum and iron in an aluminum-steel wear couple. (Courtesy of B. Venkatraman, DMRL, Hyderabad, India.)

Fracture Feature Analysis 43

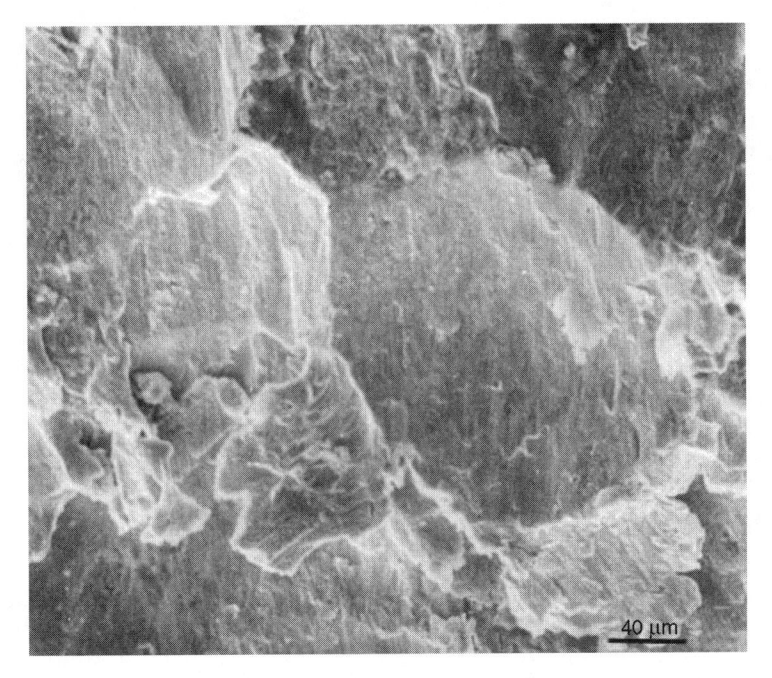

FIGURE 3.17
Solid particle erosion damaged surface showing characteristic craters and lips.

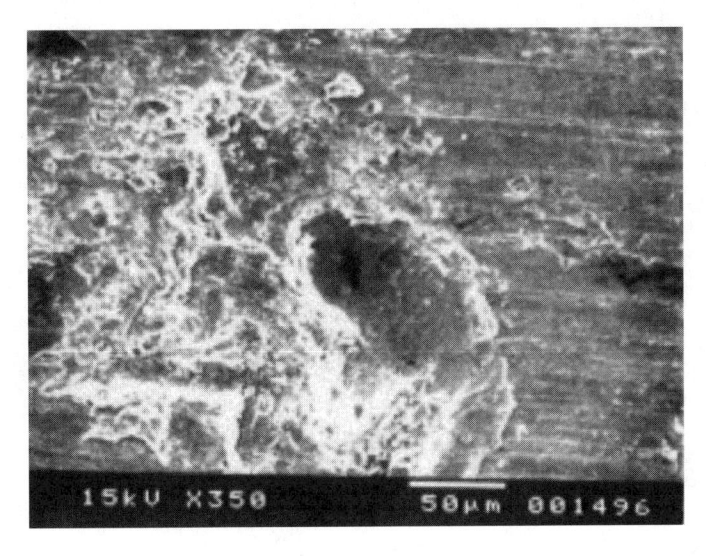

FIGURE 3.18
Fretting wear damage in a Ti-6Al-4V contact couple. (Courtesy of B. Venkatraman, DMRL, Hyderabad, India.)

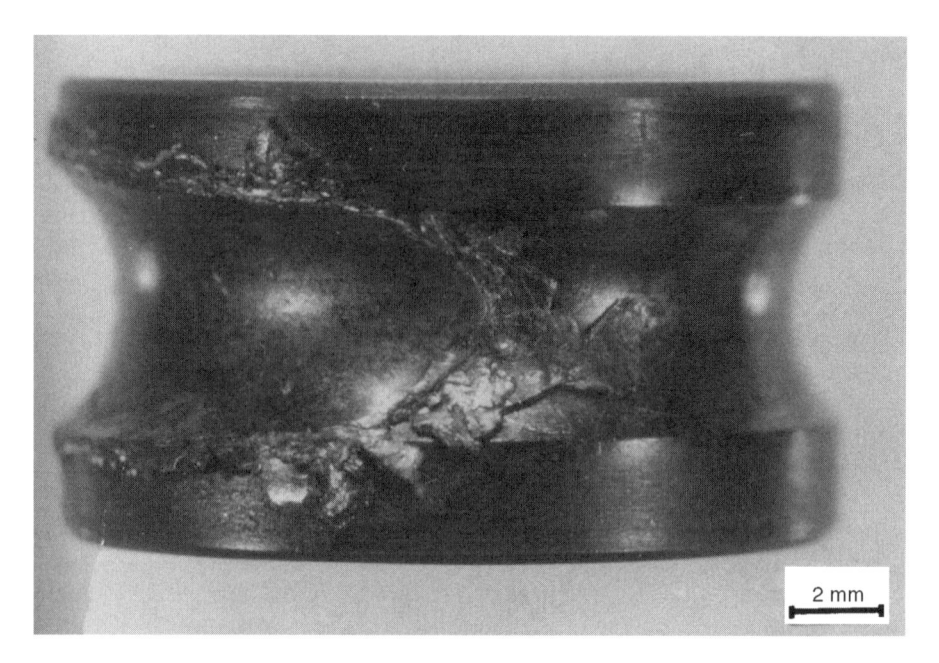

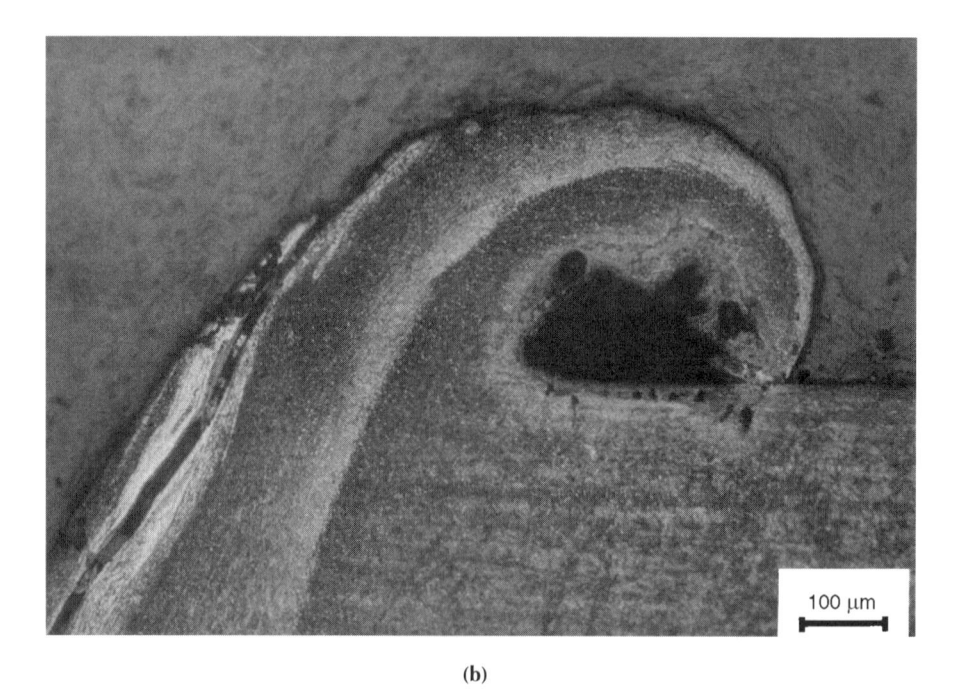

FIGURE 3.19
(a) Damaged inner race of the bearing. (b) Microstructural changes at the damaged region.

Fracture Feature Analysis

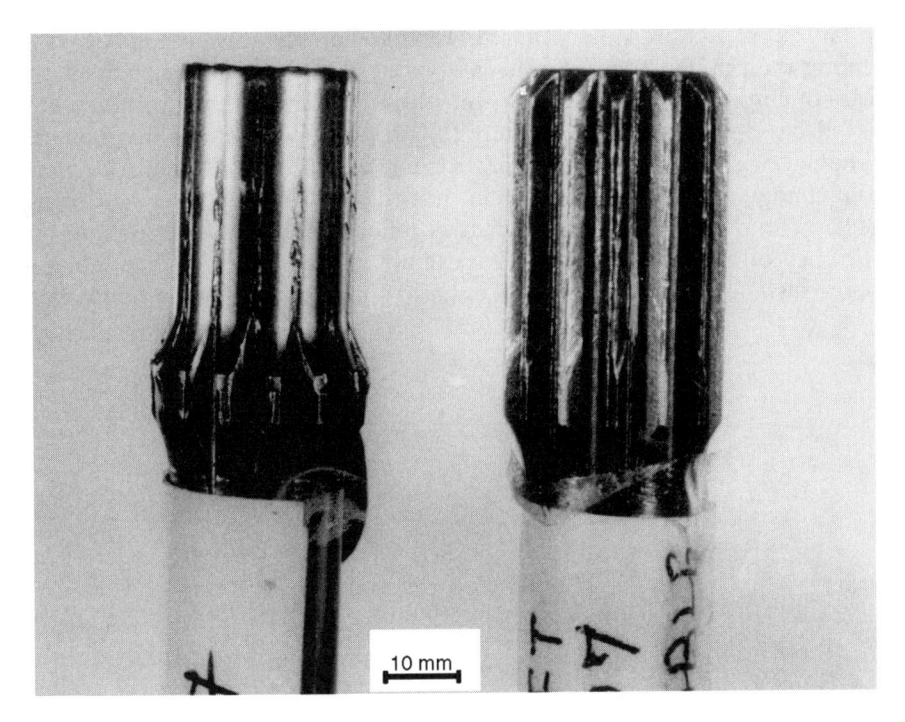

FIGURE 3.20
Worn-out splines of failed quill shaft (left) and condition of splines in a shaft after satisfactory completion of service life (right).

wrong materials and materials that were not processed per standards and (c) initiation and operation of other degradation mechanisms from the primary damage caused by wear. The extent of damage caused by frictional heat can be illustrated with the following example. The malfunctioning of the rotor of a turbo starter was traced to severe damage caused to the bearing just after 40 starts, even though the specified overhaul life was 1500 starts (Figure 3.19a). Metallurgical examination of the stripped bearing revealed severe damage to the inner race and balls, both of which were made from 1C-1Cr AISI 52100 steel. The microstructural changes (Figure 3.19b) suggested that the bearing was overheated due to starvation of lubrication resulting in scooping of the inner race material by the balls.

The failures caused by nonconformance of components to the stipulated standards can be illustrated by the following example. The splines of a quill shaft of the high-pressure fuel pump of an aeroengine wore out completely during trial run itself. The quill shaft was supposed to be made of 0.16C-4.2Ni-1.5Cr steel and the splines were specified to be carburized for imparting wear resistance. Metallurgical investigation revealed that while the quill shaft was made from the specified grade of steel, the splines were not carburized as specified. The condition of the splines of the quill shaft that failed during the trial run and that of splines of the shaft that completed service life are shown in Figure 3.20.

Many methods are now available to engineer surfaces to combat wear. Among the popular methods are (a) overlaying hard coatings such as carbides of chromium, tungsten, etc., nitrides of titanium and chromium, etc., and electrodeposition of chromium; (b) surface modifications involving (i) a change of chemistry such as carburizing, nitriding, boronizing, etc., and (ii) a change of microstructure as in induction hardening, laser hardening, etc.; and (c) overlaying coatings such as Teflon®, etc., to reduce friction. The efficiency of a system in resisting wear not only depends on the inherent wear resistance of the components but also on the efficacy of the lubrication.

4

Deficiencies in Design and Material Selection

4.1 Introduction

The efficiency of an engineering system is dependent on the complex interplay between various parameters starting with conceptual design and extending to product conformity and predefined performance standards. The five doctrines to be followed in the design and development of a functional system to ensure its satisfactory performance are given in Figure 4.1. Deficiency and laxity in any of these critical stages will result in nonconformance of the product to stipulated performance standards.

The origin of most premature failures can be traced back to lapses in either design or manufacture or operational stages of the system. The distribution of failures of an engineering system over its entire design life follows the typical "bathtub" profile (Figure 4.2). The failure rate of a product at the beginning of its life is high and progressively decreases with time and reaches a steady state with increasing utilization. The high initial failure rate is attributable to gross deficiencies in component design and manufacture or bad assembly procedures. Failures in the steady-state failure regimen, which represent the effective life of the product, occur mainly due to nonconformance of the operational and maintenance conditions to the design requirements. Sudden spurts in some of the operational variables such as mechanical loads, temperature, strain rate and loading cycles, etc., are the causes for failures in the steady-state failure regimen. The failed population increases dramatically beyond the useful life due to damages caused by environmental degradation, wear, corrosion, fatigue, etc. By periodic overhaul and replacement of life-expired parts during this regimen, it is possible to extend the life of the system.

Many things can go wrong at all stages of system development and usage, including design, material selection, manufacture, operation, and maintenance. It is prudent for the experts concerned with each of the activities to list the defects that can arise and take effective steps to eliminate them or at least minimize their incidence.

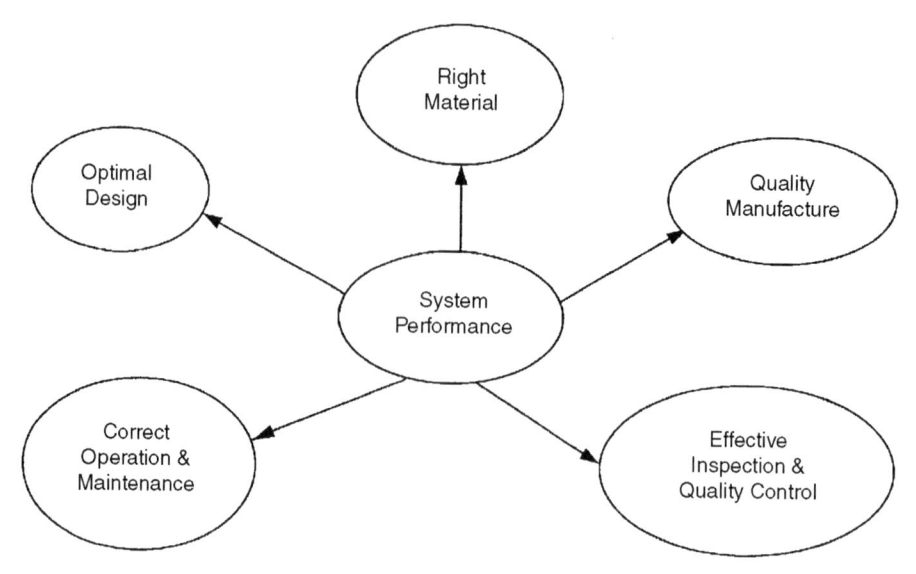

FIGURE 4.1
Critical issues in the design and development of a functional system.

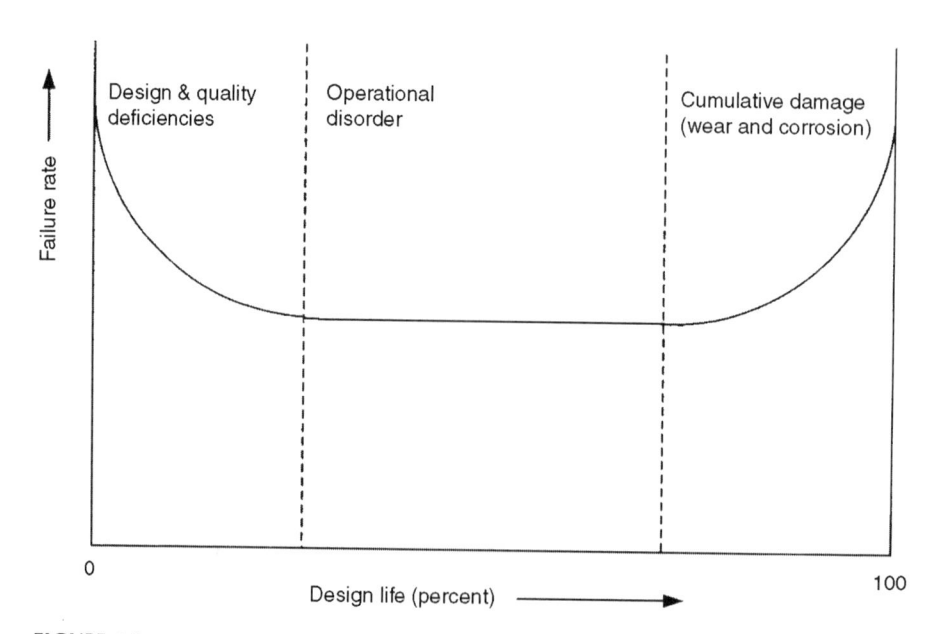

FIGURE 4.2
Distribution of failures in an engineering system over its entire design life.

Deficiencies in Design and Material Selection

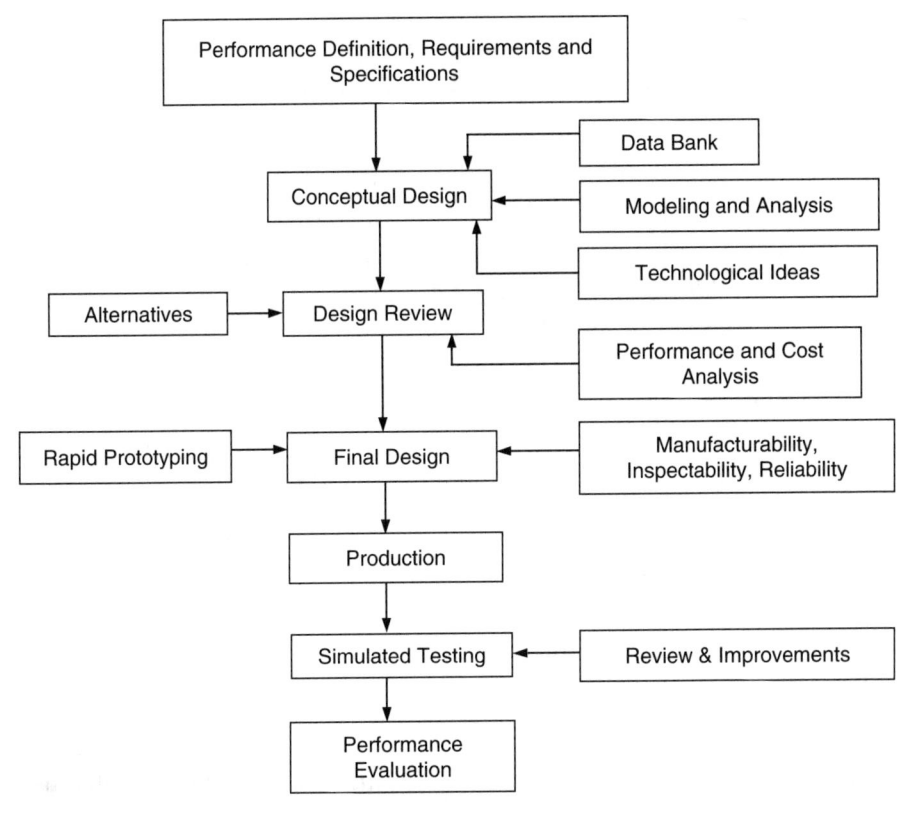

FIGURE 4.3
Stages in designing an engineering system.

4.2 Design Concepts and Concerns

During *ab initio* development of a product, the designer considers aspects such as stress pattern and its distribution, deformation state, temperature and time of loading, material degradation mechanisms, and probable failure modes. Major stages and technological inputs considered at each stage in designing an engineering system are listed in Figure 4.3.

With recent advances in modeling, simulation techniques and rapid prototyping, it is possible to validate the design *a priori* without getting into the nuances of expensive and extensive component testing. However, it is almost impossible for the designer to visualize the myriad defects that are introduced during manufacture and operation of the product that seriously affect the design life. While attention is paid to most of the defects that are likely to arise during production, the incidence of stress raisers and unfavorable residual stresses generally escape attention. A good number of service failures

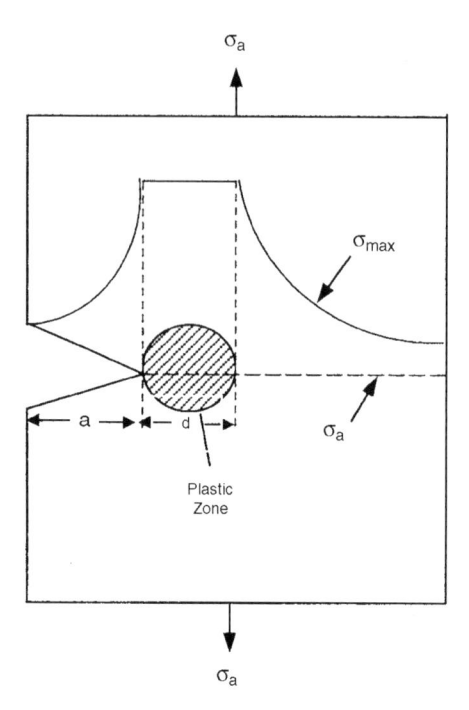

FIGURE 4.4
Stress distribution ahead of a notch.

are attributable to these two factors, despite ensuring conformity to all other design requirements.

4.2.1 Stress Raisers

Local stress, due to a stress raiser, is many times greater than the applied nominal stress, as shown schematically in Figure 4.4. The effect of the stress raiser is usually expressed in terms of stress concentration factor K_t as defined below:

$$K_t = \frac{\sigma_{max}}{\sigma_{applied}}$$

If it is assumed that the material exhibits linear elastic deformation behavior, then

$$K_t = 2\sqrt{\frac{a}{r}} \qquad (1)$$

where a = notch depth and r = notch tip radius.

Deficiencies in Design and Material Selection

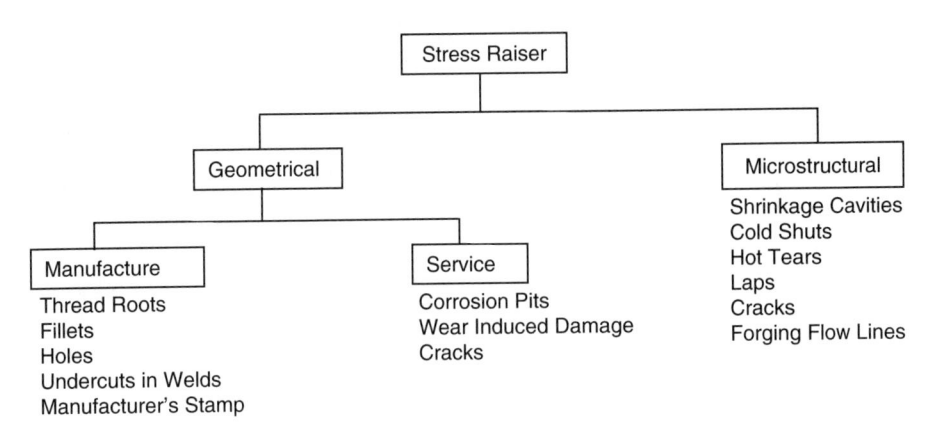

```
                          ┌──────────────┐
                          │ Stress Raiser │
                          └──────────────┘
                    ┌────────────┴────────────┐
              ┌───────────┐            ┌────────────────┐
              │ Geometrical│            │ Microstructural│
              └───────────┘            └────────────────┘
           ┌────────┴────────┐          Shrinkage Cavities
    ┌───────────┐       ┌─────────┐     Cold Shuts
    │ Manufacture│       │ Service │     Hot Tears
    └───────────┘       └─────────┘     Laps
     Thread Roots        Corrosion Pits  Cracks
     Fillets             Wear Induced Damage  Forging Flow Lines
     Holes               Cracks
     Undercuts in Welds
     Manufacturer's Stamp
```

FIGURE 4.5
Commonly observed stress raisers and their origin.

The stress concentration factor can easily be derived from elastic theory in the case of brittle materials. In very complicated cases, the factor K_t can be computed theoretically by finite element analysis or by direct measurement using photoelastic techniques. Many structural failures, particularly in impact and fatigue modes, occur due to the presence of stress raisers. In the case of ductile materials that exhibit elastic-plastic deformation behavior, if the stress increment due to the presence of the stress raiser exceeds the yield stress, the area ahead of the stress raiser undergoes plastic deformation, resulting in reduction of the peak stress. Redistribution of stress due to plastic deformation causes reduction in the stress concentration factor. The stress concentration factor increases with increasing strength and decreasing ductility in ductile material for a given geometry of the component and stress raiser.

The commonly observed stress raisers and their origin are shown in Figure 4.5. From Equation 1, it is clear that the stress concentration factor increases with increasing notch depth and decreasing notch tip radius. Circular holes, too, act as stress raisers and the stress concentration factor increases with decreasing hole diameter. If the hole is tiny or the notch root radius is very small, the stress concentration factor is about 3, which means that the maximum stress experienced is three times the nominal stress. Depending on the geometry of the component and the stress raiser, the stress concentration factor generally lies in the range of 1.2 to 3.0. Typical variation of stress concentration factors for round bars with shoulder fillets and section size changes as a function of fillet radius are shown in Figure 4.6. It is obvious that the stress concentration factor increases with decreasing fillet radius and increases with increasing section size difference. The maximum stress always occurs in the smaller part near the fillet. Similar graphs are available in various engineering handbooks for different component geometries and stress raiser configurations that can be used during design and failure analysis.

Stress raisers play a very significant role in initiating fatigue failures. The stress concentration factor for fatigue loading K_f is given as

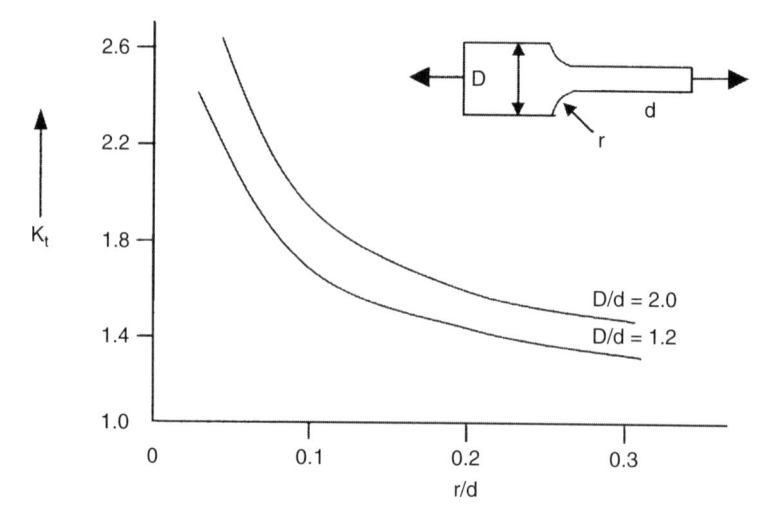

FIGURE 4.6
Variation of stress concentration factor for a round bar [1].

$$K_f = \frac{\text{Effective fatigue stress}}{\text{Nominal fatigue stress}}$$

The stress concentration factor under cyclic loading conditions is generally lower than that under static or impact loading conditions. By following the basic principles of best manufacturing practices and total quality management, it is possible to eliminate the incidence or minimize the effect of stress raisers. There are many practical guidelines to reduce the intensity of stress concentrations. Providing generous fillets at re-entrant corners, specifying better surface finish at regions of high stress, increasing the section thickness around holes, flushing the machined surfaces with liberal fillets and proper finishing of as-cast or forged surfaces of forgings and castings are some of the examples. Many structural failures can be avoided if adequate attention is paid to the existence of stress raisers and measures are taken to minimize their influence during the design stage.

The role of stress raisers in promoting premature failures can be best explained with the following examples. A fourth-stage compressor disc made from Ti-6Al-2.5Mo-1.5Cr alloy developed a circumferential crack in the hub region after serving for just 160 h, compared to the expected life of 1500 h. Failure analysis on the disc indicated that the crack had multiple origins, all of which initiated and propagated along the machine marks of about 10 µm deep. Since titanium alloys are notch-sensitive, the machine marks acted as stress raisers and promoted initiation of a fatigue crack under alternating stresses whose magnitude was lower than the design values. The disc, the crack at the machining grooves on the hub surface and the fracture surface are shown in Figure 4.7.

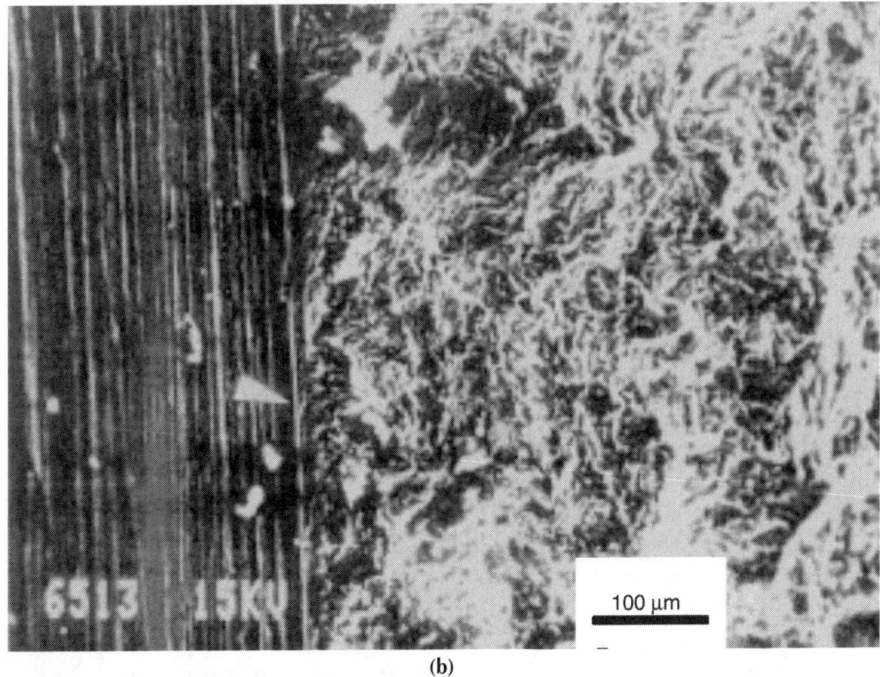

Deficiencies in Design and Material Selection

FIGURE 4.7
(a) Compressor disc. (b) Fractograph showing fracture origination from machine marks. (c) Crack along a deep machine mark. (Courtesy of A.M. Srirama Murthy, DMRL, Hyderabad, India.)

(c)

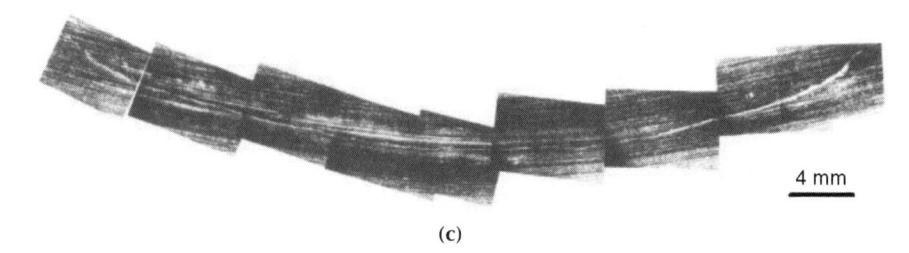

FIGURE 4.7 (continued)

In another case, a large number of drill rods, used by the mining industry, failed within a few minutes of operation (Figure 4.8). The drill rods were made from 1.0C-1.0Cr- 0.25Mo steel and were in the normalized condition with lamellar pearlitic structure instead of tempered martensitic structure. All the failures were at the shank–collar fillet. The sharp fillet, which acted as a stress raiser, aided by a microstructure with poor notch toughness (lamellar pearlite exhibits lower notch toughness than tempered martensite) was responsible for failure.

In some cases, advantage is taken of the notch sensitivity of a material and the stress concentration effect of notches in the design of components to promote functional failure along the notches provided. An example is the development of a cluster bomb. A cluster bomb is expected to fragment into a large number of pieces on detonation of the explosive inside the bomb. A large number of notches are intentionally provided in the bomblet body to promote fragmentation (Figure 4.9a). A bomblet body, investment cast to shape using 0.38C-1.5Ni-1.5Cr-0.25Mo steel in normalized condition to obtain lamellar pearlitic structure, failed to generate an adequate number of fragments. The erratic behavior was found to be due to improper notch geometry; the depth was low and the radius was high (Figure 4.9b), both of which reduced the stress concentration effects, contrary to the requirements.

4.2.2 Residual Stresses

Designers generally ignore the role of residual stresses in inducing premature failures. Residual stresses can originate at various stages of metallurgical process or can be induced by mechanical operations. The sources of residual stress generation during a component manufacture are depicted in Figure 4.10.

Depending on the stress state and level, the residual stresses can have beneficial or detrimental effects on the component's performance. The effective stress (σ_e) experienced by the component is related to the applied stress (σ_a) and the nature of residual stress. If the residual stress is compressive (σ_c) and the applied stress is tensile, then the effective stress σ_e is given by

$$\sigma_e = \sigma_a - \sigma_c$$

(a)

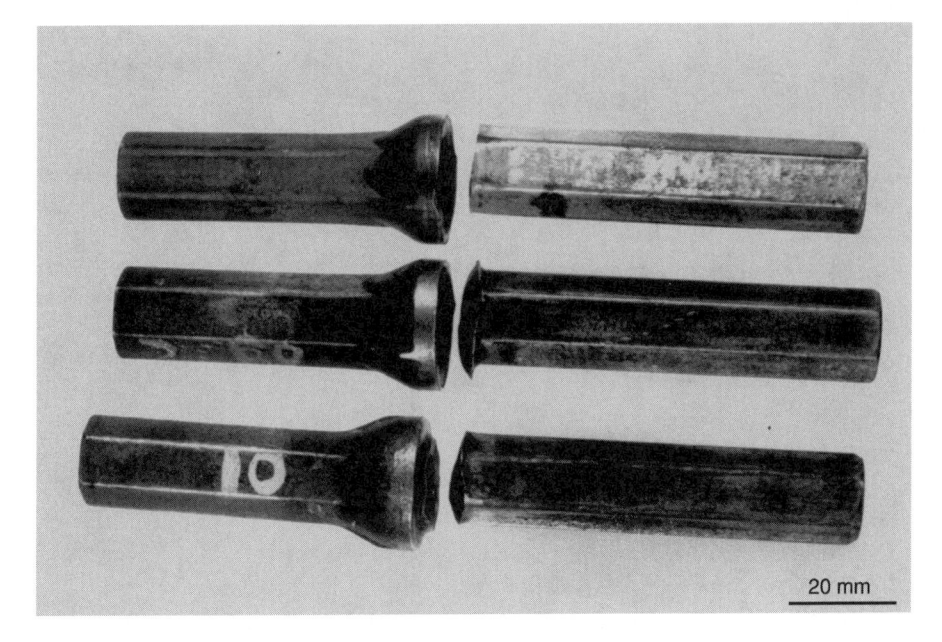

(b)

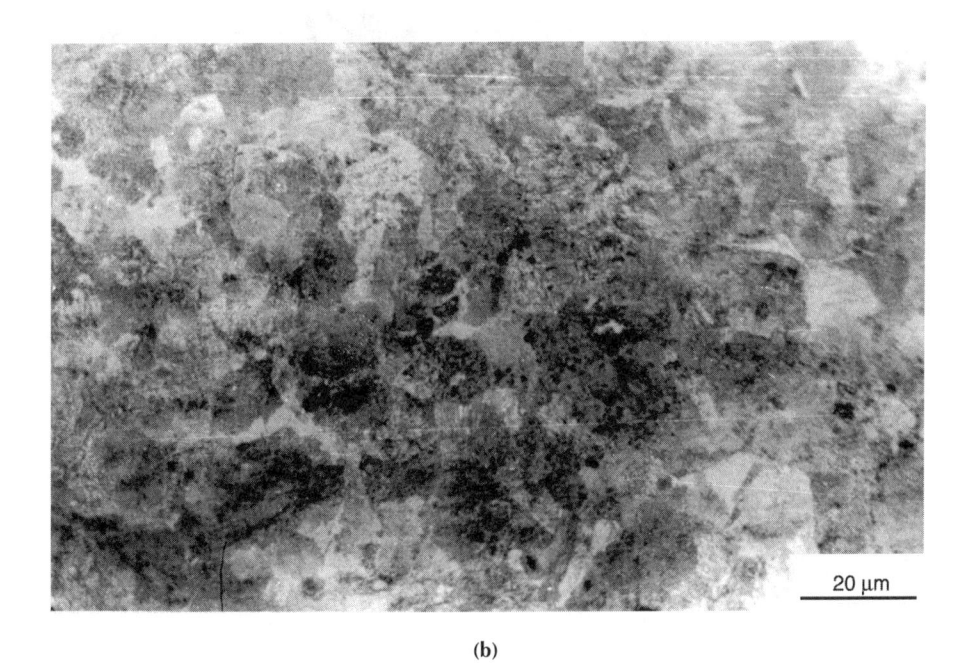

FIGURE 4.8
(a) Drill rods failed at shank-collar fillet. (b) Microstructure of the failed drill rod consisting of lamellar pearlite.

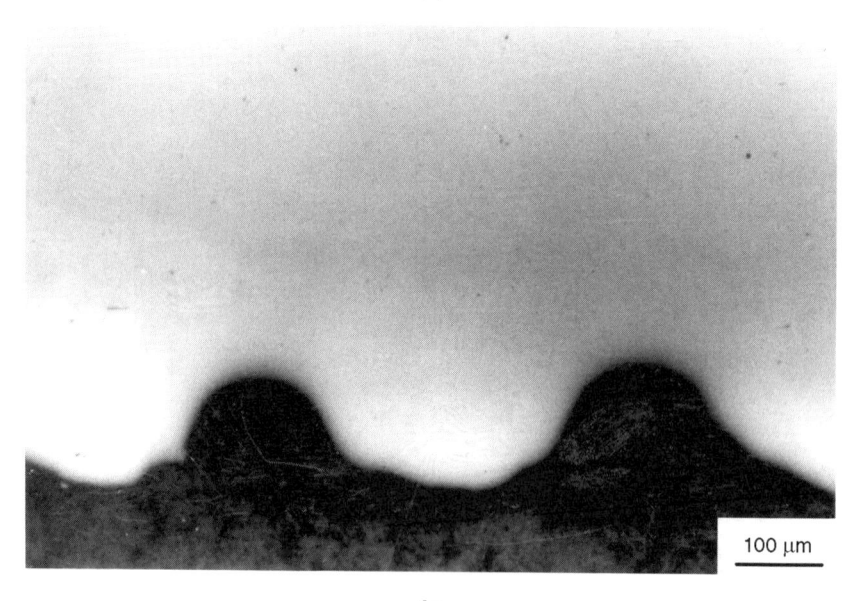

FIGURE 4.9
(a) Notched bomblet body. (b) Shallow and rounded notches on the ineffective bomblet.

Similarly, if the residual stress is tensile (σ_t) and the applied stress is also tensile, then the effective stress is calculated as

$$\sigma_e = \sigma_a + \sigma_t$$

Thus, when the component is subjected to tensile stress, the effective stress experienced by the component will be (a) lower than the applied stress if

```
                    ┌─────────────────────────┐
                    │ Sources of Residual Stress │
                    └─────────────────────────┘
                       │                    │
         ┌─────────────┘                    └──────────────┐
         ▼                                                  ▼
  ┌──────────────┐                                  ┌──────────────┐
  │ Metallurgical │                                  │  Mechanical  │
  └──────────────┘                                  └──────────────┘
```
Metal forming operations   Machining and grinding
Casting   Low plasticity burnishing
Heat treatment   Shot peening
Surface hardening   Laser shock peening
Surface coating   Interference fitments
Welding   Autofrettage

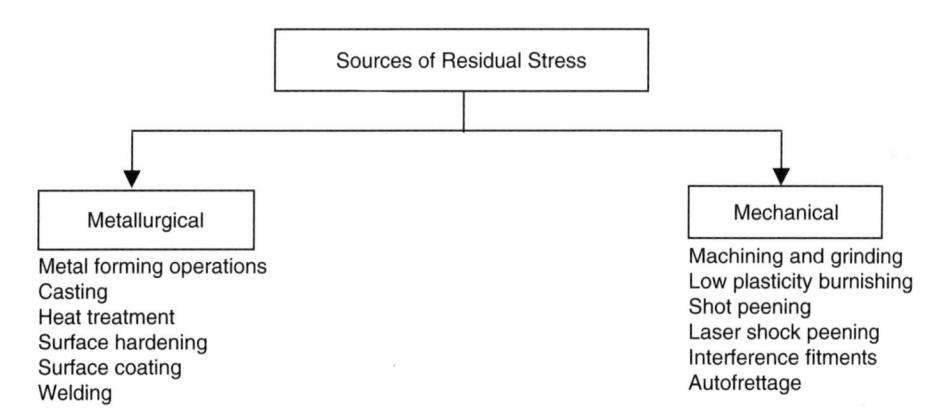

FIGURE 4.10
Various mechanical and metallurgical processes that induce residual stresses.

the residual stress is compressive and (b) higher than the applied stress if the residual stress state is tensile. The additive tensile residual stress element can cause yielding or promote crack initiation, thereby reducing the life of the component. Failure modes with cracks originating at free surfaces, such as fatigue and stress corrosion cracking, are favored by tensile residual stresses.

The nature of induced residual stresses is dependent on the manufacturing process. Some of the processes, such as heat treatment, surface hardening and surface coatings, generate residual stresses that can be either compressive or tensile, depending on how effectively the process parameters are controlled. For example, if boronizing of a steel component results in the formation of an FeB layer, the residual stresses are tensile; if the coating is of the Fe_2B type, the residual stresses are compressive. To derive the benefits of boride coating, it is necessary to control the activity of boron in the boronizing mixture to promote formation of only an Fe_2B, and not an FeB, layer.

Heat treatment is one important step wherein significant levels of residual stresses are induced and lead to severe distortion and cracking. These stresses arise due to (a) the thermal expansion/contraction in parts with varying cross sections during heating and cooling and (b) dilation or contraction as a result of phase transformations that occur during the cooling cycle. The level of residual stresses increases if the volume of the transformed phase is larger than that of the parent phase. Typical residual stress distribution in deep hardening case carburizing steel bar in the heat-treated condition before and after case carburization is schematically shown in Figure 4.11.

Welding also induces residual tensile stresses due to solidification and thermal contraction of the weld bead, leading to formation of cold cracks. The problem can be overcome by preheating the base material in the case of steels. By judicious choice of the filler material, welding parameters and pre- and postweld thermal treatments, the problems can be overcome to a great extent.

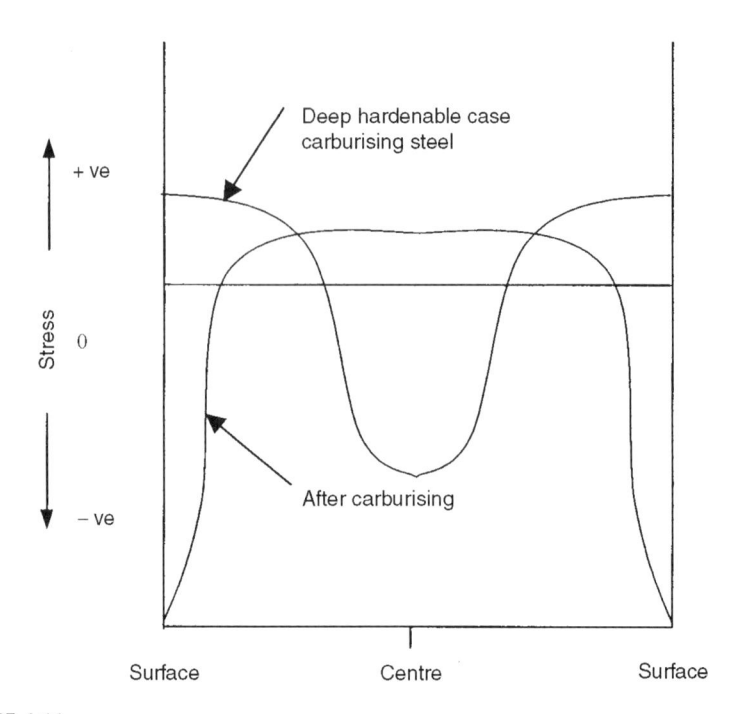

FIGURE 4.11
Residual stress distribution in a deep hardening case carburizing steel bar before and after case carburization.

Most of the surface treatments, if carried out correctly, result in residual compressive stresses in the component surfaces. Quite a few methods were developed to exploit cold work-induced compressive stresses. Various types of surface plastic deformation methods such as shot peening, laser shock peening and low plasticity burnishing are a few examples. The residual stress distribution as a function of depth below the surface in these methods is schematically shown in Figure 4.12. The process variables in each of these processes should be controlled so that the magnitude of compressive stress is as high as possible and that the induced compressive stress extends to as great a depth as possible. If the depth is too low, it is likely to be removed during final stages of component finishing and there is always the danger of exposing the layers with residual tensile stresses. Care must therefore be taken to avoid excessive material removal during any finishing operation to retain the surface layer with compressive residual stresses to ensure better life.

Autofrettage and interference fitment are two other popular methods by which residual compressive stresses are induced on the desired component surfaces. In both these methods, a small fraction of the component volume is plastically deformed, while the rest of the volume undergoes elastic deformation. On release of the deformation load, the elastically deformed region relaxes and imposes compressive stresses on the plastically deformed region.

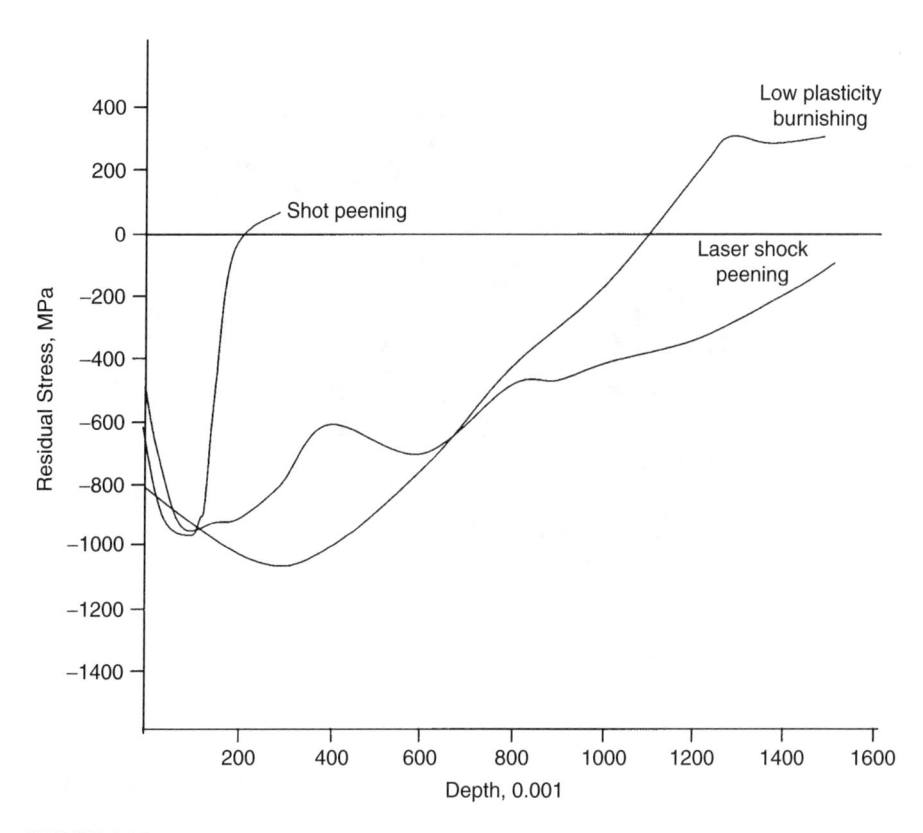

FIGURE 4.12
Residual stress distribution in some surface plastic deformation methods [2].

The life of gun barrels is significantly improved by the compressive stresses induced at the bore surface due to autofrettage. There are numerous examples of shrunk-fitted assemblies in the engineering industry.

A cold rolling mill roll made of 0.8C-2.9Cr-0.25Mo steel in quenched and tempered condition with an induction-hardened surface failed by subsurface initiated spalling in less than one tenth of its specified life. The failure was attributed to the formation of alternate bands with varying microstructure (Figure 4.13) due to improper induction hardening, the presence of over 20% retained austenite in the induction-hardened layer and high residual tensile stresses (450 MPa on the roll at unspalled location).

The measurement of residual stresses, contrary to popular belief, is fairly simple and accurate. Among the many methods reported, the nondestructive X-ray diffraction (XRD) technique is most popular. Easy-to-operate portable XRD machines are now available. Hole drilling and measuring the strain redistribution is the destructive technique used quite extensively for the estimation of residual stresses.

It is essential to eliminate these residual stresses to improve component performance. There are many methods available to relieve residual stresses

FIGURE 4.13
Failed of cold rolling mill roll indicating alternate bands.

and the choice of the correct method is dependent on the source of the residual stresses. Stress relief treatment, involving slow heating and cooling at relatively lower temperatures where no phase transformation occurs, is an extensively used method. Vibration-induced stress relief is another method that is gaining popularity even though the scientific principles are not clearly understood.

4.3 Material Selection

The benefits of a good design are easily lost if the right material is not chosen to perform the specified functions. It is generally possible to identify more than one material satisfying the broad functional requirements. Final selection is done by considering each virtue of the material based on its relative importance in determining the performance of the component. While selecting a material, it is important to consider the following aspects:

1. Functional capability to withstand the imposed service loads and operating environment
2. Ease of manufacture to produce the components in the desired shape without introducing undesirable defects

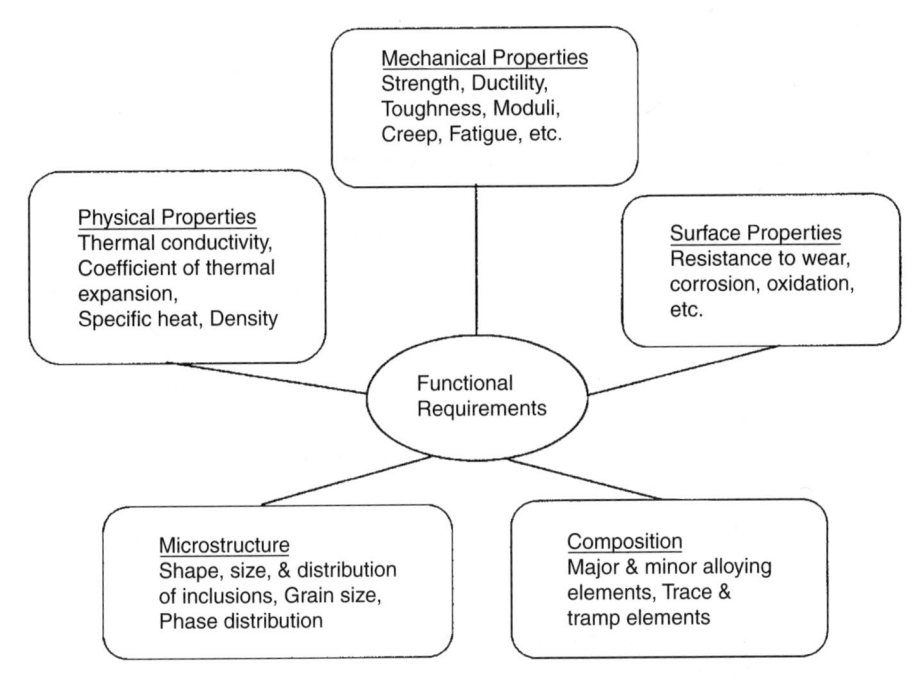

Deficiencies in Design and Material Selection

FIGURE 4.14
Functional properties of materials to be considered during material selection.

3. Ease of production in large numbers
4. Ensured inspectability to guarantee freedom from defects larger than those stipulated by the designer.
5. Readily available and cost-effective raw materials and manufacturing facilities.

To meet the various functional requirements, the designer has to carefully consider the ability of the candidate material to withstand the rigors of service conditions up to a predetermined life. The important properties that are to be considered are shown in Figure 4.14.

Even in modern times, designers consider high strength to be an all-important requirement, relegating all other material properties to insignificance. Many designs are still based on properties derived from tensile strength, based on certain empirical relationships, and designers do not realize their inadequacy. Most of them do not take into account the dependence of the derived properties on the interplay between composition, microstructure, properties and fracture behavior. This tendency is more prevalent in developing countries due to the lack of adequate design experience and the exorbitant cost of testing for the generation of necessary data.

Due consideration is to be given to the following facts while selecting metallic materials:

- Yield strength of most metallic materials depends on the temperature and rate of loading.
- Contrary to the assumption usually made, the yield strength in tension and compression need not be the same.
- High-cycle fatigue strength derived from tensile strength using empirical relationships is unreliable.
- Materials with high ductility exhibit better low-cycle fatigue resistance than those with high strength.
- Thermal properties play a significant role in dictating the life of components subjected to thermal cycling.
- Uncertainties exist in predicting long-term creep behavior by extrapolation of short-term creep data based on empirical relationships.
- The fracture behavior depends on the microstructure of the material, which in turn depends on the composition and thermomechanical processing.
- The effect of inclusions varies depending on their morphology and chemistry. While nondeformable alumina type inclusions drastically reduce the fatigue life in steel roller bearing elements, enveloping of these with deformable sulphides increases the fatigue life.
- Surface damages caused by wear, corrosion, oxidation, carburization, decarburization, etc., increase the probability of crack initiation at free surfaces.
- Under unfavorable service conditions, a ductile material can become brittle due to ductile-brittle transition, strain aging, temper embrittlement, hydrogen embrittlement, etc.

The danger of using strength as a primary material selection parameter can be illustrated by the following example. An anchor bolt with a welded base plate and stiffeners was to be used as a foundation bolt of a heavy-duty rolling mill. According to the requirements, the steel should have a minimum tensile strength of 600 MPa. Keeping in view the weldability requirements, a weldable, low-carbon, high-strength structural steel should have been selected. Instead, the contractor selected a 0.5% C steel whose weldability is poor, to meet the strength requirements, and welded the stiffener attachments, resulting in the formation of brittle untempered martensite in the HAZ with a hardness of 700 HV. The bolts broke, resulting from a crack initiating at the HAZ, while they were being unloaded from the truck, in a brittle crystalline mode. Part of the failed bolt and the HAZ microstructure are shown in Figure 4.15.

Deficiencies in Design and Material Selection

40 mm

(a)

FIGURE 4.15
(a) Failed anchor bolt. (b) Fracture surface. (c) Untempered martensite in the HAZ.

It is not uncommon to encounter failures because of nonconformity of the composition of the material used to the stipulated standard. A typical example is the failure of a 30-mm high-explosive shell during assembly itself. The details of the steel specified and that actually used for the production of the 30-mm high-explosive shell are given in Table 4.1.

(b)

(c)

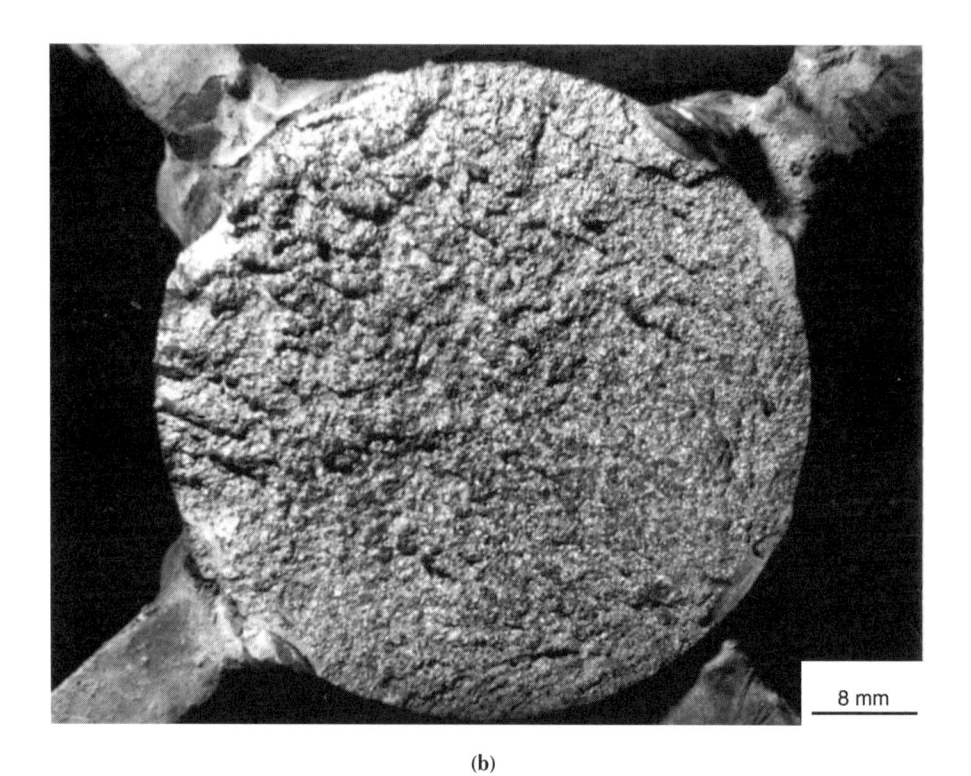

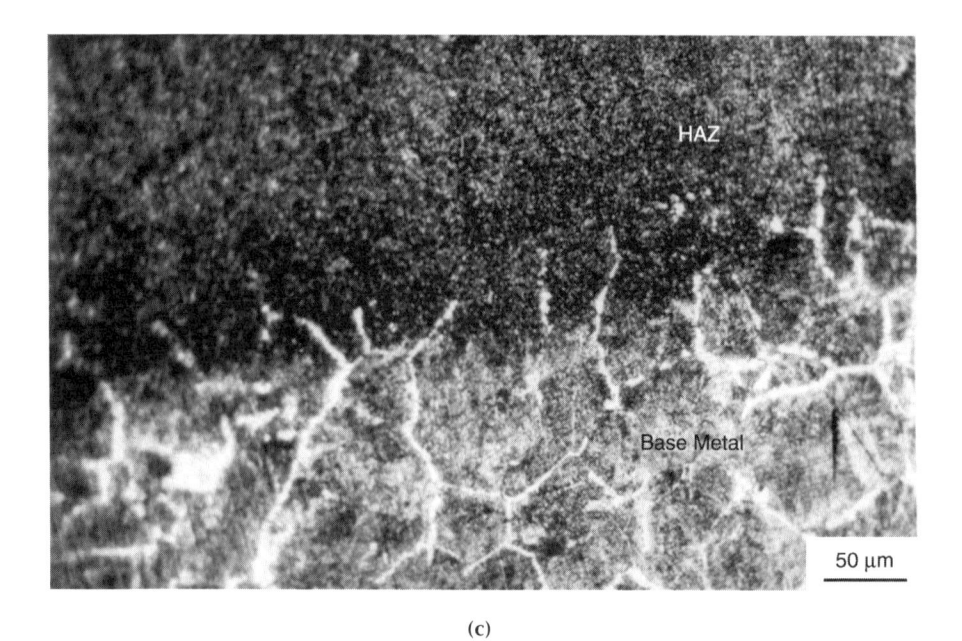

FIGURE 4.15 (continued)

TABLE 4.1
Details of the material specified and actually used for 30-mm, high-explosive shells

Material	Composition					Heat treatment	Hardness (HV)
	C	Si	Mn	Cr	Mo		
Specified	0.32–0.39	1.1–1.4	0.8–1.1	1.1–1.4	—	Heat to 840°C, quench into salt bath at 210°C, hold for 3 h, air-cool	480
Used	0.50	0.17	0.85	0.91	0.2	As above	605

Because of the differences in the composition of the two steels, the response to the specified heat treatment was different. Instead of the required tough lower bainitic structure, the steel developed a mixture of bainite and martensite with higher strength and lower toughness, resulting in failure of a number of shells in a brittle mode during insertion of the explosive pellets at 20 bar pressure (Figure 4.16).

Another case of failure due to usage of improper steel is the failure of starter motors used in tractors. Starter motor drives, specified to be made from case carburizing steel, were made of a medium carbon steel but were carburized and heat-treated as specified for the original steel. The details of composition and hardness are given in Table 4.2.

During case carburizing, the carbon content of the steel, which was already high in the steel used, further increased to such levels that about 10% austenite was retained in the case, as a result of lowering of M_s and M_f temperatures well below those expected of the specified steel after carburizing. The retained austenite promoted development of longitudinal cracks during grinding, which in turn caused the final circumferential failure (Figure 4.17).

(a)

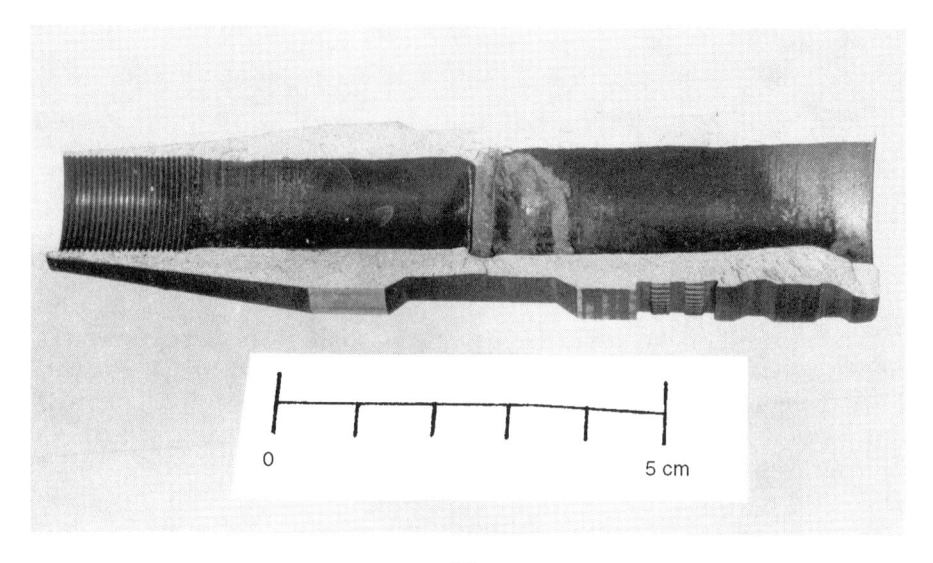

(b)

FIGURE 4.16
(a) High-explosive shell failed in brittle mode. (b) Microstructure showing a mixture of bainite and martensite in the shell.

TABLE 4.2
Composition and hardness of steel used for starter motor drives

Material	Composition						Heat treatment	Hardness (HV)	
	C	Si	Mn	Ni	Cr	Mo		Core	Case
Specified	0.18–0.23	0.20–0.35	0.7–0.9	0.4–0.7	0.40–0.60	0.15–0.25	Carburize at 850°C, oil-quench, temper at 180°C, 1 h	415	700
Used	0.49	0.51	0.85	1.48	1.07	0.26		750	570

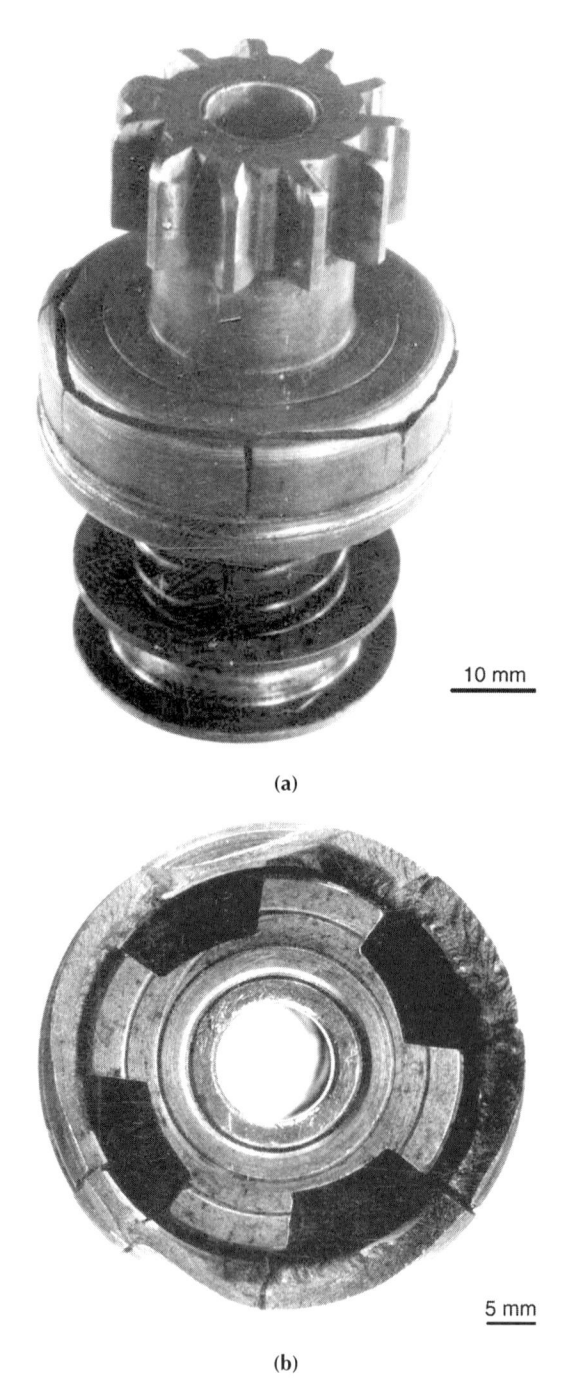

FIGURE 4.17
(a) Starter motor drive. (b) Fracture surface.

The above examples illustrate the dangerous consequences of nonconformity of materials used to the specified material standards. If alternate materials are to be used for reasons of nonavailability and economic viability, it is essential to assess the processibility and heat treatment of substitute material and define process procedures to ensure trouble-free production and service performance.

5
Manufacturing Defects

5.1 Introduction

Despite perfect design and correct material selection, many components fail prematurely. A great majority of these failures are attributed to the presence of defects generated during manufacturing. Most of the defects essentially develop because of the abuse of science during manufacturing. Failures are triggered when the manufacturing defects are either greater in number or larger than those considered during design.

There are a large number of process stages in the production of components ranging from melting of the alloy to finish machining. The major manufacturing stages and the defects generated at each stage are shown pictorially in Figure 5.1. Each process stage is a potential defect creator.

5.2 Melting and Teeming Defects

The primary step in the manufacture of a component is alloy preparation and production of sound ingot. There are many complex alloys for which correct compositional balance has to be maintained to avoid formation of deleterious phases during solidification and their retention during subsequent hot working. It is important to recognize that even though the composition of the alloy is well within the specification limits, an improper balance of critical elements can still be a cause for concern.

The significance of composition balance can be best described taking the example of 17Cr-2Ni martensitic stainless steel. The influence of ferrite stabilizing elements is included in the chromium equivalent (Cr_{eq} = Cr + 2Si + 1.5Mo + 5V + 5.5Al + 1.75 (Nb + Ti) + 0.75W (in wt%)) and that of the austenite stabilizing elements is considered in the nickel equivalent (Ni_{eq} = Ni + Co + 0.5Mn + 0.3Cu + 25N + 30C (in wt%)). If the Cr_{eq} is higher than the optimum, stringers of δ ferrite appear in the product, affecting the transverse ductility and corrosion resistance. If the Ni_{eq} is higher than the desired

```
┌─────────────────────────┐      ┌─────────────────────────┐
│ Finishing Operations    │      │ Melting & Teeming       │
│ Defects due to cleaning,│      │ Composition deviations  │
│ Machining, Coating      │      │ Shrinkage defects       │
│                         │      │ Segregation             │
└─────────────────────────┘      └─────────────────────────┘
```

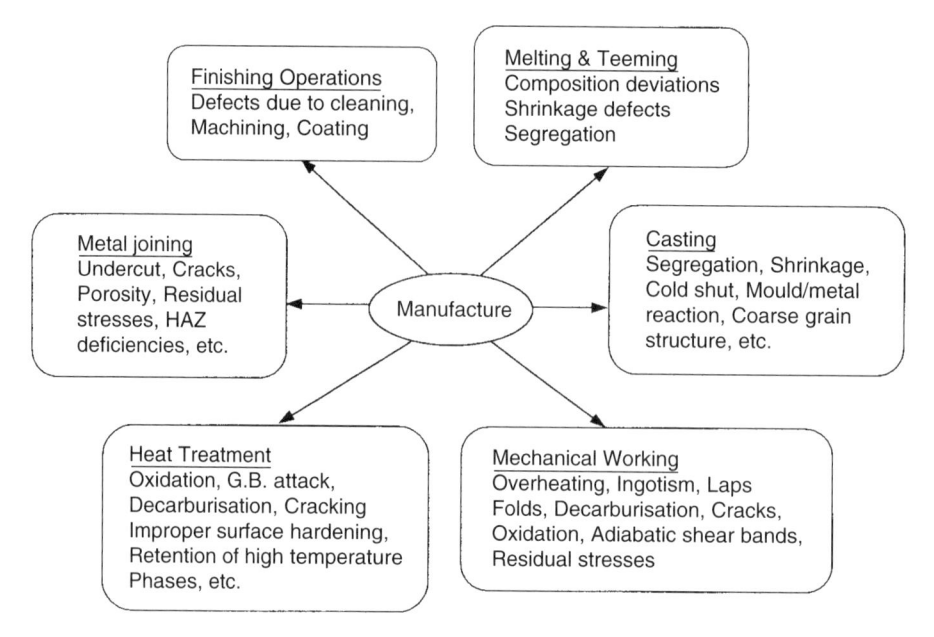

Diagram: Central node "Manufacture" with arrows to/from six surrounding categories:

- **Finishing Operations**: Defects due to cleaning, Machining, Coating
- **Melting & Teeming**: Composition deviations, Shrinkage defects, Segregation
- **Metal joining**: Undercut, Cracks, Porosity, Residual stresses, HAZ deficiencies, etc.
- **Casting**: Segregation, Shrinkage, Cold shut, Mould/metal reaction, Coarse grain structure, etc.
- **Heat Treatment**: Oxidation, G.B. attack, Decarburisation, Cracking, Improper surface hardening, Retention of high temperature Phases, etc.
- **Mechanical Working**: Overheating, Ingotism, Laps, Folds, Decarburisation, Cracks, Oxidation, Adiabatic shear bands, Residual stresses

FIGURE 5.1
Defects generated during manufacture.

level, the austenite to martensite transformation does not proceed to completion during hardening, leading to retention of up to 40% volume fraction of austenite. Such high amounts of retained austenite not only result in lower yield strength, but also in warpage, cracking and dimensional changes upon transformation to martensite.

The design and control of composition is extremely important in titanium alloys. The role of each alloying element in the stability of α and β phases is of paramount importance in maintaining the correct composition balance to achieve the desired volume fractions of the phases. Improper control of the alloying elements results in the formation of embrittling α_2 and ω phases. The interstitial elements O, N, C and H play a significant role and need to be precisely controlled along with major alloying elements to ensure desired microstructure and mechanical properties. The modern practice is to reduce the concentration of interstitial elements to as low a level as possible by judicious selection of raw materials and employing triple vacuum melting. The importance of interstitial elements in titanium can be gauged by the fact that reducing the oxygen level by one third in commercially pure (CP) titanium doubles the fracture toughness in a corrosive environment.

Similarly, the composition control is recognized as a critical quality parameter in many superalloys and its importance can be illustrated with the example of CM247 LC (Ni-8.0Cr-0.5Mo-3.2Ta-0.8Ti-9.5W-5.5Al-9.2Co-1.5Hf). The tendency for the formation of topologically close-packed phases (TCP) such as σ, μ and laves is dependent on the composition of the alloy. These phases reduce resistance to fracture, and their formation tendency is estimated by computerized calculations. A convenient and routinely employed method is to calculate the average electron hole number (N_V) for the alloy, considering the atom percentage and individual electron hole number for each of the elements present in the alloy. The alloy composition is balanced to ensure that the N_V is less than 2.15, thus ensuring freedom from the undesirable phases.

The control of trace and tramp elements is as important as the compositional balance. The trace elements segregate preferentially to certain microstructural features and embrittle the material. A classic example is the temper embrittlement of quenched and tempered steels due to segregation of Sn, As, Sb and P to prior-austenite grain boundaries, promoting intergranular fracture. It is also well known that diffusion of nitrogen atoms to dislocations retards their mobility in low-carbon steels and promotes strain age embrittlement, which seriously affects the formability. Similarly, a small amount of Sb embrittles copper.

Persistence of nonuniform distribution of alloying elements, popularly known as segregation, is a cause of concern. The larger the gap between liquidus and solidus, the greater is the segregation. The types of segregation are as follows:

1. *Microsegregation*: Solute-rich interdendritic regions and banded structures are typical examples of fine scale microsegregation, caused during solidification.
2. *Macrosegregation*: Essentially arises due to solidification-induced concentration gradients. Macrosegregation manifests in the formation of dendritic pattern, ring pattern, freckles and white spots.
3. *Gravity segregation*: The chemical inhomogeneity is caused by the differences in densities of the solutes. As a result, the ingot bottom is richer in high-density elements.

Macro etching is a good technique to reveal all types of chemical inhomogeneities. Increasing the cooling rate during solidification can minimize the extent of segregation. Techniques such as argon or helium or liquid metal cooling are routinely employed to improve homogeneity in large ingots.

The path-breaking technological improvements made during the 20th century in melting techniques (VIM, VAR, ESR, EBM) and analytical procedures (XRF, AAS, ICP) have led to development of alloys with precise control over the alloying elements and the undesirable trace elements. Still, many failures occur because of improper chemistry, not because of technological limitations but because of callousness of personnel.

A few failures are caused by the presence of large slag pockets and solidification defects such as secondary pipe, shrinkage cavities, etc. These defects are retained in the ingot due to inadequate ingot top and bottom discard. These volume defects on hot deformation develop into sheetlike defects, leading to separation along the defect-metal interface. Delamination in AISI 4130 steel sheet made by the ESR technique due to a slag pocket in the ingot is shown in Figure 5.2.

FIGURE 5.2
A layer of slag in AISI 4130 steel plate. (Courtesy of S. R. Sahay, DMRL, Hyderabad, India.)

5.3 Casting Defects

The types of casting defects and their causes and remedies are well documented in the literature. Blow holes, pin holes, shrinkage cavities and defects caused by metal-mold reaction are the important defects that initiate fracture. The terms *blow holes*, *pin holes* and *porosity* are used interchangeably by some failure analysts without realizing that the cause and correction of each of these defects are different. Blow holes, generally large and spherical in shape, are caused by entrapped gases whose solubility decreases with decreasing temperature. Shrinkage cavities are very small and either round or irregular in shape and develop during the final stages of solidification. Typical blow holes and interdendritic shrinkage cavities are shown in Figure 5.3.

Effective degassing of liquid metal reduces the tendency for blow hole formation. Ensuring availability of a sufficient liquid metal pool during solidification by providing adequate risers and maintaining the superheat in the liquid metal by the use of exothermic compounds, mold heating, and ceramic insulation sleeves are some of the popular techniques employed by foundries to minimize the incidence of shrinkage defects. Defects such as mold material embedment on the casting surfaces and surface porosity induced by a green sand mold–metal interaction can be minimized by correct selection of mold materials and preheating the molds.

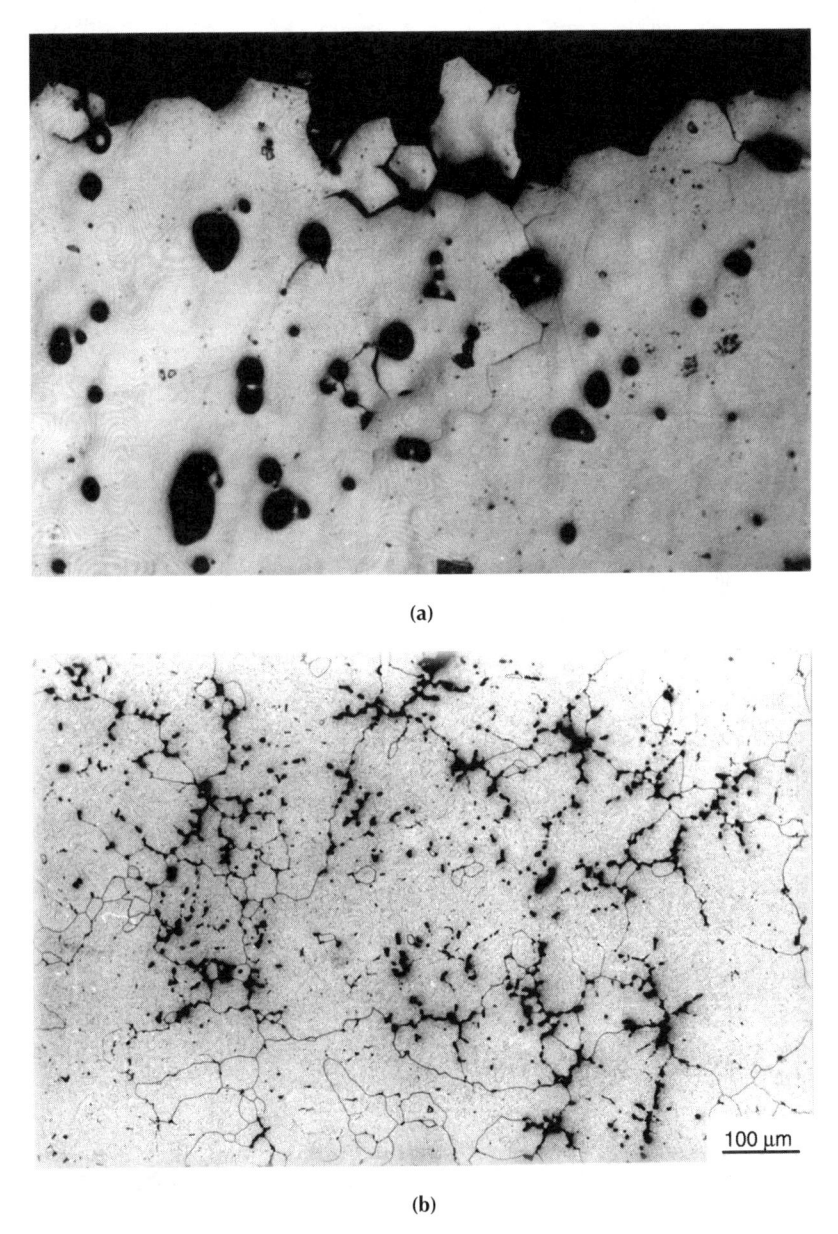

FIGURE 5.3
(a) Blow holes. (b) Interdendritic shrinkage cavities in Al-11% Mg casting.

Chemical inhomogeneity caused by segregation of solute to the last-to-solidify regions of the casting is an important defect that accounts for a significant number of failures of castings. Chemical inhomogeneity results in different strength levels at different locations for a given heat treatment and undesirable residual stress patterns. Segregation can be eliminated to a

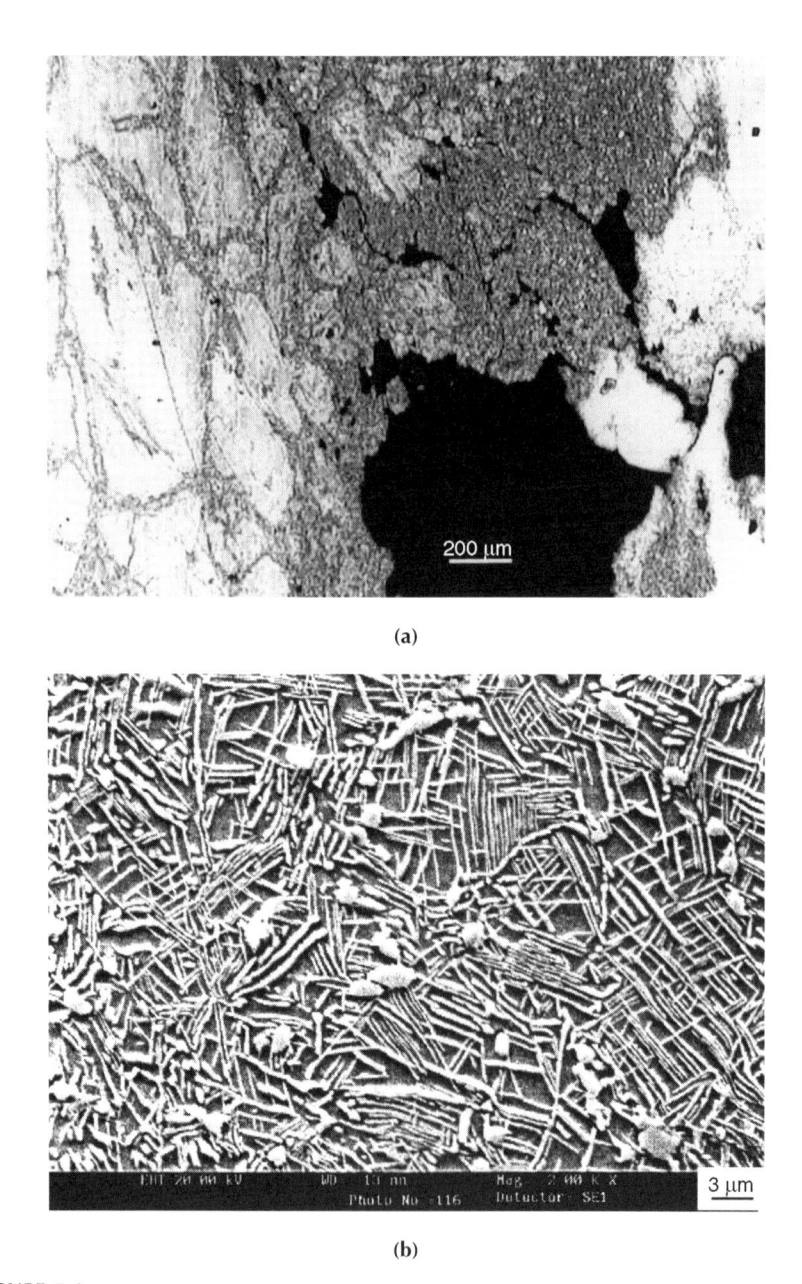

FIGURE 5.4
(a) Freckle defect associated with a cavity in an Inconel 718 forging. (b) Acicular and globular phases in the dark regions of figure (a). (Courtesy of R.V. Krishnan, NAL, Bangalore, India.)

great extent by proper homogenization treatment. A freckle defect (Nb-rich region) in an Inconel 718 superalloy forging that was inherited from the ingot is shown in Figure 5.4.

5.4 Metal Working Defects

Numerous defects of differing severity are created during thermomechanical working of ingots. These defects can broadly be classified into two categories: (1) inherited defects that have their origin in the ingot and (2) generated defects that develop during various metal working operations and thermal treatments.

5.4.1 Inherited Defects

These defects already exist in the ingot/billet prior to the metal working. Some of the inherited defects, under favorable conditions, are healed during hot deformation. Typical examples are (a) welding-up of defects such as blow holes and porosity free from oxidized surfaces, (b) minimization of chemical inhomogeneity as a result of improved diffusivity due to synergistic effects of deformation and high temperature, (c) grain refinement of the coarse-grained cast structure because of dynamic recrystallization during hot deformation and (d) overall microstructural refinement caused by controlled deformation at the selected finishing temperature.

There are a few defects that are aggravated during metal working. Many surface defects in ingots and billets such as tears, seams and cracks, if not properly dressed and removed prior to hot working, will develop into major flaws during hot working. These defects cannot be healed because of the oxidation of the exposed surfaces of the defects during ingot/billet heating. Internal defects such as slag pockets entrapped in the liquid metal during solidification and large deformable inclusions develop into sheetlike defects after metal working. These sheetlike defects are particularly dangerous as they can lead to delamination. For example, the presence of a sheetlike slag defect in a forged compressor blade made from ESR quality martensitic stainless steel (0.12C-12Cr-2W) resulted in delamination of the blade, as shown in Figure 5.5.

Yet another defect encountered, particularly in forgings, is the persistence of dendritic structure, a defect known as *ingotism*. Inadequate reduction during forging result in the persistence of dendritic structure. Failures caused by the presence of dendritic structure with attendant chemical inhomogeneities have been increasing due to the increasing usage of continuous cast billets of the smallest possible size as feedstock for forging. The relatively poor toughness associated with dendritic structure impairs service performance of the components. For example, pins of the screw coupling used to couple railway cars were forged from 0.5C–1.0Mn continuous cast steel billet. The pins failed to meet the specified impact toughness requirements in spite of imparting correct heat treatment due to the persistence of dendritic structure, as shown in Figure 5.6.

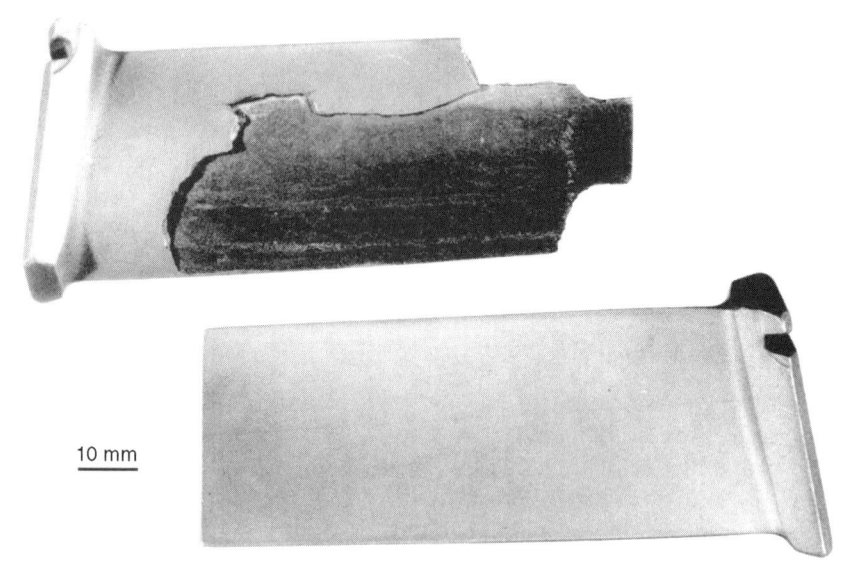

FIGURE 5.5
Delamination in a compressor blade forging due to a sheetlike slag defect.

5.4.2 Generated Defects

A number of defects can originate during hot working operations such as forging, rolling, etc. Central burst caused by improper forging conditions, formation of cracks due to fast cooling because of large thermal gradients between the surface and center of the stock, surface damages such as oxidation and decarburization, formation of laps and seams, incipient melting and grain boundary oxidation are some of the examples. Strain localization during deformation leading to formation of adiabatic shear bands and cracks, an uncommon defect, noticed in a forged compressor blade of an aeroengine made of 0.12C-12Cr-3W martensitic stainless steel is shown in Figure 5.7.

Overheating and burning are two other common defects generated during hot working of metals. With a view toward taking advantage of lowered flow stress with increase in temperature, working at temperatures close to the solidus temperature is not uncommon. Depending on how close the hot working temperature is to the solidus, either overheating or burning may take place. Overheating occurs when working temperatures are high, resulting in excessive grain coarsening. The damage caused by overheating can be corrected by reheating to a lower temperature and faster cooling. Normalizing treatment corrects the damage caused by overheating in steels. Burning is a more serious and irreversible damage caused by unintentional heating close to solidus temperature of either the alloy or the low melting phases present in the alloy. The evidences of burning are incipient melting and excessive grain boundary oxidation. Such microstructural damages

Manufacturing Defects

FIGURE 5.6
(a) Screw coupling assembly. (b) Persistence of dendritic structure in the forged pins.

result in drastic reduction in toughness of the components. Burnt surfaces can be easily recognized from their rough appearance, as shown in Figure 5.8.

Typical evidence of grain boundary oxidation in an Al-11% Mg alloy casting used in an undercarriage lever assembly of an aircraft is shown in Figure

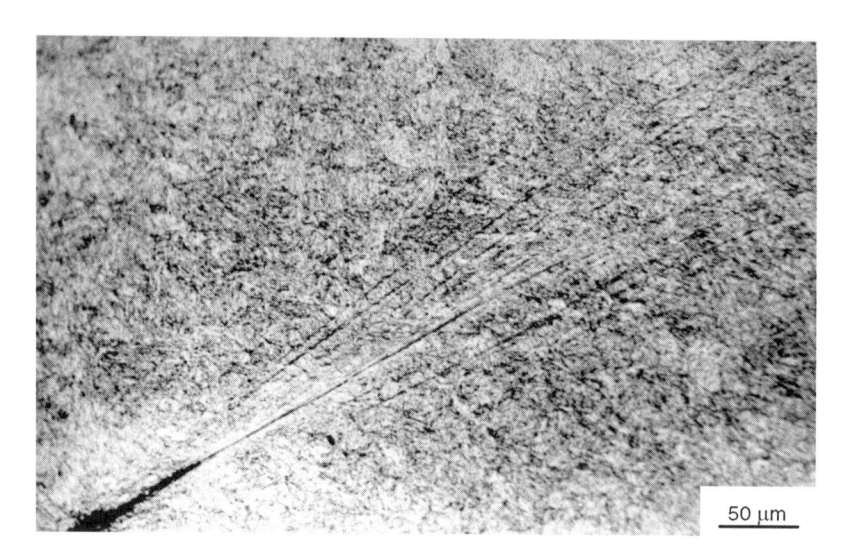

FIGURE 5.7
Adiabatic shear bands and cracks in a forged compressor blade of martensitic stainless steel.

5.9. The grain boundary oxidation occurred due to the use of a forging temperature that was higher than permissible.

Nonmetallic inclusions, depending on their type, morphology and distribution, can create defects during metal working. Hot tears caused by low melting constituents such as iron sulphides in steels, lead inclusions in brass, etc., are still encountered. Their incidence, however, is somewhat rare because of the use of practices such as modifying the chemistry of the low melting inclusions to increase their melting point. The severity of a given inclusion depends on the type of product being produced. While inclusions of about 5 µm in size do not affect the quality of thick wrought products, they induce necking and fracture in fine wires, causing loss in the quality and productivity. Severe necking caused by a Cu_2O inclusion in oxygen-free electrolytic copper (OFEC) wire of 150 µm diameter is shown in Figure 5.10.

The region of flash in a forging is always susceptible to initiate cracks. The important reasons are that (a) the flash region, being thin, cools much faster than the rest of the forging, leading to formation of unfavorable residual stresses at the trim line, and (b) trimming of the flash exposes *end grains* that are oriented in transverse direction to the main grain flow direction. A large grain boundary area coupled with residual tensile stresses promotes initiation of cracks in fatigue and stress corrosion modes. A typical case of intergranular stress corrosion cracking originating at the fine trim line of an Al-4Cu-1Si-1Mg alloy forging is shown in Figure 5.11.

It is common to observe cracks in hot-worked products if adequate care is not taken during heating the input stock and cooling the product. These cracks essentially develop as a result of thermal shock. Too fast a heating leads to higher surface temperatures compared to those of the core, leading to the development of high thermal stresses due to the constraint imposed

Manufacturing Defects

(a)

FIGURE 5.8
(a) Complex low-alloy steel forging with drastic changes in the section size. (b) Surface roughness developed due to burning.

(b)

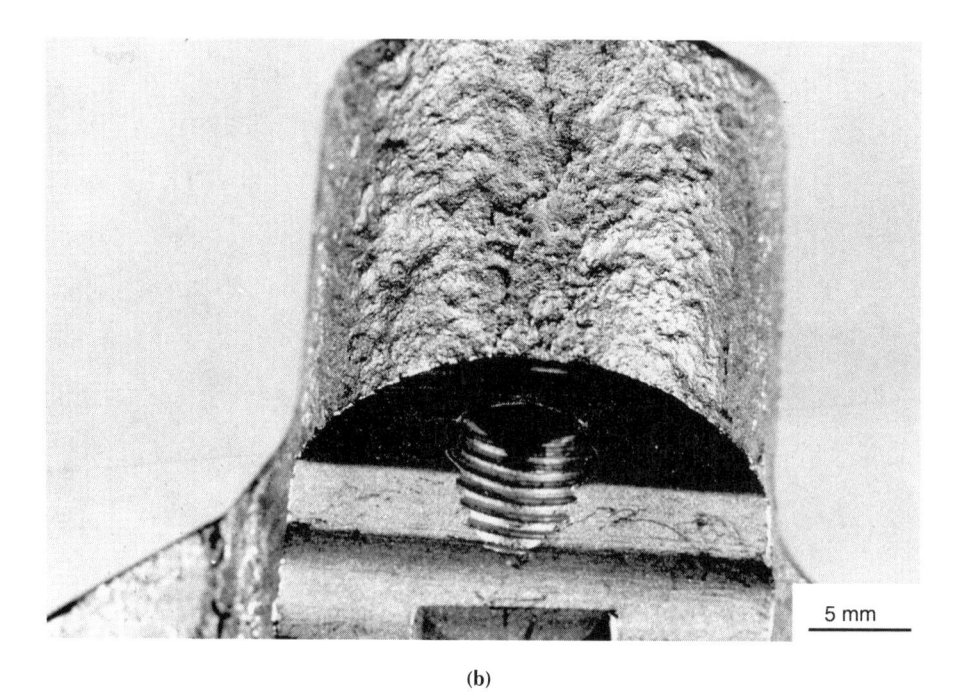

FIGURE 5.8 (continued)

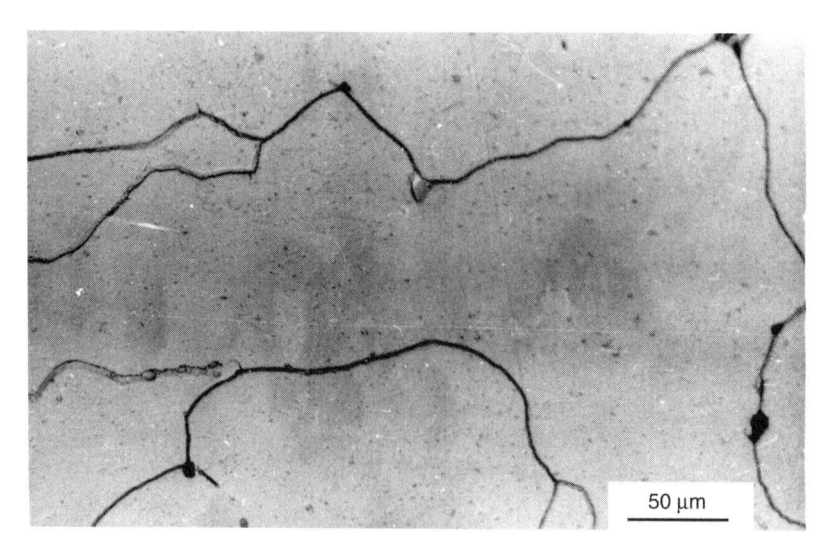

FIGURE 5.9
Coarse oxidized grain boundaries in an Al alloy casting.

by the core on the thermal expansion of the surface. If fast cooling rates are imposed on the product, the surface layers cool faster and restrain the contraction of the slowly cooling core. In either case, the thermal stresses can

Manufacturing Defects

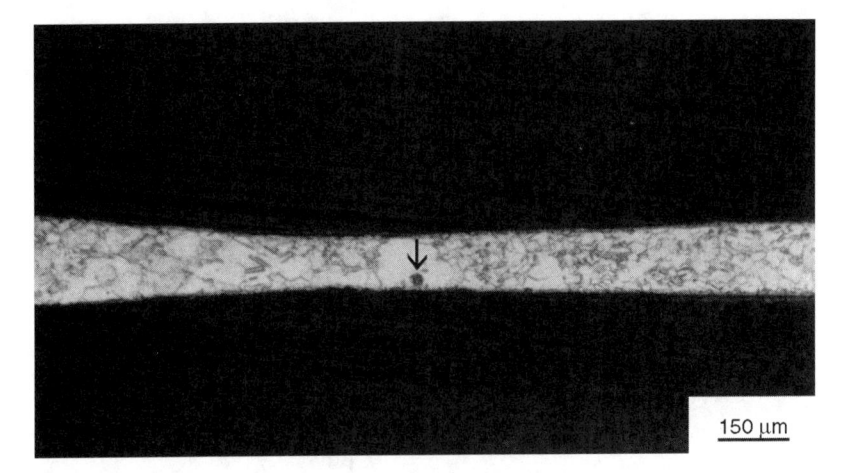

FIGURE 5.10
Necking induced by an oxide inclusion in a Cu wire 150 μm in diameter.

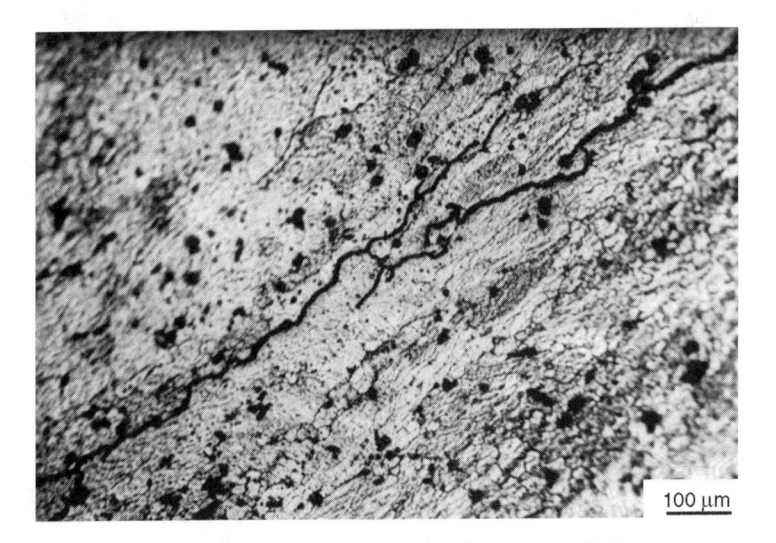

FIGURE 5.11
SCC at the trim line of an Al-Cu-Si-Mg alloy forging.

be of such magnitude to cause extensive cracking in the components. The problem is aggravated if there are gross changes in the section thickness of the product, as shown in the following example. Wedge forgings made of 0.45C-1.54Ni-1.15Cr-0.20Mo steel used in the assembly of the track of an infantry combat vehicle failed due to cracks that developed at the head-stem junction due to fast cooling after forging, as shown in Figure 5.12.

Failures due to microstructural banding are not uncommon. Chemical inhomogeneity and segregation of trace elements are primarily responsible for the development of banded microstructures during hot working. Banding causes significant reduction in transverse strength and toughness, fatigue strength and

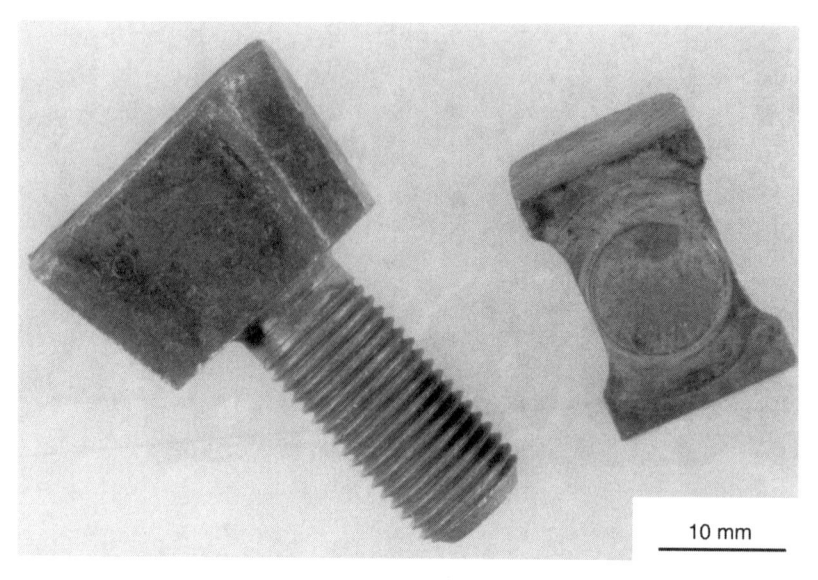

FIGURE 5.12
Failure of a track wedge at the head–stem junction due to fast cooling.

corrosion resistance because of the presence of alternate layers of microstructural features each with a different solute concentration and mechanical properties. Banding of carbon-rich and carbon-lean layers caused by phosphorus segregation in many low-alloy steels, banding of delta ferrite in stainless steels, carbide banding in high-carbon tool steels and beta-phase banding in many alpha-beta alloys are some of the examples. In many cases, it is possible to minimize banding by proper thermal treatment after hot working. Normalizing of low-alloy steels, homogenization of tool steels at temperatures close to the hardening temperature, etc., are effective. In some cases, it is not possible to eliminate alignment of a particular microstructural constituent during working, such as δ-ferrite in stainless steels. It is desirable to avoid formation of such phases during solidification by proper control of composition and choice of melting route. Typical examples of banding in a low-alloy steel and δ-ferrite banding in a martensitic stainless steel are shown in Figure 5.13.

Carbide banding in tool steels can cause damage to the fine features of the tool. The dislodgement of the splines in a thread roll made of AISI D2 steel along the carbide bands is shown in Figure 5.14. When the splines were oriented perpendicularly to the carbide bands, the tool life improved significantly.

5.5 Heat Treatment Defects

Of all the manufacturing stages in the production of metallic components, heat treatment is by far the most critical and also the most abused step. A

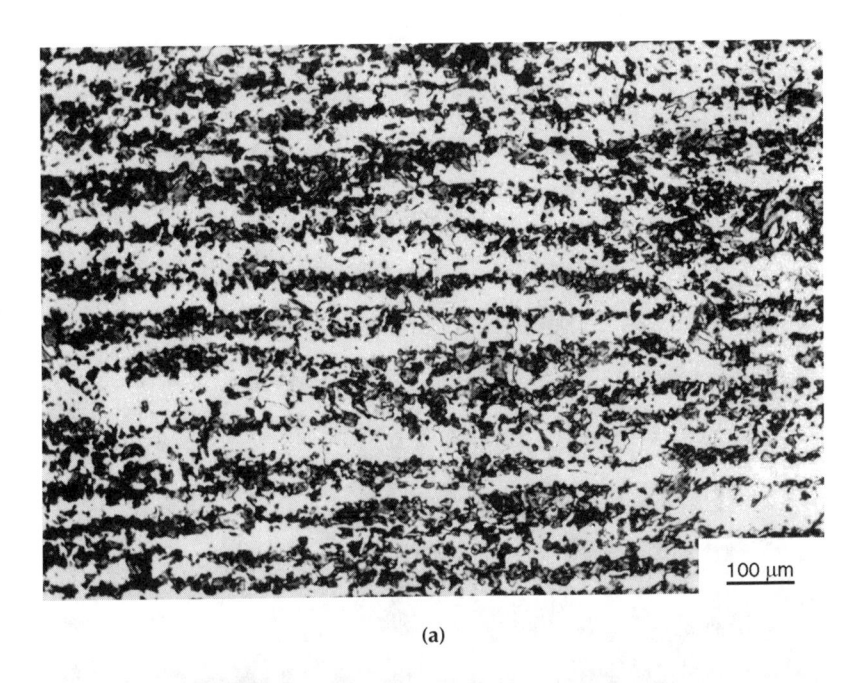

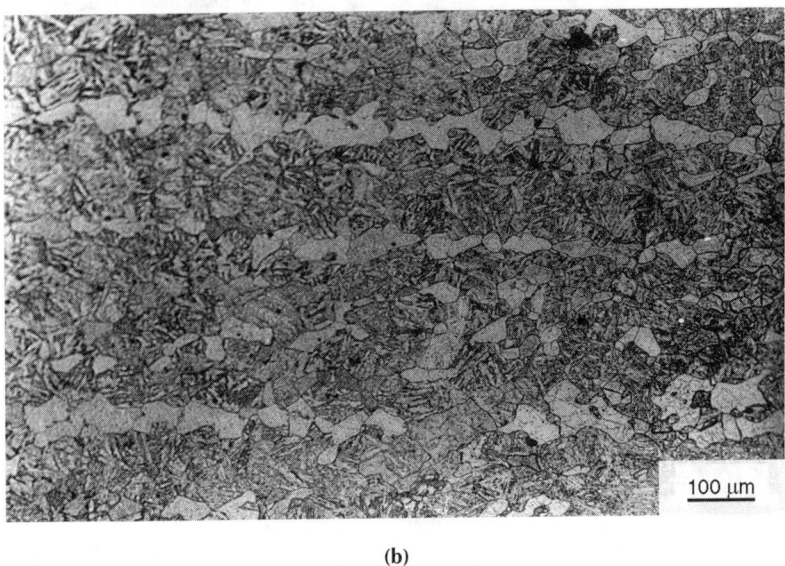

FIGURE 5.13
(a) Banding in AISI 4140 steel. (b) δ-Ferrite banding in martensitic stainless steel.

large number of problems arise because of the technical incompetence of the personnel engaged in heat treatment and a few because of shortcuts practiced. The problem is acute in developing countries where heat treatment shops are generally managed by "versatile" mechanical engineers who are not technically trained on metallurgical practices to define heat treatment

(a)

FIGURE 5.14
(a) Failed splines parallel to carbide bands in a thread roll. (b) Undamaged splines cut perpendicularly to carbide bands. (c) Carbide banding in AISI D2 tool steel.

schedules and understand the cause and effect of process deviations. The problem is aggravated by the nonconformance of materials used to the material standards and heat-treating them as per the procedures defined for the specified material. Even if the scientific capabilities of the heat treatment

(b)

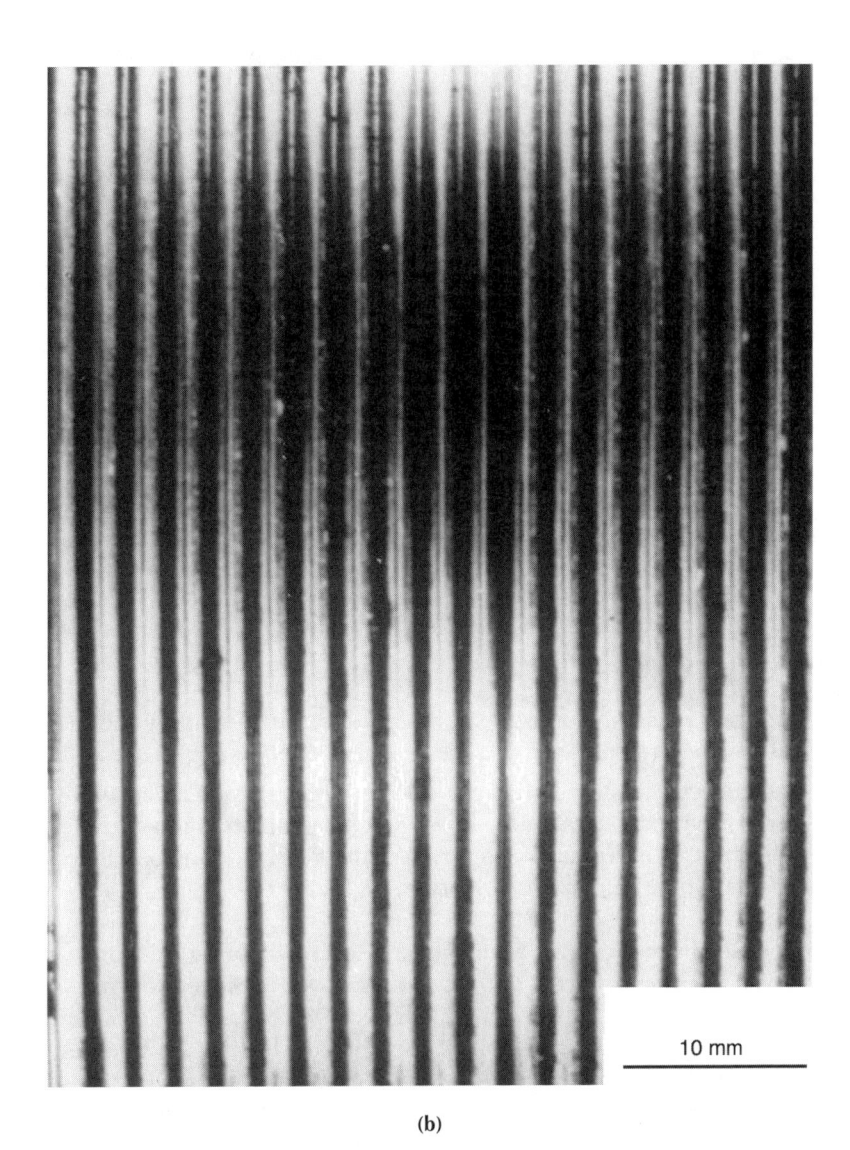

FIGURE 5.14 (continued)

personnel are of a high order, certain defects may still arise during heat treatment mainly due to nonadherence to defined procedures.

The parametric variables in the heat treatment are (a) heating rate, (b) furnace atmosphere, (c) temperature, (d) time at temperature, (e) ruling section, (f) cross-sectional variations, (g) quenching, (h) cryogenic treatment, (i) tempering/aging temperature and time and (j) cooling rate after final stage. If adequate care is not taken in evolving an appropriate heat treatment schedule based on metallurgical principles, defects can develop at each of these stages. The defects generated can be classified into two categories:

(c)

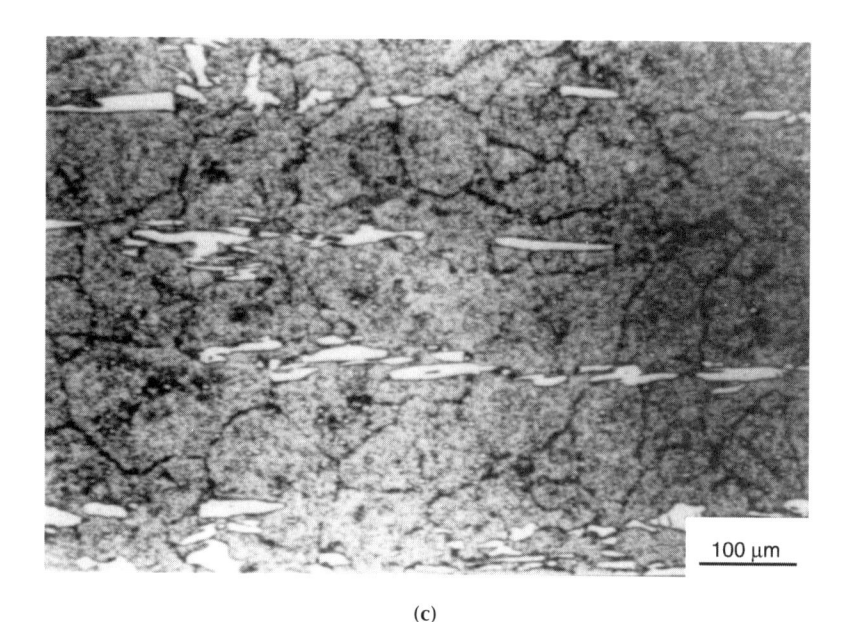

FIGURE 5.14 (continued)

1. Surface defects, including oxidation and decarburization and residual tensile stresses
2. Bulk defects, including grain coarsening, incipient melting, incomplete transformation, retention of undesirable phases, formation of undesirable phases, embrittlement, development of cracks, unfavorable residual stress distribution and distortion and warpage

Depending on the furnace atmosphere used for heating the components during heat treatment, certain surface changes occur. Since most of the heat treatments are carried out in electrical furnaces with air as the environment, oxidation and loss or gain of elements at the surface are the two damages that can occur. Taking heating of steel as an example, the steel surface reacts with the oxygen in the furnace atmosphere, leading to the formation of oxide scale on the surface. The thickness of the oxide scale depends on, in addition to the temperature and time of exposure, the composition of the steel and the oxide and its porosity. In general, the oxide scale in itself is not deleterious if the component dimensions are not altered beyond the acceptable limits. Since grain boundaries are the most favorable paths for oxygen diffusion, the depth of damage caused by oxidation is always greater than the thickness of the scale. Grain boundary oxidation, however shallow it may be, is dangerous as oxidation makes grain boundaries weaker. Innumerable high tensile fastener failures at thread roots in fatigue mode are attributable to grain boundary oxidation. For example, steel fasteners made of 0.4C-1.5Ni-1.0Cr-0.25Mo steel in quenched and tempered condition have failed in fatigue due to the formation of an oxide network along the prior-austenite grain boundaries

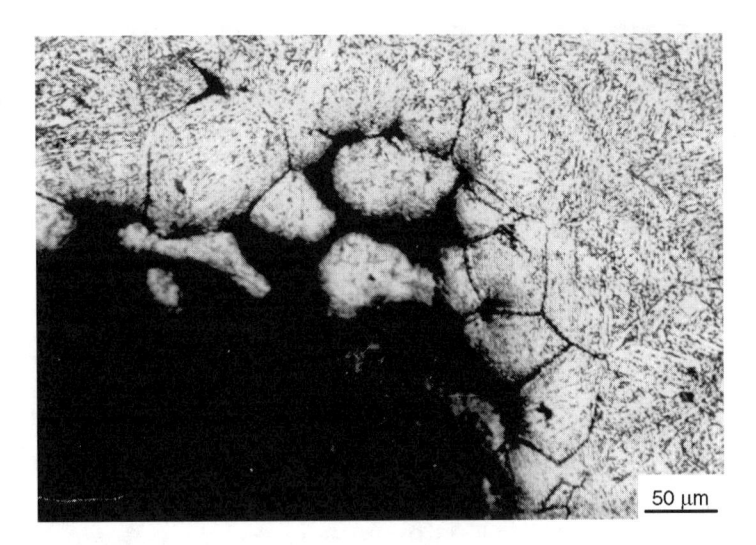

FIGURE 5.15
Grain boundary oxidation in a steel fastener causing fatigue failure.

during hardening treatment, which promoted initiation of fatigue cracks. The thread root and the grain boundary oxidation are effective stress raisers to initiate fatigue crack (Figure 5.15).

Decarburization is another type of surface damage frequently noticed in heat-treated steel parts. Both oxidation and decarburization occur concurrently. Decarburization results in ferrite formation; ferrite, being soft, is detrimental to surface-specific properties. The extent of damage caused by decarburization is dependent on the carbon activity and the alloy type and content, which dictate the diffusivity of carbon. In general, the higher the carbon content and lower the concentration of elements that form stable carbides, the greater the tendency of decarburization. Because of the fact that grain boundary diffusivity is higher than bulk diffusion rate, formation of ferrite is observed along prior-austenite grain boundaries to depths far greater than the depth of the total decarburized layer. Decarburization need not always result in complete ferrite formation but can lead to a noticeable reduction in carbide population at the decarburized surface, as generally observed in heavily alloyed tool steels. Decarburization causes a drastic reduction in tensile strength, fatigue strength and wear resistance of the components. The presence of a decarburized layer at the surface of a low-alloy steel (AISI 8660) component that was responsible for premature fatigue failure of the component is shown in Figure 5.16.

Packing the components with oxygen-gettering materials such as cast iron chips, coke, etc.; the use of protective coatings on the components (e.g., borax coating on steel components and glass coating on Ti-alloy components); and the use of protective atmospheres and neutral salt baths, etc. are some of the shop floor practices used to minimize oxidation and decarburization during heat treatment.

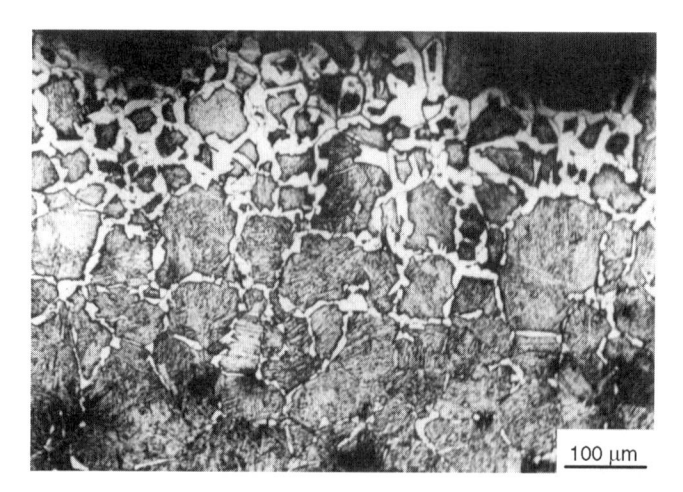

FIGURE 5.16
Decarburized surface of an AISI 8660 steel component.

Damages to the material during heat treatment are more serious than the surface damage. Employment of as high a temperature and as short a time as possible to achieve the desired result is a common practice at many heat treatment shops to improve productivity. Too high a temperature can lead to grain coarsening, which in turn results in inferior strength and toughness. Another reason for the undesirable grain growth is the heat-treater's practice of clubbing components made of different steels to add up to the furnace charge quantity and heating them to a predecided temperature on the premise that the differences in actual heat treatment temperatures are marginal. The "marginal" difference in temperature is sometimes as high as 100°C. If steel A, whose defined austenitization temperature (T_a) is 850°C, and steel B, whose T_a is 950°C, are together austenitized at 950°C, the austenite grains in steel A will coarsen significantly. In one incident, a batch of 0.35C-1.0Cr-0.25Mo steel components was austenitized (as a made-up charge along with other components) at 930°C instead of at 860°C. This resulted in significant grain growth and inferior tensile and impact properties (Table 5.1), leading to premature failure of the components.

Incipient melting during heat treatment is a serious concern in the case of Ni-based superalloys, particularly in castings. For example, airfoil castings made of advanced Ni-based superalloys such as CM 247 LC are to be given solution treatment at temperatures typically 10 to 15°C less than the incipient melting temperature to dissolve as much coarse γ-γ' eutectic as possible in order to maximize the mechanical properties. There is every possibility for incipient melting to occur when operating at such a close proximity to the incipient melting temperature either due to marginal inaccuracies in furnace controls or due to local drop in incipient melting point as a result of excessive microsegregation, causing irrecoverable damage to the component serviceability. A typical case of incipient melting of γ-γ' eutectic nodules in superalloy CM 247 LC is shown in Figure 5.17.

TABLE 5.1

Effect of higher austenitization temperature on the structure and properties of a low-alloy steel

Treatment	Grain size ASTM no.	YS (MPa)	UTS (MPa)	% El	% RA	CVN (J)
Q&T: defined 860°C, OQ 540°C, 2 h, AC	8	908	1180	17	57	42
Q&T: practiced 930°C, OQ 540°C, 2 h, AC	4	846	1157	12	43	30

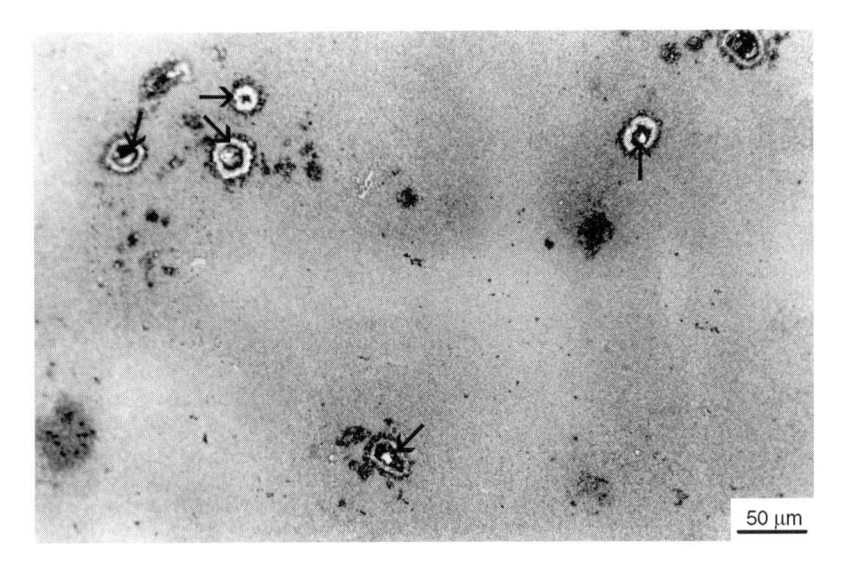

FIGURE 5.17
Incipient melting in CM 247 LC superalloy.

The final microstructure that forms after a given heat treatment depends on the inherent metallurgical characteristics of the material to transform from the high-temperature phase to the low-temperature phase with correct phase distribution. If a quench-hardenable steel is considered as an example, the characteristics of the high-temperature austenite depend on the composition of the steel, austenitization temperature and time. The transformation of austenite completely to martensite on quenching depends on the hardenability, the martensite start (M_s) and finish (M_f) temperatures and the quench rate.

In the case of high-alloy steels, the selection of austenitization temperature plays a major role in developing the desired structure and properties. If the austenitization temperature is too low, most of the carbides remain undissolved, leading insufficient carbon levels in the austenite. Consequently, the martensite formed upon quenching will have lower strength. On the other hand, if the austenitization temperature is too high, all the carbides get dissolved, resulting in the formation of austenite, rich in carbon and alloying elements. Consequently, the M_s and M_f temperatures are lowered, leading to the retention of excessive quantities of austenite after quenching. Therefore, the austenitization temperature should be judiciously selected to take into austenite a preselected volume fraction of carbides to achieve nearly complete martensitic transformation on quenching.

A further consideration in the selection of austenitization temperature is the size of carbides in the initial microstructure. At a given austenitization temperature, the larger the size of carbides, the lesser the extent of dissolution. The consequences of disregarding the above consideration are illustrated in the following example. A 0.2C-16Cr-2Ni (AISI 431) steel was produced at two firms, "M" and "I." Both producers have employed identical heat treatments: 1000°C/75 min, oil quench followed by two-stage tempering, the first at 650°C for 60 min and the second at 600°C for 60 min, followed by oil quench. The steel produced at firm M failed to meet the minimum strength requirements, while the steel produced at firm I met all the requirements. Microstructural examination of the heat-treated samples from both firms (Figure 5.18) showed that the steel produced at firm M contained large, undissolved carbides, while the steel produced at firm I contained fine, dissolved carbides. This indicates that the austenitization temperature employed was not adequate to cause sufficient carbide dissolution in the steel produced at firm M as a result of very coarse carbides in the initial microstructure leading to insufficient carbon levels in the austenite and subsequent formation of low-strength martensite. Thus, the selection of an appropriate austenitization temperature should be based on the size of carbides in the microstructure of the material to be austenitized.

Retained Austenite

A major problem in the quenched and tempered steels, which is not given adequate importance, is the incidence of high levels of residual austenite. The retained austenite, if stable, enhances the toughness significantly. The deleterious effects of instability lead to (a) variations in strength as a result of progressive transformation of austenite, (b) an increased tendency to distortion, warpage and cracking caused by its subsequent transformation with attendant volume changes, (c) promotion of tempered martensite embrittlement due to transformation of interlath austenite to carbide films and (d) development of cracks during grinding as a result of its transformation under the influence of pressure and temperature experienced during grinding.

The amount of retained austenite in a quenched steel depends on the interdependence of the composition, hardenability, austenitization temperature, quench rate and tempering parameters. The stability of austenite occurs due to chemical stabilization caused by solute partitioning to austenite during

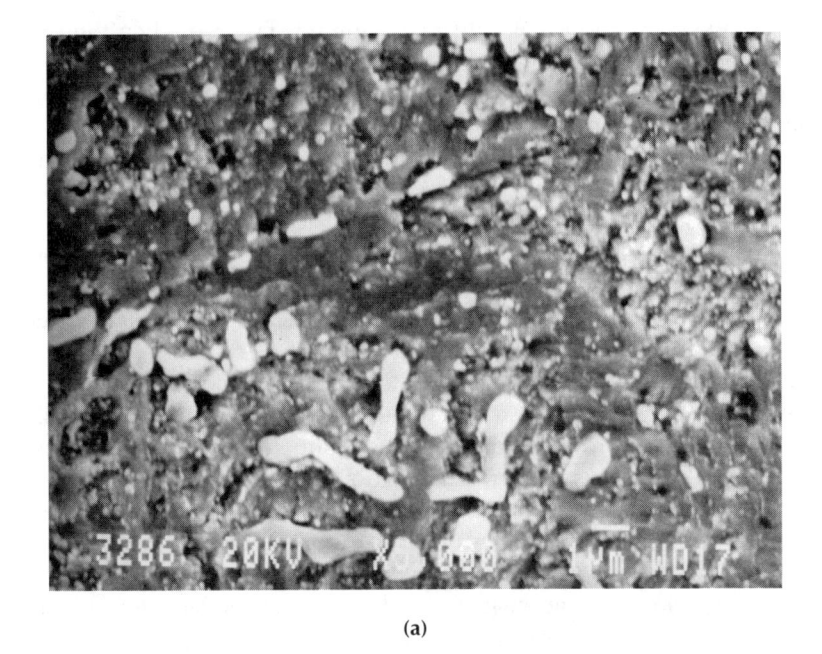

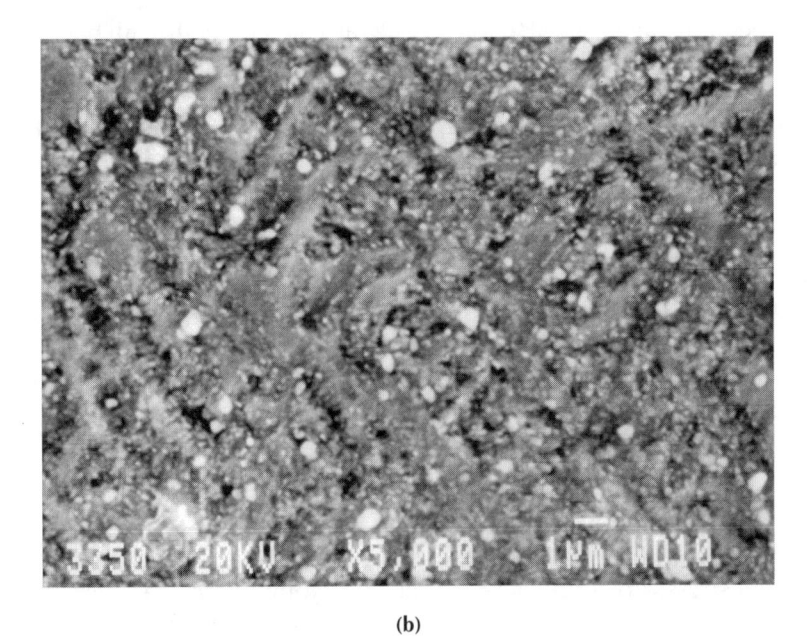

FIGURE 5.18
Microstructure of quench and double-tempered AISI 431 steel [3]. (a) Large, undissolved carbides and precipitated carbides in the steel produced at firm M. (b) Fine, dissolved carbides and precipitated carbides in the steel produced at firm I.

cooling between M_s and M_f. Solute partitioning is favored when the cooling rate between M_s and M_f is low, resulting in an austenite rich in solute content, driving its M_s and M_f temperatures lower. When the M_f falls below ambient temperature, transformation of austenite to martensite does not proceed to completion, leading to retention of untransformed austenite. Mechanical stabilization is also caused by the accommodation of transformation strains by plastic deformation of residual austenite.

The hardening temperature, time and quench rate have a bearing on the quantum of retained austenite after quenching. In general, (a) the higher the hardening temperature and time, the greater the amount of carbides going into solution and, hence, the higher the amount of retained austenite and (b) the slower the quench rate, beyond the critical cooling rate for the steel, the higher the chemical stabilization of austenite and the higher the residual austenite content. In order to transform the retained austenite, it is necessary to employ either cryogenic treatment or double tempering. In double tempering, the first tempering is carried out at a higher temperature at which the stable austenite can be made unstable by promoting diffusion of solute elements out of it, and the second tempering done at a lower temperature is aimed at tempering the martensite formed from the unstable austenite during cooling after the first tempering. The retained austenite, usually distributed along the martensite lath boundaries, transforms to carbide when tempered in the temperature range of 250 to 350°C. The carbide films at the lath boundaries in association with segregation of P and N are responsible for the loss of toughness; this phenomenon is known as *tempered martensite embrittlement* (TME). Whenever low tempering temperatures are to be employed to get high strength levels, care must be taken to avoid tempering in the TME range.

The problems caused by retained austenite are more acute in case-carburized steels and high-alloy steels. Because of the common belief that higher carbon content results in greater hardness and wear resistance, there is a tendency to overcarburize. Too much carbon in the case leads to retention of large quantities of austenite in the case with a significant reduction in case hardness. The high amount of retained austenite in a case-carburized, low-alloy steel due to high carbon content that has resulted in formation of cracks in the case during grinding is shown in Figure 5.19.

5.5.1 Internal Stresses and Quench Cracks

Internal stresses in hardened steel parts consist of thermal stresses and transformation stresses. Thermal stresses develop in rapidly cooled parts due to nonuniform cooling. Under conditions of rapid cooling the surface layers cool faster than the core, resulting in the formation of tensile stresses at the surface and compressive stresses at the core. The transformation stresses are caused by the solid-state transformation of austenite to martensite, which involves volumetric expansion. In rapidly cooled components, the austenite at the surface quickly transforms to undeformable martensite, while the

Manufacturing Defects

FIGURE 5.19
Retained austenite (light) in a case-carburized steel.

transformation progresses gradually to the inner layers. This leads to the development of tensile stresses at the surface and compressive stresses at the center. The internal stresses reach a maximum level in deep-hardening steels during transformation of austenite to martensite at the core regions because of the difficulty in accommodating transformation stresses by the nondeformable martensite initially formed at the surface. The resultant tensile stresses at the surface of the quenched steel component, because of the thermal and transformation stresses, are primarily responsible for distortion, warpage and quench crack formation. To avoid the formation of quench cracks, extra care must be taken while quenching components with (a) large variation in section size, (b) sharp re-entrant corners and (c) stress concentrations such as small holes, punch marks, thread roots, etc.

Formation of quench cracks invariably leads to premature failures. For example, the track clamps of an infantry combat vehicle forged from 0.45C-1.5Cr steel, water-quenched from 850°C and tempered at 200°C, were found to have cracks at the junction of the lateral threaded hole and the inner surface. The intergranular fracture observed over the entire peripheral region of the cross section was found to be due to formation of quench crack (Figure 5.20). The problem was overcome by employing oil quenching instead of water quenching for hardening.

5.5.2 Tempered Martensite Embrittlement and Temper Embrittlement

The toughness of the brittle martensite formed after hardening is increased by tempering. The toughness generally increases with increasing tempering

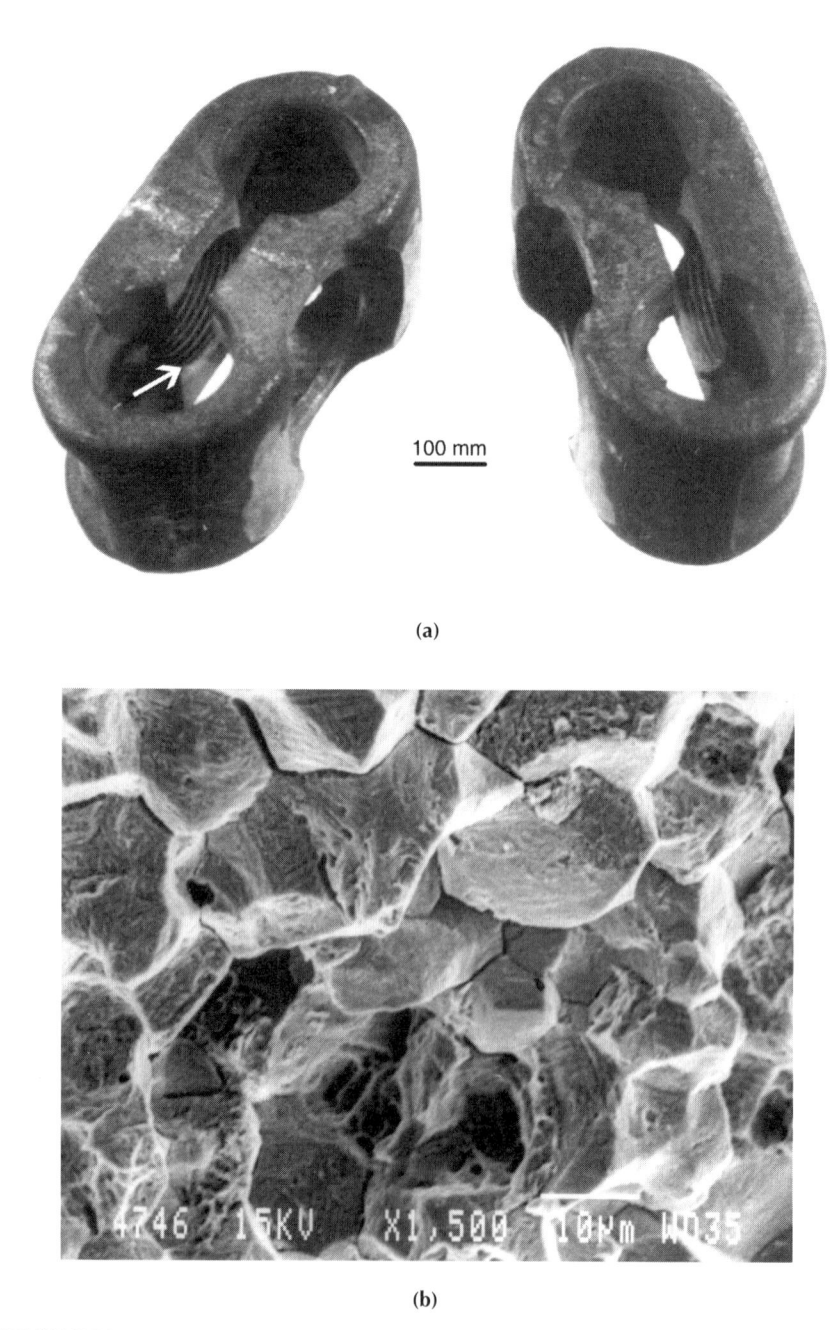

FIGURE 5.20
(a) Track clamps. (b) Intergranular fracture due to quench cracks.

temperature with an accompanying decrease in strength. However, when some grades of steels are tempered in a susceptible temperature range, loss of toughness is observed. The variation of impact toughness with tempering

Manufacturing Defects

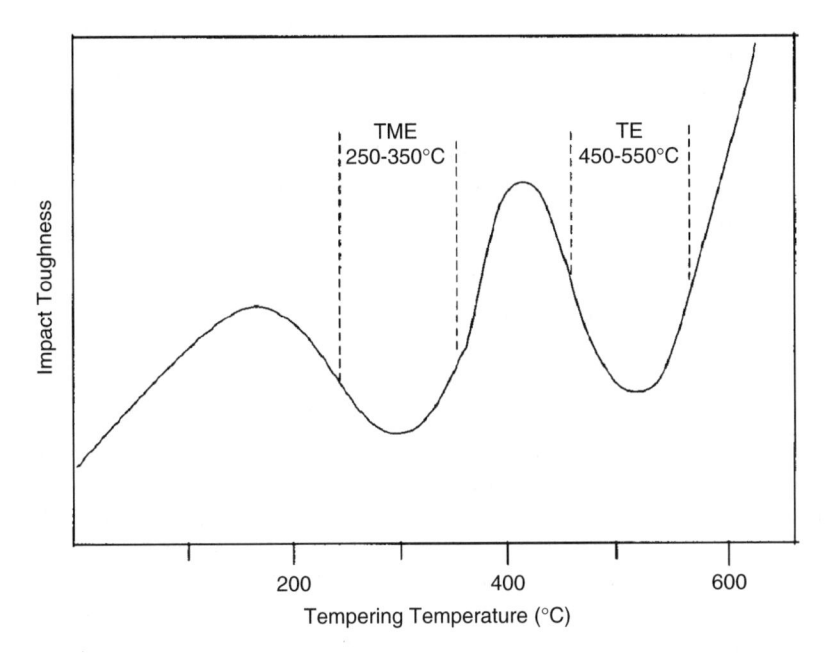

FIGURE 5.21
Variation of impact toughness with tempering temperature in quenched and tempered low-alloy steels.

temperature in quenched and tempered low-alloy steels is shown schematically in Figure 5.21.

The loss of toughness in high-strength hardened and tempered steels when tempered between 250 and 350°C was known as "blue brittleness" in the past and has now been rechristened as TME in recognition of the mechanism responsible for this phenomenon. It is now well accepted that TME occurs due to transformation of retained austenite films at the lath boundaries to cementite films, which act as potent crack nucleation sites and also aid the propagation of cracks along the lath boundaries. If cryogenic treatment can transform the retained austenitic after quenching, most problems associated with TME can be overcome. If it is not possible to completely eliminate retained austenite after quenching, one way of minimizing TME is to add alloying elements that can retard carbide nucleation and growth. Si, Al, Mo, and W are found to be beneficial in shifting TME temperatures to over 400°C. Another way to get over TME is to produce ultrapure steels by reducing the levels of deleterious elements.

The loss of toughness in certain low-alloy steels slowly cooled after hardening and tempering in the temperature range of 450 to 550°C is called *temper embrittlement* (TE). TE is known to be related to the segregation of certain impurity elements such as P, As, Sb and Sn to prior-austenite grain boundaries. The interactions between alloying elements and impurity elements (Ni-Sb, Ni-P, Ni-Sn, Mn-Sb etc.) lead to trace element segregation at prior-austenite grain boundaries, resulting in lowering of grain boundary cohesion and thereby promoting intergranular fracture. It is possible to minimize TE by resorting to the following:

- Reducing the level of impurity elements in the steel by using scrap free from the trace elements and employing vacuum melting and/or electroslag remelting techniques to further reduce their concentration
- Adding alloying elements such as Ti and Mo that have stronger interaction with impurity elements and hold the impurity elements away from the grain boundaries
- Avoiding tempering in the susceptible temperature range
- Employing faster cooling in the susceptible temperature range to minimize segregation of impurity elements

The efficacy of some of these solutions can be assessed by the following example. Balancing gear springs of a field gun made of AISI 8660 steel failed during trials (Figure 5.22a). The springs were austenitized at 900°C for 50 min, quenched in oil, and tempered at 500°C for 75 min and air-cooled. The electric arc-melted steel contained 200 ppm of S, 250 ppm of P, 90 ppm of As, and 60 ppm of Sn in addition to the specified alloying elements. The Ni-Cr-Mo steel is highly prone to TE when tempered at 450 to 525°C. The failure occurred in intergranular mode as a result of segregation of the trace elements to the prior-austenite grain boundaries (Figure 5.22b). Coarse grain size of the spring material and slow cooling after tempering further aggravated the problem.

The problem was overcome by the following corrective actions:

- The springs were made of the same steel with virgin scrap with low concentration of the tramp elements.
- The austenitization temperature was reduced to 850°C and the tempering temperature was lowered to 400°C.
- The desired strength levels were achieved by increasing the tempering time at 400°C, calculated based on tempering parameter.
- The springs were cooled in oil after tempering instead of air cooling.

5.6 Defects Generated in Finishing Operations

Normally, a cast or forged component goes through many finishing operations before it is cleared for assembly. From the defect generation point of view, among the many finishing operations, the notable ones are (a) pickling, (b) sand blasting, (c) machining and (d) surface coatings.

Sand blasting is an effective method to clean the component surfaces free from adherent scale inherited from the last high-temperature operation in

Manufacturing Defects 99

(a)

FIGURE 5.22
(a) Failed Q&T AISI 8660 steel spring. (b) Intergranular fracture caused by temper embrittlement.

the manufacturing sequence. If adequate care is not taken in selecting the impinging particle-related variables such as velocity, particle size, angularity, friability, etc., significant damage can be caused to the surface of the component. Too high a velocity and fragmentability of the particles result in embedding of the broken particle tips in the metallic surface, leading to the formation of a thin composite layer consisting of the embedded particles in the metal matrix. This damaged surface condition affects corrosion resistance and fatigue strength. For example, in an investigation on martensitic stainless steel compressor blade failures, some angular particles were observed below the Ni-coated surface that might have been embedded during the sand blasting operation prior to Ni coating. These were confirmed to be silica sand particles by EPMA analysis (Figure 5.23) and were found to be responsible

(b)

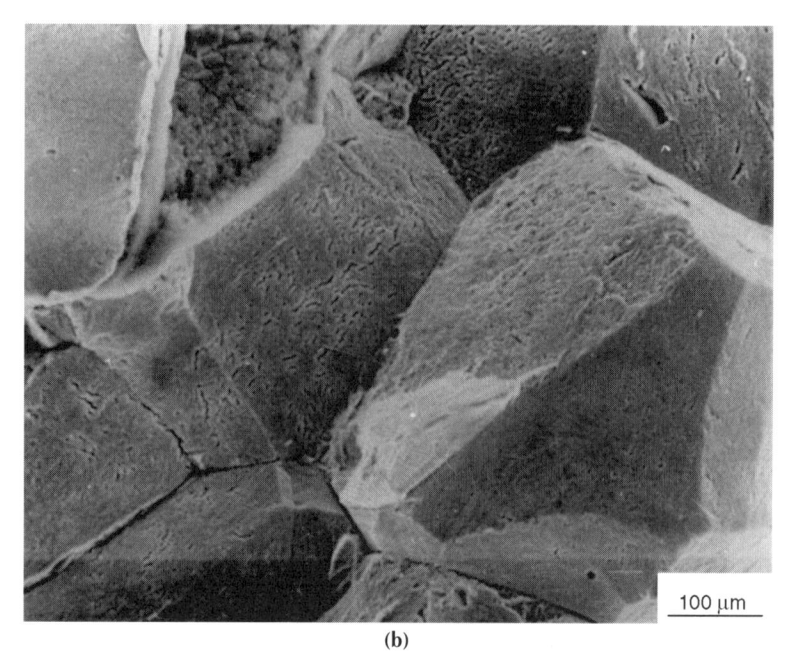

FIGURE 5.22 (continued)

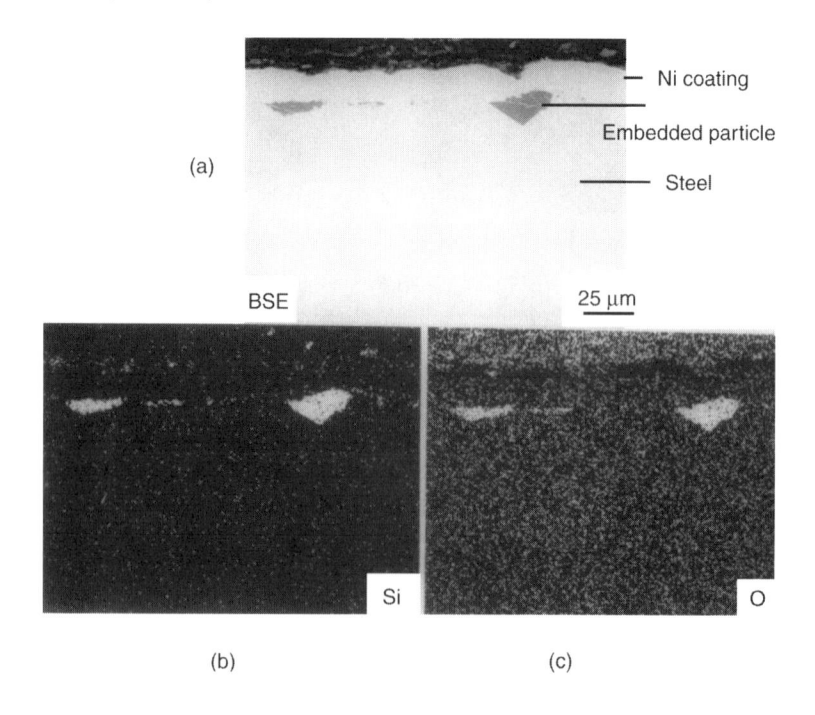

FIGURE 5.23
(a) BSE image showing the embedded particles below the Ni-coated surface. (b) X-ray image of Si. (c) X-ray image of O.

for the failures by initiating fatigue cracks. The problem can be obviated by correct choice of sand blasting parameters and by the use of a less friable erodent such as steel grit or steel shot.

The major defects that originate during machining are machining grooves, grinding burns and cracks. Machining grooves are extremely dangerous in critical rotating parts, with life limited by fatigue. Because of their stress-raising effect, machining defects are the preferred sites for fatigue crack initiation. A large number of service failures occur due to improper machining.

Formation of grinding burns and cracks are common in quenched and tempered high-strength alloy steels. Grinding burn is the result of microstructural changes caused by the high temperature generated at the contact surface between the work piece and grinding wheel due to the ineffectiveness of the coolant. The burnt regions, which are relatively softer, appear dark when the ground surface is etched with 5% Nital (5% HNO_3 + 95% ethanol) and can easily be recognized. Because of the ease with which fatigue cracks can initiate in the softer areas of the burnt region, fatigue life is dramatically reduced.

The occurrence of failures due to grinding cracks is more common than the failures due to grinding burns. The problem is acute in tool and die steels and case-hardened steels, which are heat treated to very high hardness levels and invariably contain some retained austenite. Grinding cracks develop mainly due to transformation of retained austenite either under the influence of the pressure exerted by the grinding wheel or local temperature rise or the combined effect of pressure and temperature. To minimize formation of cracks during grinding, in addition to judicious choice of operational variables it is essential to minimize the retained austenite, either by subzero treatment or by multiple tempering treatments. Grinding cracks, like quench cracks, are invariably intergranular and are oriented perpendicularly to the grinding direction. The role played by grinding cracks in initiating failures can be illustrated with the following example. The breech block of a rifle was made out of 0.15C-3Ni-0.8Cr-0.2Mo steel, case-carburized at specified locations, hardened and tempered and finish ground to drawing dimensions. The case contained a very high amount of retained austenite and its hardness was 418 HV instead of the specified minimum of 575 HV. The grinding operation led to the formation of shallow intergranular surface cracks, which in turn caused the failure of the breech block after firing a few rounds in quasi-cleavage mode. The component and the fracture features are shown in Figure 5.24.

The major problem encountered in pickling and electroplating of high-strength steel components is hydrogen embrittlement (HE). HE is caused either by the diffusion of hydrogen atoms to the dislocations inhibiting their mobility or its segregation to micropores in molecular form and formation of high-pressure hydrogen bubbles and their implosion. HE is observed at low strain rates and low temperatures, which retard diffusion of hydrogen atoms from the dislocations and the dislocation pinning effect is sustained. The threshold hydrogen concentration for embrittlement depends on the

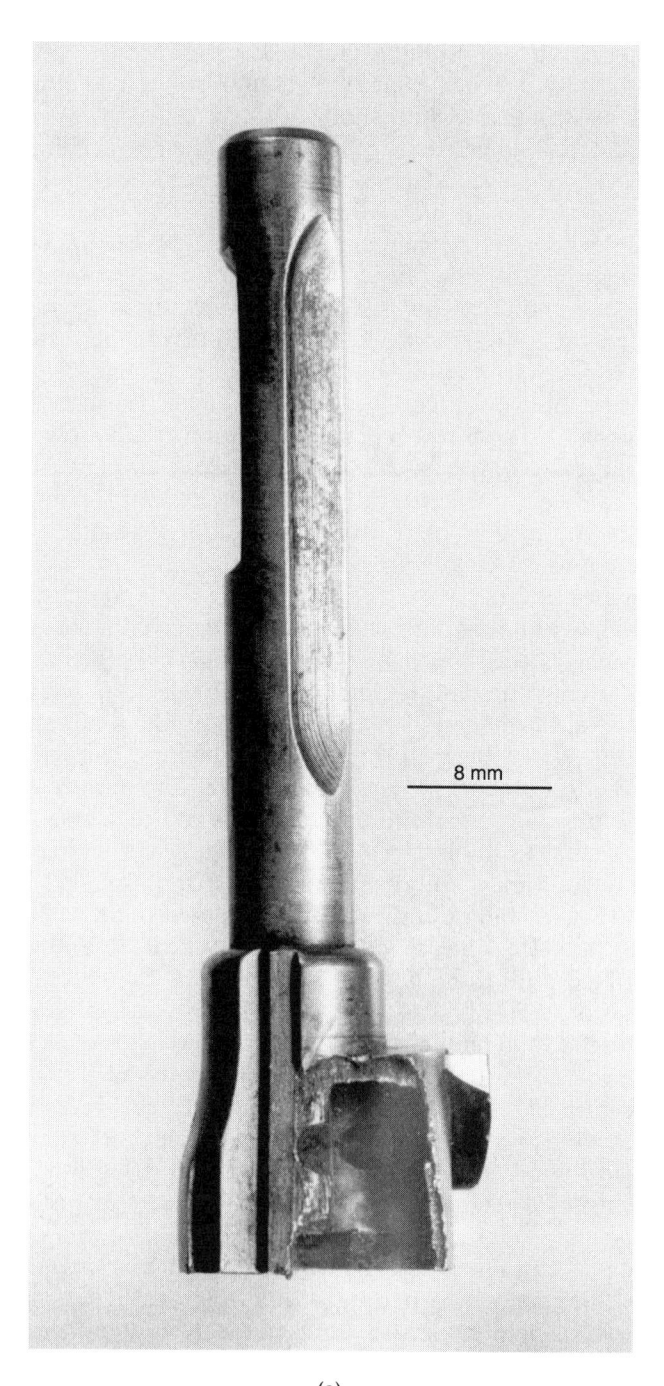

(a)

FIGURE 5.24
(a) Failed breech block. (b) Intergranular failure caused by grinding crack. Final fracture is in quasi-cleavage mode.

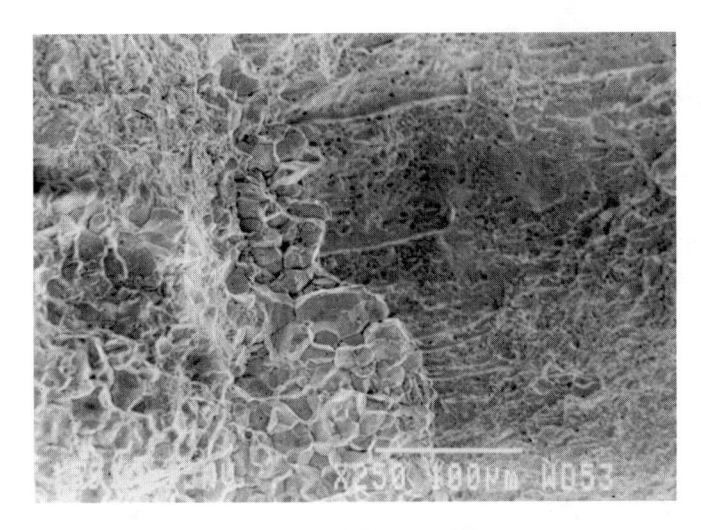

(b)

FIGURE 5.24 (continued)

strength levels, lattice strains and residual stresses. The greater the strength, the lower the tolerable hydrogen content. Usually a maximum of 2 ppm is specified for all high-strength steels.

Generally, the fracture due to HE in high-strength steels occurs along either prior-austenite grain boundaries or along the martensite lath pocket boundaries; sometimes both modes are in operation. The hydrogen in the lattice can easily be diffused out by suitable baking treatment between 200 and 250°C for times commensurate with section size and thickness of overlay electrodeposited coatings. Certain titanium alloys also suffer HE. Interstitial hydrogen and formation of brittle hydride precipitates are considered responsible for HE in titanium alloys. While both α-β and β alloys are susceptible, the loss of ductility due to HE is more pronounced in β alloys. Reduction of hydrogen levels by careful selection of raw materials and melting route is the most practical solution to avoid HE in Ti-alloys.

The role played by hydrogen in premature failures of components protected by electroplated coatings can be illustrated by the following example. The studs used in a turbo pump failed after 3 months of assembly. The studs were made from 0.35C-1.5Ni-1.0Cr-0.2Mo steel, quenched and tempered to 400 HV hardness, cadmium-plated and baked at 200°C for 2 h. The delayed failure shown in Figure 5.25 was predominantly in intergranular mode mixed with ductile failure features characteristic of delayed failure due to hydrogen embrittlement. Increasing the baking time to 12 h considering the low diffusivity of hydrogen across the relatively impervious cadmium coating solved the problem.

The examples cited indicate that innumerable defects originate during various stages of manufacture and all of them affect the component performance and reliability to some degree, depending on their severity. While it

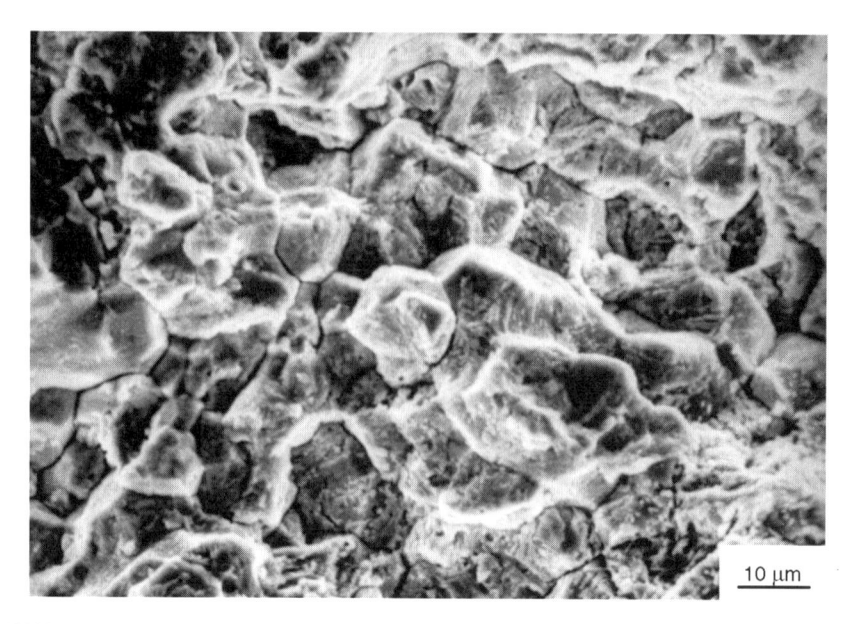

FIGURE 5.25
Intergranular fracture due to hydrogen embrittlement in Cd-plated Ni-Cr-Mo steel.

is to be acknowledged that it is impossible to produce defect-free components, every possible step should be taken to ensure conformance to the defect acceptance limits and to minimize defects. If the occurrence of certain defects beyond the acceptance limits is unavoidable, their influence on component performance has to be carefully evaluated by the designer to reestablish functional capability of the system.

6
Operational and Maintenance Defects

6.1 Introduction

A high percentage of failures occur as a result of defects developed due to abuse during operation of the system and human errors during mandatory maintenance. Depending on the actual operating conditions during the service life of an engineering system, the design life can decrease or increase as shown in Figure 6.1.

Operating any system beyond the design limits causes irrecoverable damage to the components due to unintended exposure to overload, overheating, corrosion and wear. Consequently, failures due to overload conditions; stress rupture, and creep and thermal shock caused by overheating; corrosion damages such as pitting corrosion, hot corrosion, stress corrosion cracking; and wear, erosion and fretting damages caused by wear and tear are commonly encountered. The most important life-limiting factors emerging from operational abuses are listed in Figure 6.2.

The primary damages are predictable and can be corrected during overhaul and maintenance. A more serious problem is the secondary damage, often in the form of initiation of fatigue cracks, originating at the defects formed as a result of structural and microstructural changes induced by any of the primary operational damages.

Corrosion- and wear-related failures are common occurrences due to operational and environmental conditions. An example of material degradation due to corrosion-induced damage is illustrated below. In a titanium sponge production unit, austenitic stainless steel pipes (AISI 310) through which molten magnesium flows intermittently were found to have been corroded at the inner surface. A typical corroded surface with corrosion debris, nickel-depleted layer and the substrate are shown in Figure 6.3. It is well known that molten magnesium leaches out nickel preferentially in austenitic stainless steels, resulting in severe corrosion. The problem was overcome by using austenitic (AISI 310)–ferritic (AISI 430) bimetallic stainless steel pipes, ensuring that magnesium is in contact with only the nickel-free ferritic stainless steel.

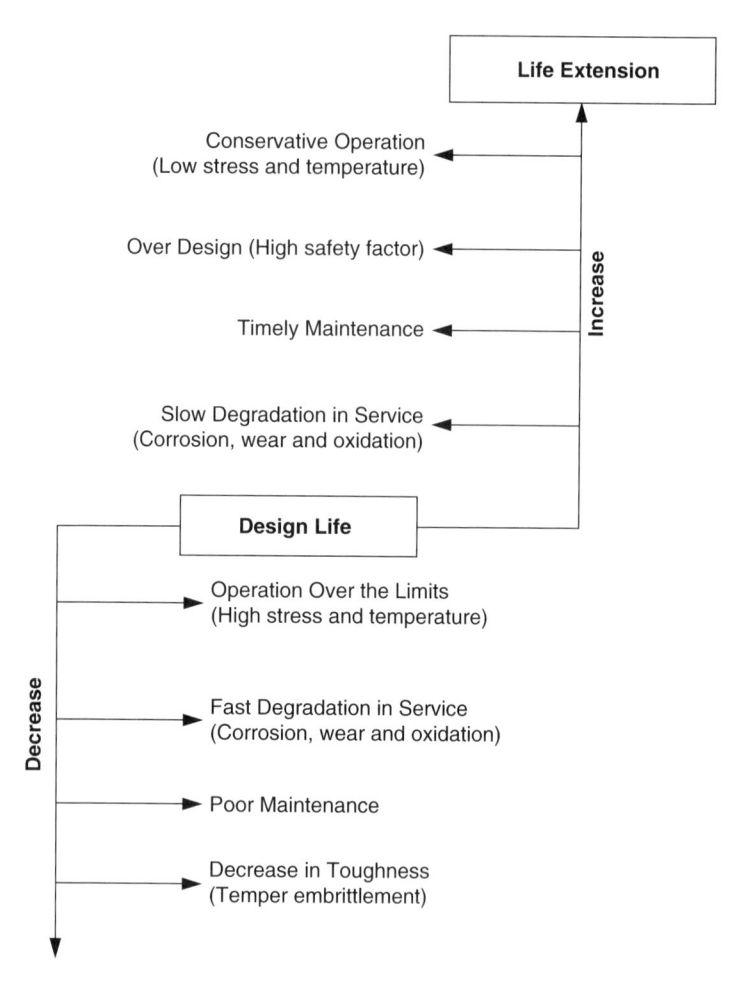

FIGURE 6.1
The variation of design life with operating conditions.

Premature failures due to wear are frequently encountered. Such failures generally occur due to operational abuses and metallurgical deficiencies associated with gross deviations in the specified manufacturing practice. The damage caused by improper operation can be illustrated with the malfunctioning of a labyrinth seal of an aeroengine that led to seizure of the shaft. The tapered labyrinth seal consists of a sleeve (0.6C-0.79 Cr steel) coated with abradable material (Ni-22Cu-8C-15BN) and ring (0.15C-2Ni-17Cr martensitic stainless steel). Metallurgical examination of the component revealed that (a) the abradable coating was completely lost, some of it deposited and fused to the grooves of the ring, (b) the projections of the ring were in direct contact with the sleeve during rotation of the shaft and (c) the sleeve and ring became fused at the contact points due to high frictional heat. The malfunctioning seal and the microstructural details are shown in Figure 6.4

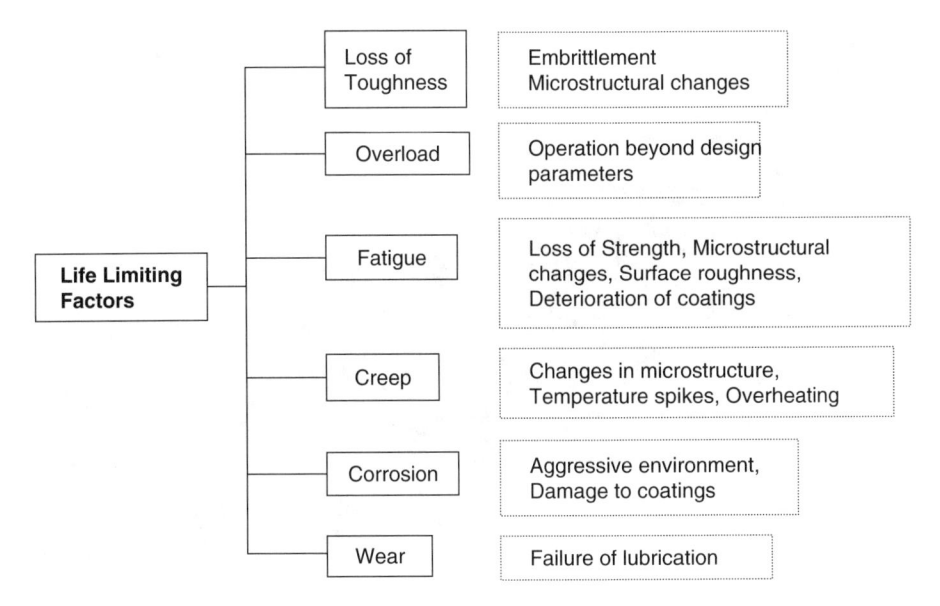

```
Life Limiting Factors
├── Loss of Toughness ──── Embrittlement, Microstructural changes
├── Overload ──── Operation beyond design parameters
├── Fatigue ──── Loss of Strength, Microstructural changes, Surface roughness, Deterioration of coatings
├── Creep ──── Changes in microstructure, Temperature spikes, Overheating
├── Corrosion ──── Aggressive environment, Damage to coatings
└── Wear ──── Failure of lubrication
```

FIGURE 6.2
Life-limiting factors due to operational abuses.

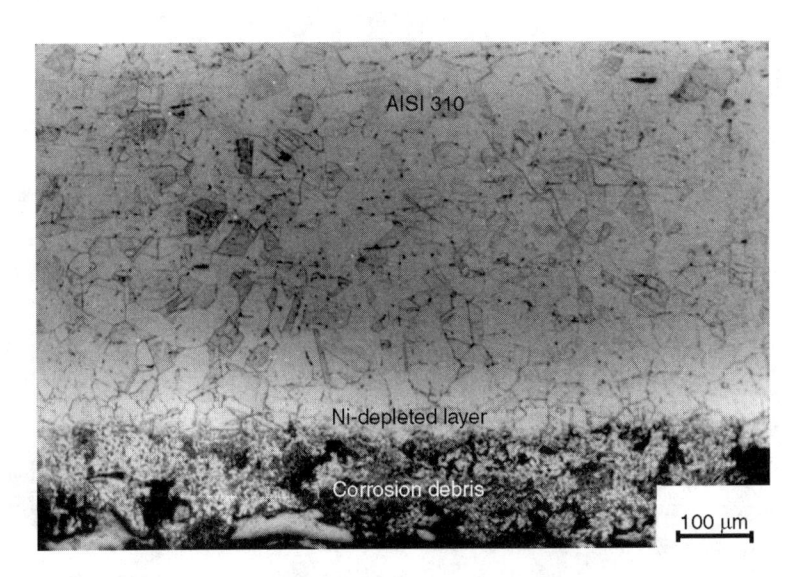

FIGURE 6.3
Microstructure of corroded AISI 310 pipe showing corrosion debris and nickel-depleted layer.

Operation of engineering systems beyond the specified limits leads to failures because of overload, excessive temperature and abnormal wear. The influence of overheating on the turbine stator ring of an aeroengine is described below. Severe surface roughness and the presence of pits and

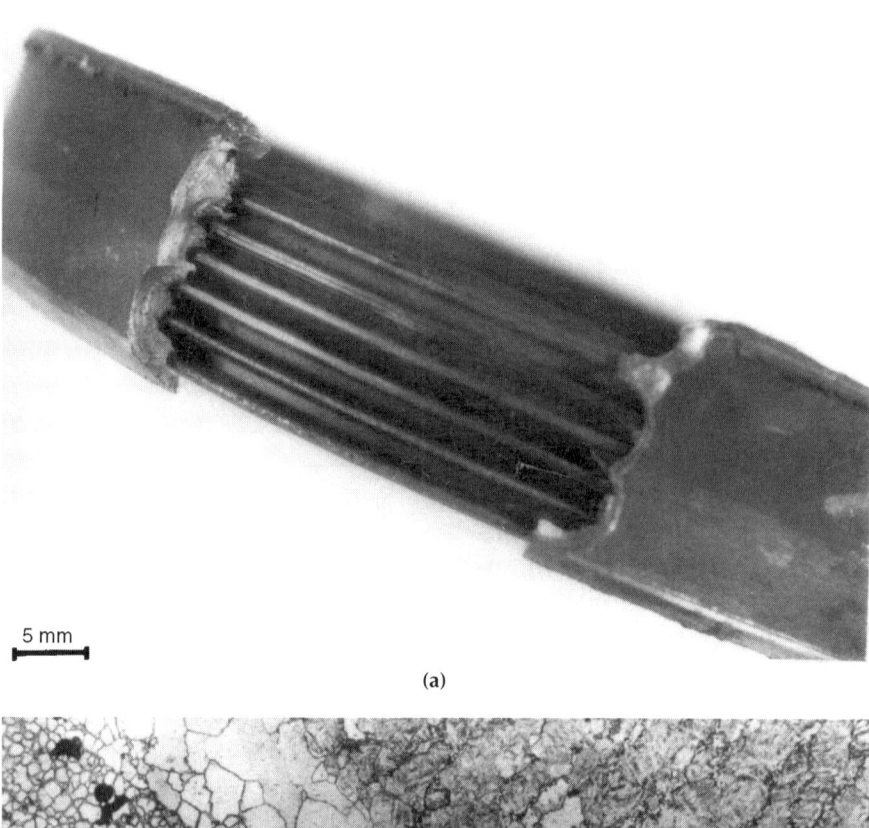

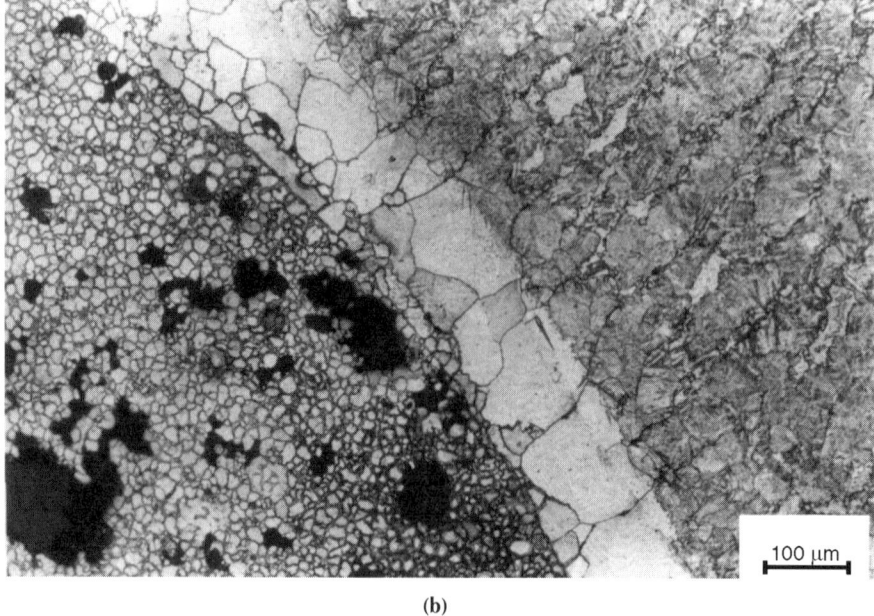

FIGURE 6.4
(a) Failed seal with partly removed sleeve showing metallic debris in the grooves of the ring. (b) Micrographs showing fused seal material in the grooves of the ring. Note decarburization and grain coarsening of the ring material. (c) Micrograph showing ring material welded to the sleeve material.

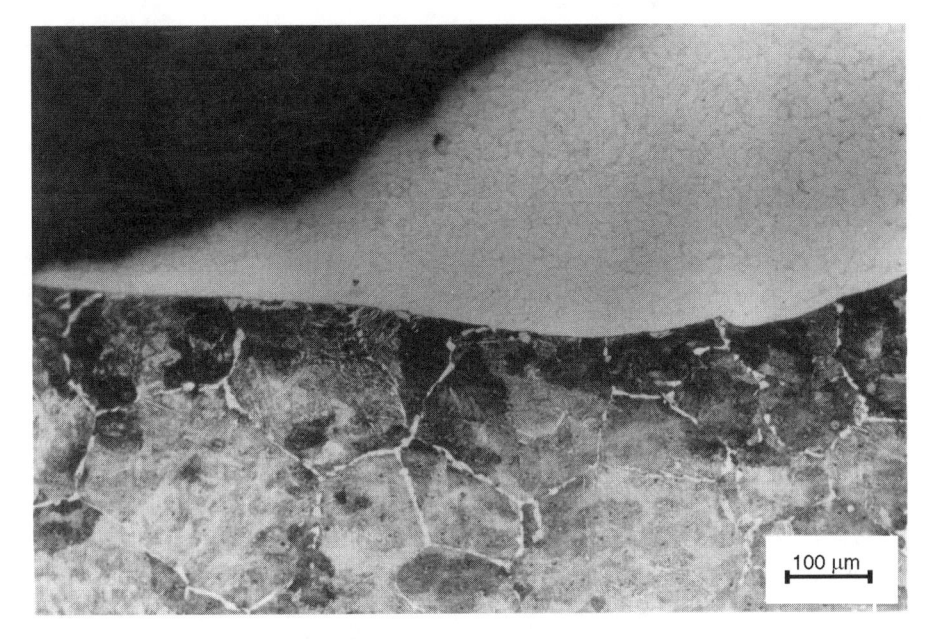

(c)

FIGURE 6.4 (continued)

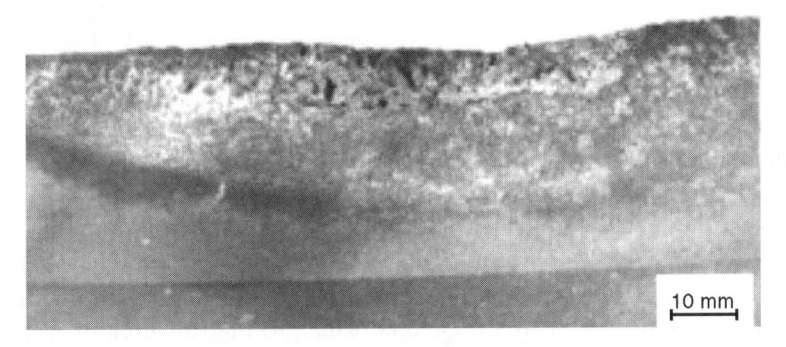

FIGURE 6.5
Photograph showing evidence of incipient melting, scaling and cavities on the overheated stator ring.

heavy scaling were observed on the stator ring when the aeroengine was strip-examined. Metallurgical examination revealed that (a) one edge of the ring was exposed to temperatures close to the melting point of the martensitic stainless steel (0.14C-10.5Cr-0.53Mo-0.25V), although the specified maximum operating temperature at the zone was 690°C, (b) microstructural and hardness changes had occurred and (c) shrinkage cavities were formed as a result of incipient melting and resolidification. The damaged portion of the turbine stator ring is shown in Figure 6.5.

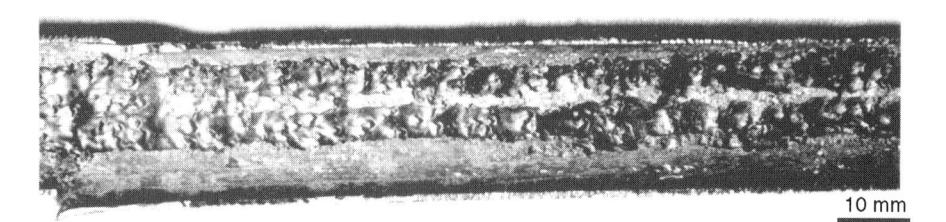

FIGURE 6.6
Erosion damage to a condenser tube caused by fly ash-laden cooling water.

The catalytic effect of operational and environmental damages in triggering primary mechanical failures is an unavoidable reality. For example, the synergistic effect of corrosion and erosion in causing premature failure is illustrated in the following example. The condenser tubes of a thermal power station made of Cu-20Zn-2Al-0.5Ni brass developed leaks. The small perforations responsible for the leaks were found to be associated with the degradation caused by the combined action of corrosion and erosion resulting in a reduction in the wall thickness. Analysis of the suspended particles in the cooling water indicated that the erosion was caused by fly ash particles. The corrosion-erosion damage to the condenser tube is shown in Figure 6.6.

Damage to upstream components and debris deposition on critical downstream components can cause serious accidents. For example, flame-out of an aeroengine was caused by the failure of the plunger of a fuel pump. The pump operates at a temperature of 105°C and pressure of 150 kg/cm^2. The failed plunger assembly and the segments are shown in Figure 6.7. On analysis, it was found that the plunger cylinder and the ball made of 1.5C-13.4Cr-0.38Mo-0.33V steel were covered with adherent splashes of Al-Si alloy. The Al-Si alloy, probably from the cylinder head, splashed onto the cylinder and blocked the orifice, resulting in bursting of the plunger under the operating pressure.

Surface damages such as dents, score marks, wear and corrosion caused by operational excesses can act as potential sites for initiation of cracks, as can be evidenced by the following example. The throttle lever connected to a fuel-metering unit of an aircraft failed during ground run after an overhaul. The lever was made of 16Cr-2Ni martensitic stainless steel. The fatigue failure of the lever originated at a deep score mark (Figure 6.8) generated due to improper handling during overhaul.

6.2 Maintenance Defects

Proper maintenance is the essence of extracting the design life from the system. During maintenance, it is necessary to measure the extent of damage caused to the components in order to assess the reusability of the system

Operational and Maintenance Defects 111

(a)

FIGURE 6.7
(a) Failed plunger assembly. (b) Plunger cylinder and balls with ring surface. (c) White metallic deposition on the surface of the ball.

after repairs and replacements. The type of maintenance, the thoroughness of inspection at each stage and the time interval between overhauls depend on the criticality of the components under defined loading conditions, and the time required for the initial crack to grow to a critical length (estimated based on fracture mechanics concepts) that can lead to catastrophic failure dictates the maintenance frequency. The checks at each level of maintenance are generally derived from known precedents. The maintenance concept generally includes (a) base level periodic inspection to ensure freedom from obvious defects that can impair the performance of the system, (b) secondary level maintenance to postpone or prevent the occurrence of an anticipated

20 mm

(b)

100 μm

(c)

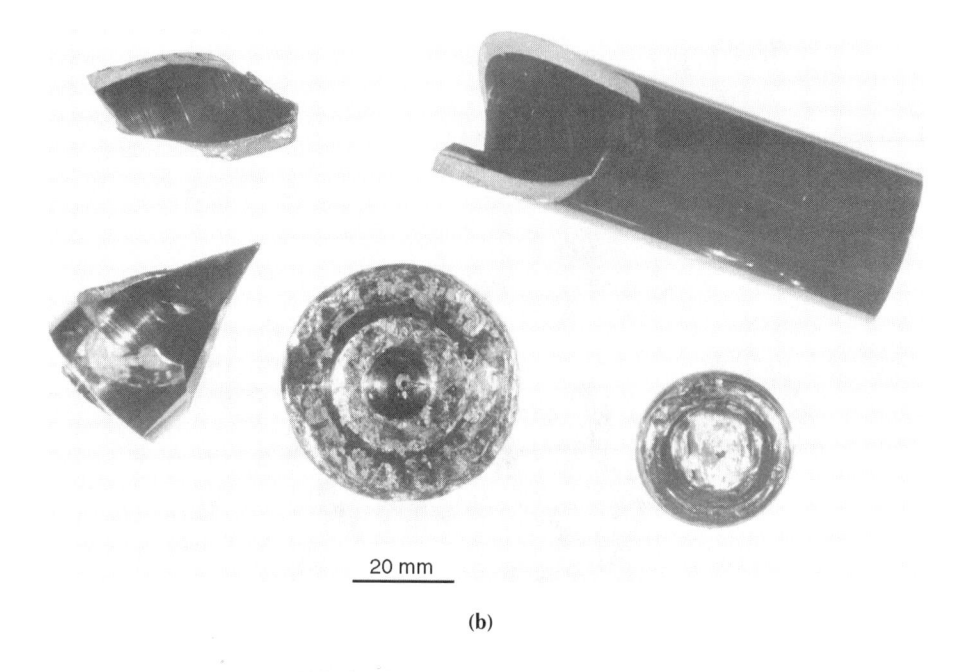

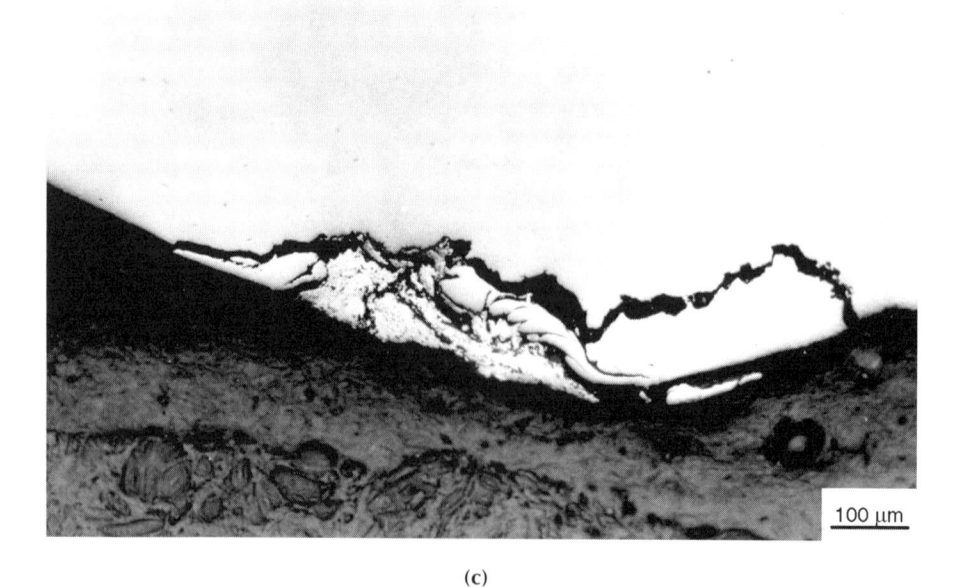

FIGURE 6.7 (continued)

failure and (c) more exhaustive breakdown maintenance to restore the functionality of the system after the failure event. The scope of various maintenance schemes and the expertise level desired under each scheme are shown below:

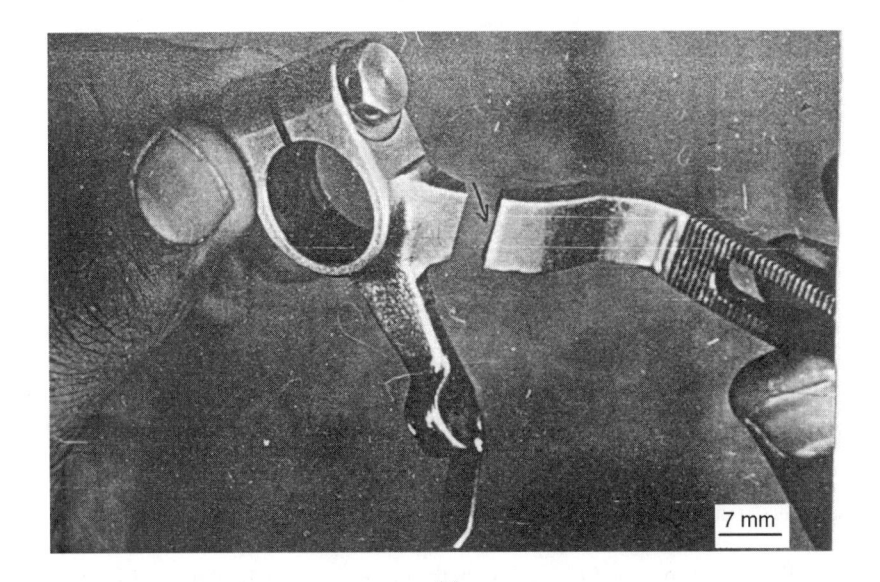

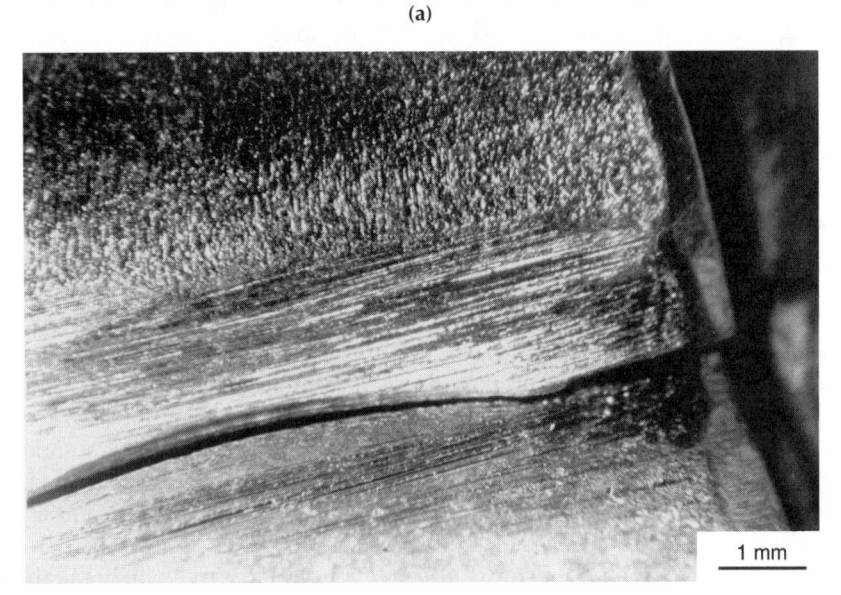

FIGURE 6.8
(a) Failed throttle lever. (b) Deep score mark at which a fatigue crack initiated.

Periodic Maintenance
   Limited expertise is adequate
   Inspection of defect-prone components carried out
   Visual inspection is generally adequate
   Carried out very frequently

Preventive Maintenance
  Aimed at preventing system breakdown
  Less frequent but more elaborate inspection
  Selective strip examination is necessary
  Exhaustive nondestructive testing
  Comprehensive expertise is needed
  Limited repair and replacement is practiced
Breakdown Maintenance
  Complete strip examination is carried out
  Comprehensive damage assessment is done
  Exhaustive repair and replacement schemes are employed
  Corrective schemes for secondary damages are practiced
  Interdisciplinary and high technological skills are required

Periodic maintenance is to ensure that the system has not suffered any damage during previous use and that it will function without failure in the next run. Preventive maintenance is intended to ensure safe performance by early detection of operational damages and to initiate corrective measures. Breakdown maintenance carried out after the failure occurs is intended to put the system back into service. Most of the sophisticated engineering systems are backed up by well-laid maintenance plans issued by the designer. Nonadherence to the specified procedures and human errors during stripping, quality checks and reassembly are responsible for many failures occurring immediately after maintenance.

The increased incidence of failures of cylinder heads of an aeroengine after reworking to remove eroded surfaces to prevent leakage of gases between the cylinder and cylinder head is an example of the things that can go wrong during maintenance. The cylinder head, an aluminum alloy casting (Al-4Cu-1.2Mg-2.1Ni-0.54Si-0.27Fe), is fixed to the cylinder with steel studs. During reworking, the studs are removed and oversized studs are fitted after retapping the stud holes. Oblong-shaped, retapped holes, the presence of broken studs left in the holes and fitting oversized studs over the original broken studs were responsible for the development of cracks at the cooling fin grooves where the thickness of the casting is the lowest. The details are shown in Figure 6.9.

A large number of defects are generated during operation and maintenance of engineering systems. While many defects are corrected during various maintenance stages, others go unnoticed or are allowed as insignificant without proper evaluation. Such defects on their own can cause premature failures or can trigger some serious failure modes. Extreme caution is to be exercised in permitting use of components with noticeable defects detected during maintenance. Operating the system within the stipulated guidelines is a sure way to ensure freedom from failures caused by operational abuses.

60 mm

(a)

FIGURE 6.9
(a) Cracked cylinder head. (b) Transverse section showing oblong stud hole. (c) A stud fitted over another broken stud [4].

Summary

Despite all our technological advancements, premature service failures of engineering components is a hard reality that causes huge direct and indirect losses to society. The driving force for failure analysis lies not just in finding the fault, but also in the necessity to learn from mistakes in order to prevent recurring failures. Failure analysis involves systematic and careful probing

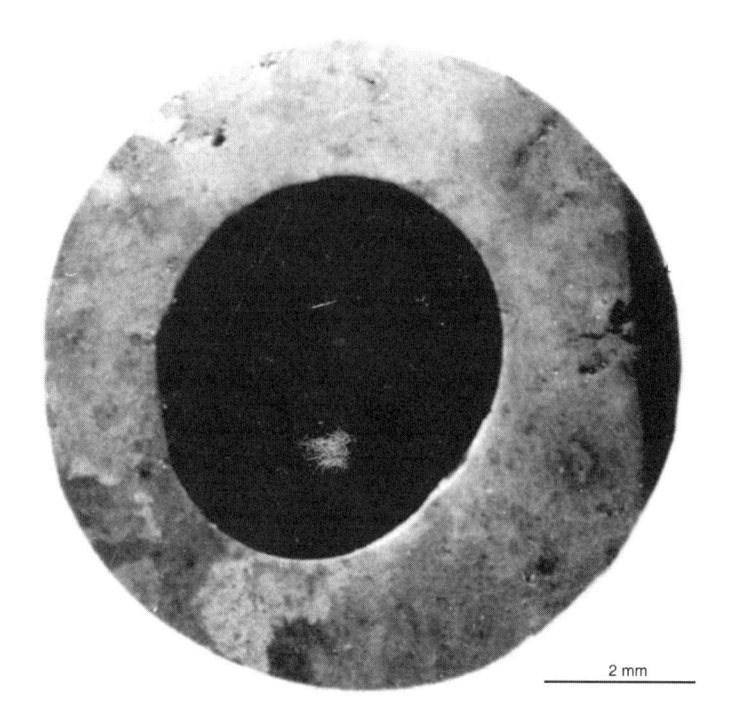

(b)

FIGURE 6.9 (continued)

into every minor detail of the failed component to find out the root cause of failure and is, as such, an extremely involved affair demanding extensive knowledge, common sense, practical wisdom, and, importantly, impeccable honesty on the part of failure analysts. It is the job of a failure analyst to suggest appropriate and practicable remedial measures, to devise suitable methodologies for their implementation in consultation with the shop floor engineers from the production organization and to reassess the effectiveness of his or her suggestions for preventing failures. However, it is the manufacturer who makes the ultimate decision whether or not to implement the remedial measures suggested by the failure analyst.

Failure analysts should gather as much information as possible about the failed component, about its service history until the failure event, and about the whole engineering system to which the failed component belongs. They should work with an open mind and should not become prejudiced by their own opinions on the probable cause of failure. They should establish all possible causes of failure and systematically eliminate all other possible causes to arrive at the single most probable cause of failure based on their own experimental results. Various tools such as FTA and FMAAM are available to help failure analysts in this regard. Careful visual examination, optical microscopy, study of fracture surfaces using SEM, and hardness measurement

Operational and Maintenance Defects

(c)

FIGURE 6.9 (continued)

are adequate in analyzing most of the failures. Fracture surfaces provide the most vital clues in identifying the root cause of failure. Fracture feature and causative factor analysis provides a systematic method of identifying the cause of failure based on the observed fracture features. A variety of advanced characterization techniques such as EPMA, AES, TEM, etc. can also be employed, if needed, to generate supportive evidence for accurately identifying the cause of failure. Failure analysts must be able to filter out misleading information and redundant experimental evidences using their own discretion. The conclusions drawn by the investigators must always be based on unambiguous experimental evidence.

Almost all premature service failures can be traced back to any one or more of the following: deficiencies in component design, wrong material selection, manufacturing deficiencies, lapses in quality assurance, operational abuse and poor maintenance. While elimination or minimization of all the above would certainly prevent most of the premature failures, the same is easier said than done. The presence of defects in metallic engineering components is a particularly common occurrence. It is necessary to assess the damage potential of a defect before defining specification limits, bearing in mind the fact that minimization of defect incidence involves extensive and expensive efforts. While every defect need not cause premature failure, adequate attention must be paid to control them within the design stipulations. Strict enforcement of quality assurance and quality control procedures is important in order to prevent defective components from getting into service. Involvement of a mature failure analyst right from the design stage in the development of an engineering component would certainly help in eliminating or minimizing the deficiencies in component design, material selection and manufacture. Similarly, respecting safe operating limits and maintenance regulations would help in preventing premature service failures.

Composition is the first step –
Place it correctly!
Cleanliness is the second step –
Assess and advance!
Processing and heat treatment are the
left and right –
Watch and proceed!
Quality assurance is the "*Agni*" –
Let the competent handle!
Maintenance is murky –
Manage it well!

Case Studies

Part 1

Failures Due to Improper Material Selection and Heat Treatment

1.1 Failure of Emitting Electrodes of an Electrostatic Precipitator

Key words: Austenitic stainless steel, Residual stresses, Stress corrosion cracking

Introduction

The electrostatic precipitator (ESP) with 27630 electrodes is used in thermal power stations to collect dust from flue gases. The electrodes, made of AISI 316-grade austenitic stainless steel wire in the form of helical springs, are stretched and assembled in ESP as shown in Figure C1.1. A number of emitting electrodes of ESP failed over a period of 45 days during commissioning of a power station, which was operated intermittently with heavy oil as the fuel [5]. The fuel oil contained 0.5% max. S and 200 ppm Cl. The operating temperature was in the range of 70 to 120°C.

Experimental Results

Examination of the emitting electrodes revealed the presence of some brown spots and a few transverse cracks. The surface corresponding to the inner surface of the helical spring was more damaged than the outer surface. The fracture surface of the failed electrodes had two distinct regions. Region I was rough and covered with a dark brown deposit and Region II was smoother, brighter and twisted (Figure C1.2). Region I showed mostly transgranular faceted fracture with a number of secondary cracks and Region II showed elongated dimples. The fracture features at the transition are shown in Figure C1.3.

A few samples containing the defects were examined. A number of pits with brown deposits were observed (Figure C1.4). A number of branching cracks propagating in mixed inter- and transgranular mode in deformed

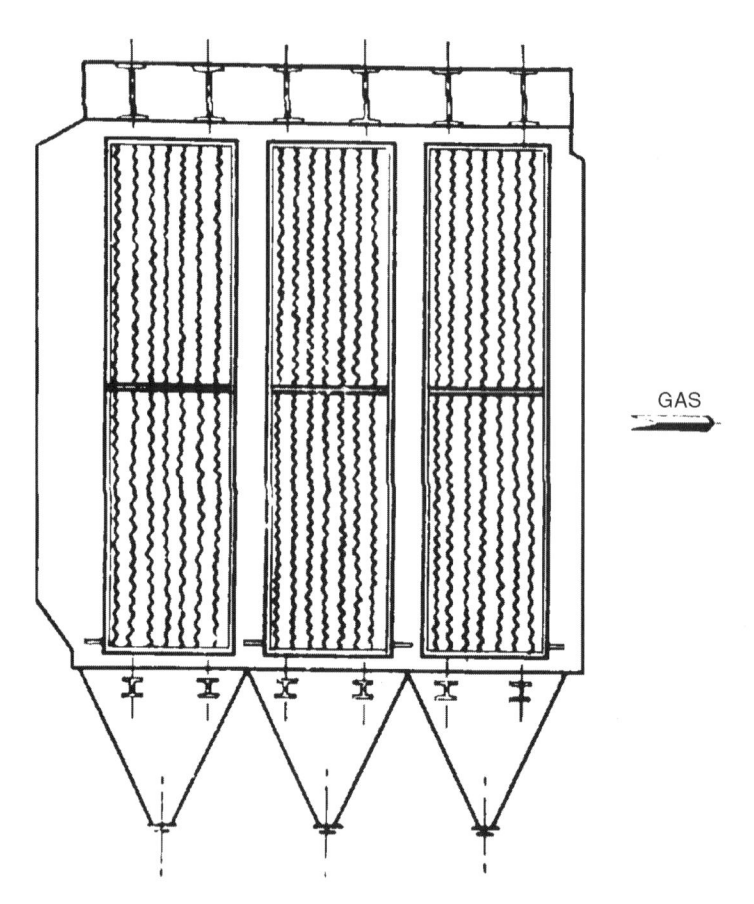

FIGURE C1.1
Emitting electrode system in the electrostatic precipitator.

austenite grain structure were also observed, along with a few yawning cracks (Figure C1.5). The brown deposit in some of the cracks on analysis by EDS was found to contain chlorine, sulfur, calcium, iron, chromium, and silicon. The chemical composition of the failed electrode wire is given in Table C1.1. The hardness of the failed electrode wires was 425 HV.

FIGURE C1.2 Fracture surface of the failed electrode wire showing two different regions.

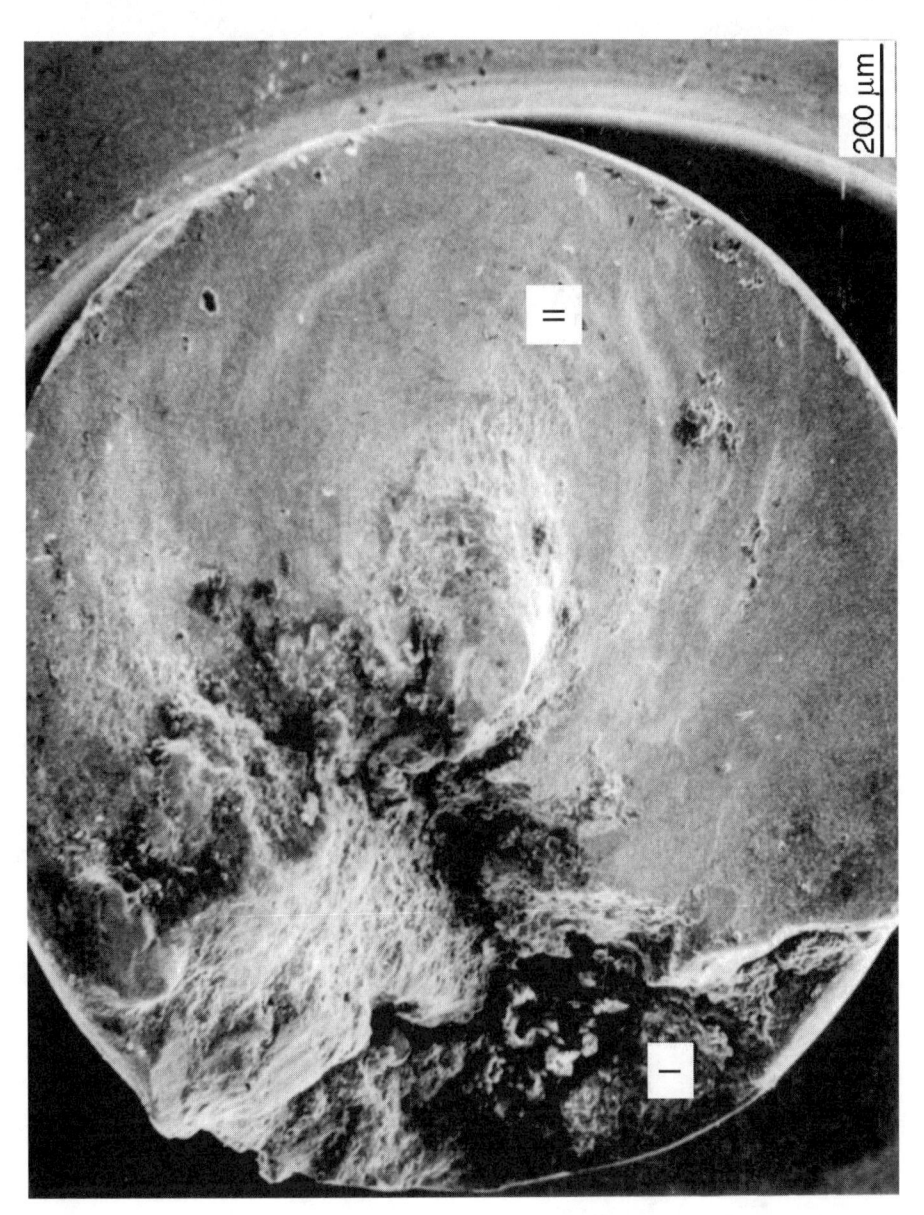

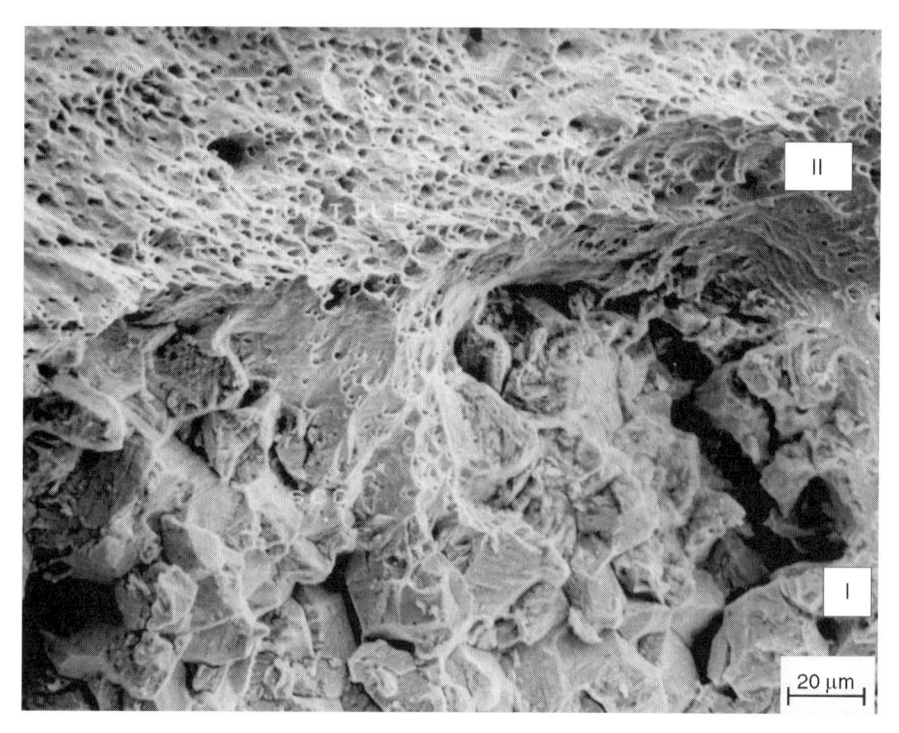

FIGURE C1.3
SEM fractograph showing the transition from the transgranular faceted fracture to the dimpled rupture.

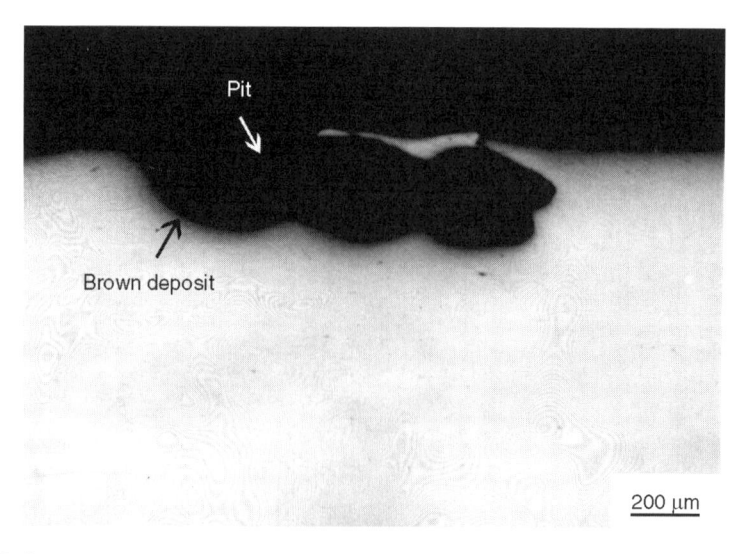

FIGURE C1.4
Photomicrograph showing a pit with brown deposit.

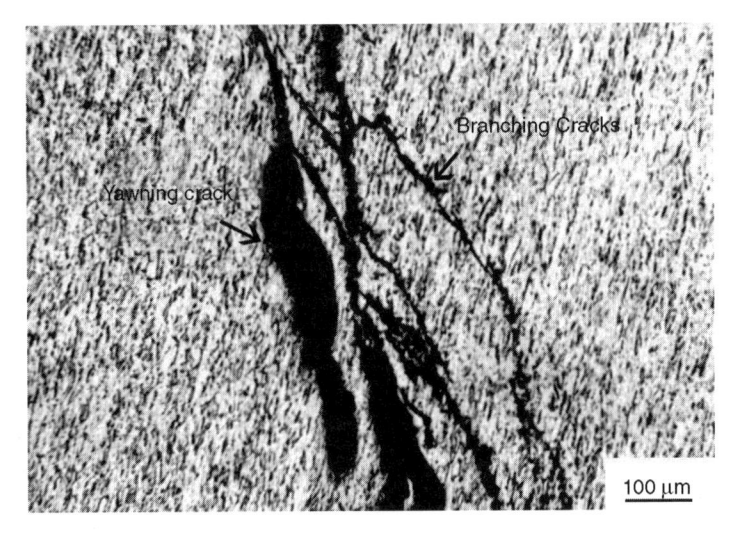

FIGURE C1.5
Micrograph showing branching cracks in mixed inter- and transgranular mode in deformed austenite grains. A few yawning cracks are also noticeable.

TABLE C1.1

Chemical Composition of the Failed Electrode Wire

Element	Wt%
C	0.05
Si	0.4
Mn	1.2
Ni	10.6
Cr	17.5
Mo	2.3
P	0.02
S	0.01
Fe	Bal

Discussion

The presence of pits, broad cracks filled with corrosion product, and sharp branching cracks suggested that the damage to the electrodes was caused by corrosion. Austenitic stainless steels, including the Mo-bearing grades, are prone to pitting and SCC in atmospheres containing chlorides at temperatures above 80°C. The presence of large numbers of branching cracks propagating in mixed mode and the fracture features lend evidence to the fact that the initial failure was due to SCC, the stress component essentially coming from the residual stresses due to cold drawing and the stress imposed due to stretching during assembly. The intermittent shutdown of the plant during commissioning might have promoted stress corrosion cracking due to stagnant conditions leading to autocatalytic effects in the chemistry of the electrolyte. The final failure of the wire was due to overload, as evidenced by dimpled rupture.

Conclusion

The failure of the emitting electrodes was due to SCC.

Remedial Actions

The most effective solution to the problem of electrode cracking in ESP is to use electrodes made of duplex "super" stainless steels of the type 22-10-05.

1.2 Failure of Impellers of a High-Pressure Water Pump

Key words: Brass, Intergranular fracture, Stress corrosion cracking

Introduction

The impeller blades of the first, second and third stages of a self-priming, multistage, centrifugal high-pressure water pump of 33 m^3/h capacity at an operational speed of 1450 rpm were damaged [6]. The pump was in service for 2 years and was used only intermittently. The impellers were specified to be cast with leaded tin bronze conforming to IS 318–1987, grade 2.

Experimental Results

The blades of I and II stage impellers were completely broken at different levels from the hub and those of the III stage impeller were badly bent and twisted. The failed I stage impeller is shown in Figure C1.6a/b. While two of the blades that failed close to the root had undamaged fracture surfaces showing coarse crystalline fracture on the convex surface, the rest of the fracture area of the failed blades and all the other blade fracture surfaces showed fibrous fracture. The section thickness at the failed region was 8 mm. The fracture surfaces of many blades were completely damaged.

The fracture surface of the blade that failed at the root showed faceted crystalline fracture covered with debris to a depth of 4 mm from the convex surface (Figure C1.7). The intergranular fracture with secondary cracks at the origin and dimpled rupture away from it are shown in Figures C1.8a and C1.8b, respectively. EDS analysis of the debris on the fracture surface

(a)

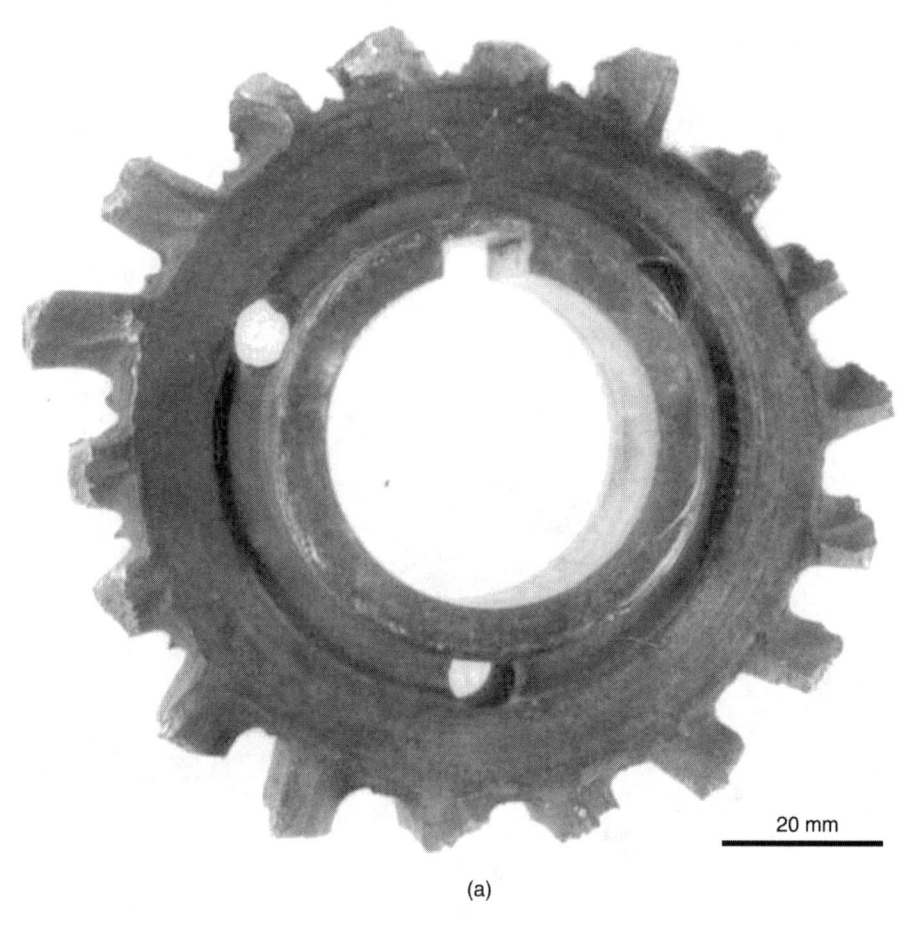

FIGURE C1.6
(a) I stage impellers with broken blades. (b) The two blades that failed close to the root (arrows). Fracture originated on the convex surface.

showed that it contained Cl, S, and Ca in addition to the elements in the material (Figure C1.9). A longitudinal section at the root of the blade was examined. General corrosion with evidence of intergranular cracking was observed (Figure C1.10). The microstructure consisted of needles of α in β matrix with dark particles of Fe-rich phase. The chemical composition of the failed impeller blade along with the specified composition is given in Table C1.2. The hardness of the impeller was 132 HV.

FIGURE C1.6 (continued)

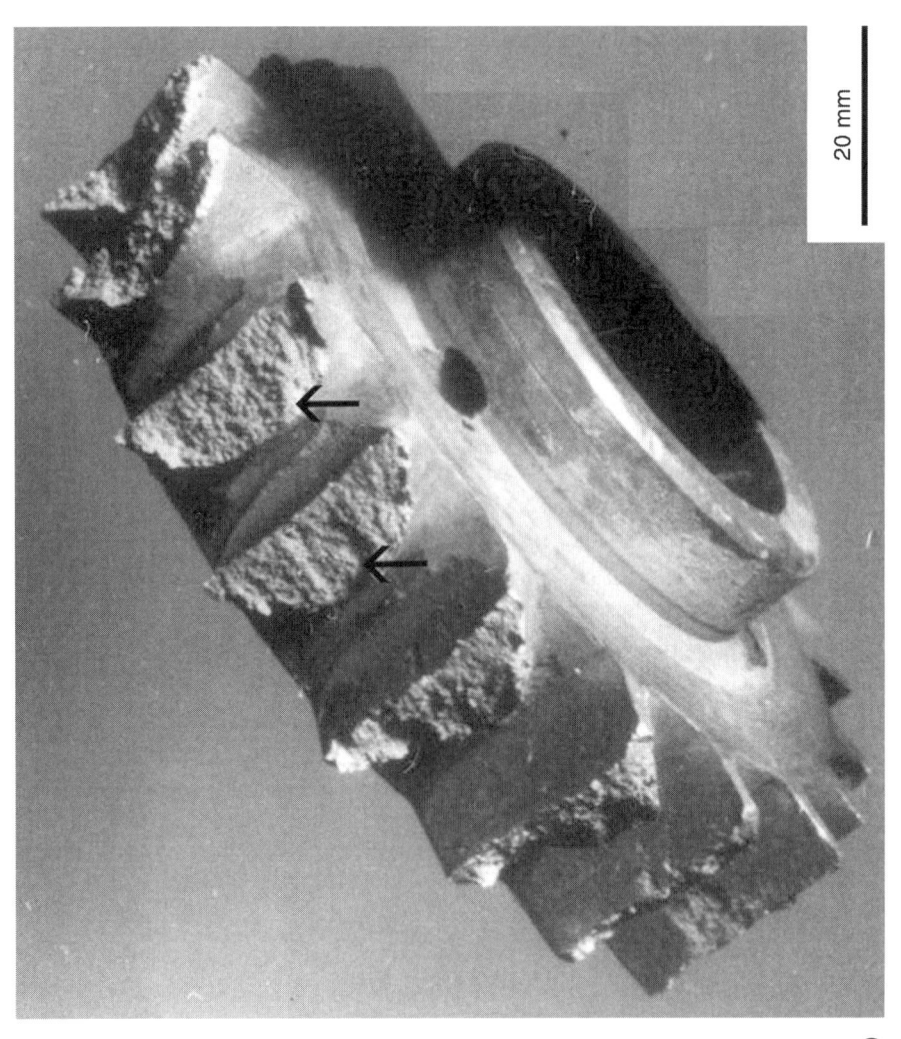

FIGURE C1.7
SEM fractograph of the blade that failed at the root showing faceted crystalline fracture features covered with debris.

(a)

(b)

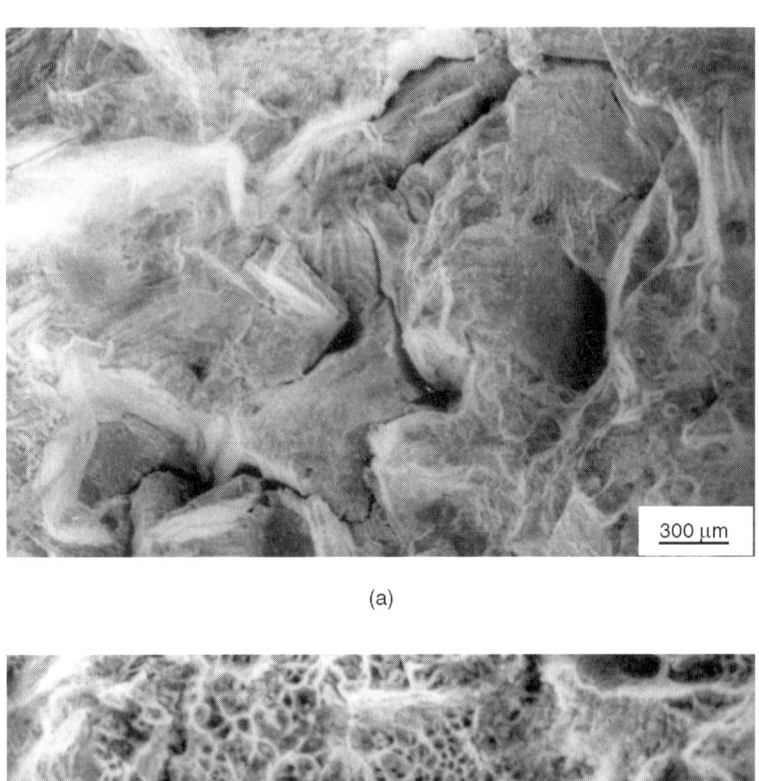

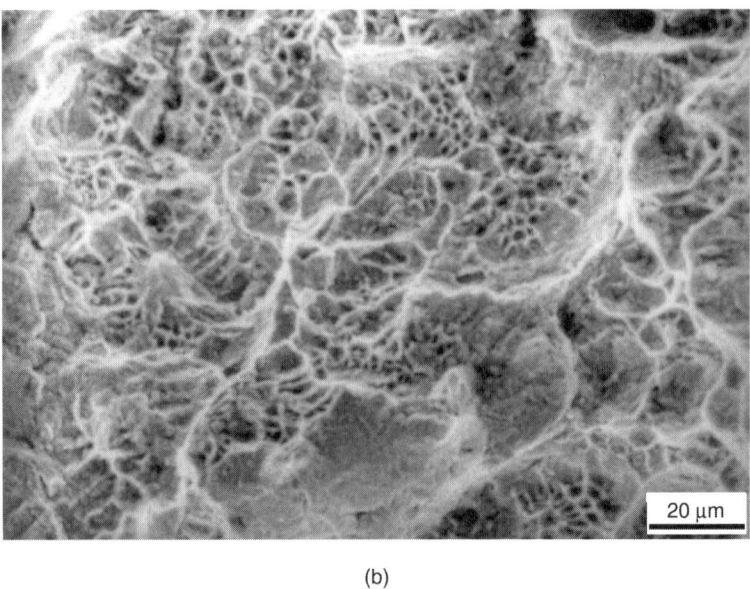

FIGURE C1.8
SEM fractograph of I stage impeller blade showing (a) intergranular fracture features and secondary cracks at the fracture origin and (b) equiaxed dimples on the remaining part of the fracture surface.

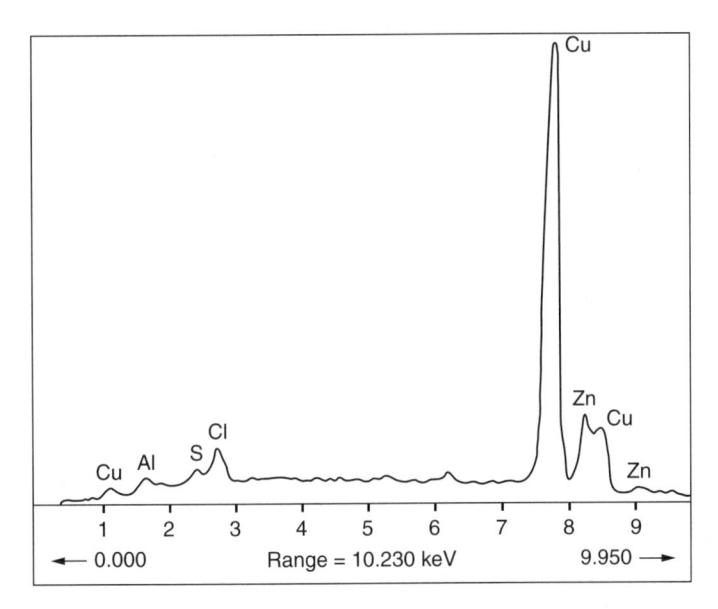

FIGURE C1.9
EDAX spectrum obtained on corrosion debris.

FIGURE C1.10
Microstructure of the blade sample showing intergranular corrosion and cracks near the fracture surface. Etchant: acidified $K_2Cr_2O_7$ solution.

TABLE C1.2

Chemical Composition of the Failed Impeller Blade Along with the Specified Composition

Elements	Sample (wt%)	Specified (wt%)
Zn	39.2	4–6
Fe	0.7	0.35 max
Pb	0.19	4–6
Mn	0.60	—
Al	1.45	—
Sb	0.16	0.3 max
Sn	—	4–6
Cu	Bal	

Discussion

The chemical composition of the impeller was not the same as the specified grade. The experimental results suggest that (a) two blades of I stage impeller failed first and remaining damages are consequential, (b) failure of the first-to-fail blades originated at the convex surface near the root, (c) initial cracking was due to SCC in intergranular mode as evidenced by corrosion debris and secondary cracks at the fracture origin, and (d) final failure occurred under overload conditions as evidenced by the dimpled rupture features on the fracture surface away from the fracture origin. The failure therefore was caused by stress corrosion due to the combination of chloride- and sulphate-containing water and the operational stresses. Stagnant conditions during the unused period of the pump might have aggravated the problem. The change of material from a more corrosion-resistant tin bronze to brass is the primary cause for the premature failure.

Conclusion

Two of the I stage impeller blades failed due to stress corrosion cracking initially and failures of the remaining I, II, and III stage impellers were consequential damages.

1.3 Failure of a Track Shoe of an Infantry Combat Vehicle

Key words: *Mn-Mo steel, Intergranular fracture, Phosphorus segregation, Tempered martensite embrittlement*

Introduction

A few batches of track shoes used in the track assembly of an infantry combat vehicle failed while conducting acceptance trials on the vehicle [7]. The track shoes were specified to be forged from electric arc-melted steel conforming to Russian grade 20ХГСНМ, machined and heat treated. The specified heat treatment consists of heating to 880°C, quenching in oil and tempering at 230°C for 3 h.

Experimental Results

The failed track shoe is shown in Figure C1.11. The failure occurred along the length of hole through which the track pin is inserted for track assembly. The fracture surface was crystalline over its entire length and was covered with dark oxide at many places at the inner surface of the hole (Figure C1.12). Fractographic examination revealed that the oxide-covered region showed mixed intergranular and transgranular cleavage fracture with a number of secondary cracks (Figure C1.13). The rest of the failure was in intergranular mode (Figure C1.14). In order to identify the cause for intergranular failure, AES analysis was carried out on an *in-situ* fractured sample and segregation of phosphorus to prior-austenite grain boundaries was identified (Figure C1.15).

During microstructural examination, long stringers of silicate inclusions along with a number of small oxide inclusions were noticed in the as-polished sample (Figure C1.16). On etching with 2% Nital, the sample showed tempered martensitic structure (Figure C1.17). The specified chemical composition and that of the failed sample are given in Table C1.3. The average hardness of the failed track shoe was 485 HV.

FIGURE C1.11
Photograph showing the failed track shoe.

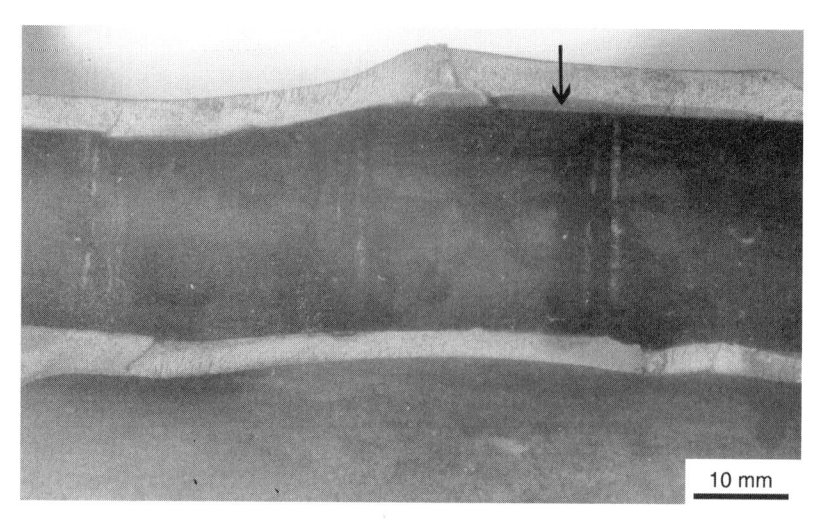

FIGURE C1.12
Fracture surface of the track shoe showing coarse crystalline fracture and dark areas near the inner surface (arrow).

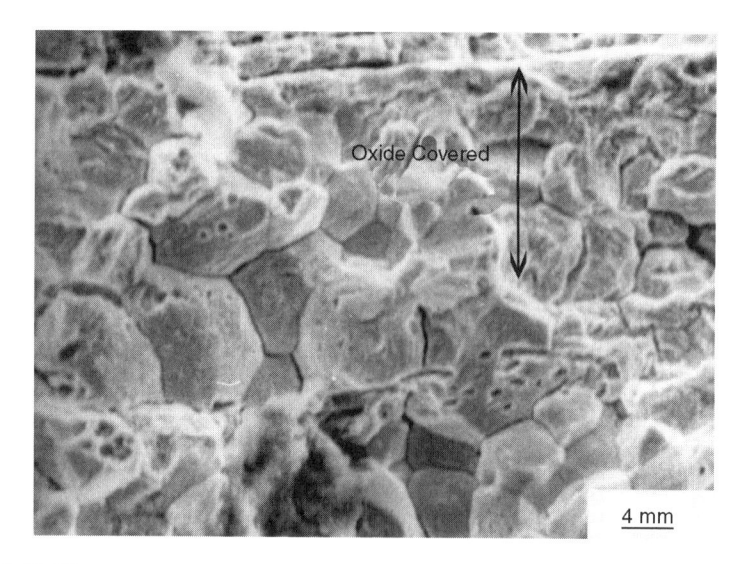

FIGURE C1.13
SEM photograph near the inner surface showing mixed transgranular and intergranular fracture with secondary cracks.

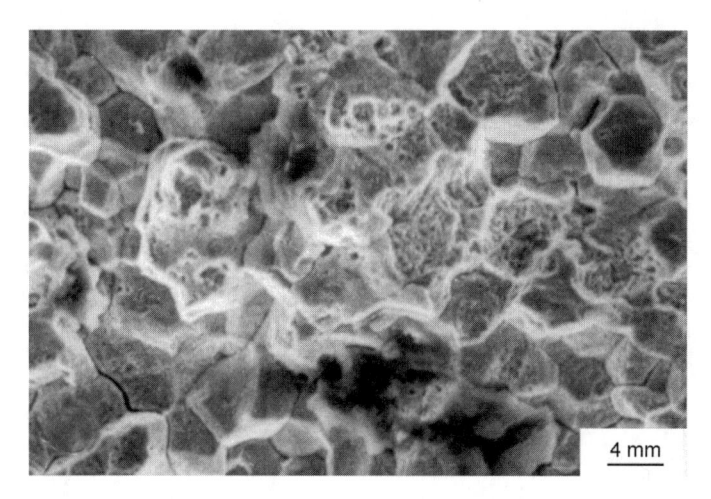

FIGURE C1.14
SEM fractograph of the fracture surface showing intergranular fracture and fine dimples on the grain facets.

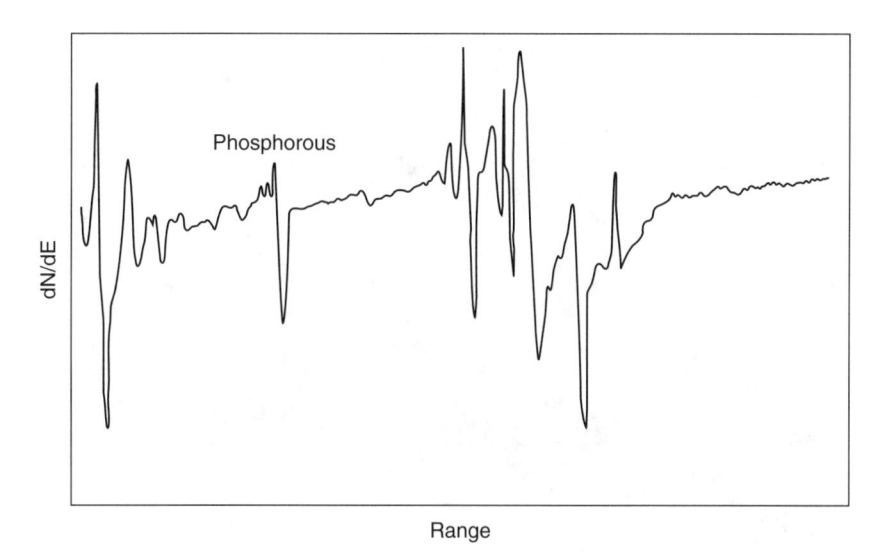

FIGURE C1.15
Auger electron spectrum of the specimen showing phosphorous peaks.

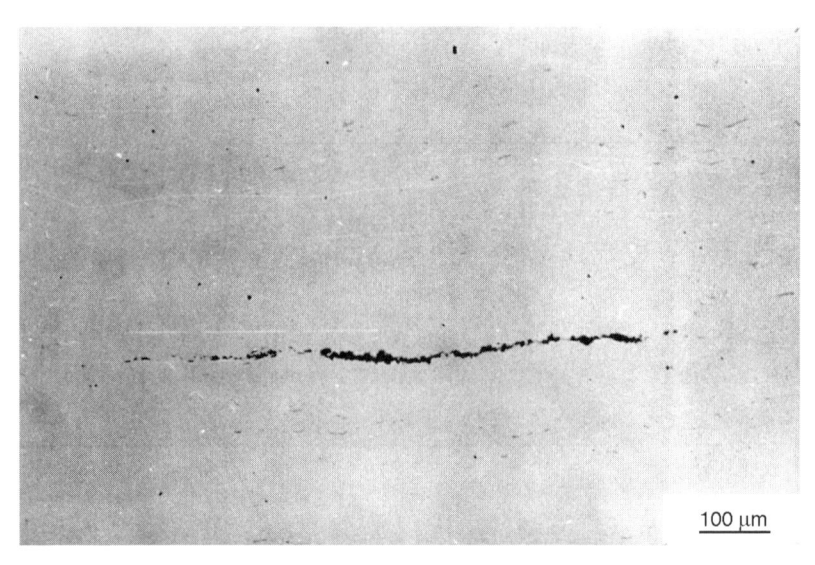

FIGURE C1.16
Photomicrograph of the as-polished sample showing a long stringer of nonmetallic inclusion.

FIGURE C1.17
Photomicrograph of the etched sample showing tempered martensite and stringers of nonmetallic inclusions.

… Failures Due to Improper Material Selection and Heat Treatment

TABLE C1.3

Chemical Composition of the Failed Track Shoe Along with Specified Composition

Element	Failed shoe	Specified
C	0.33	0.18–0.24
Si	0.20	1.20–1.50
Mn	1.44	0.90–1.20
Ni	—	0.90–1.20
Cr	0.16	0.60–0.90
Mo	0.43	0.10–0.15
Cu	0.18	0.3 max
S	0.05	0.035 max
P	0.02	0.035 max
Fe	Bal	Bal

Discussion

The test results indicated that the steel used for the track shoes did not conform to the specified grade. The extremely large population of nonmetallic inclusions in a component with such a low wall thickness is detrimental to the transverse ductility. The oxide-covered initial cracks at the inner surface of the hole should have existed prior to hardening as the formation of such an adherent oxide layer at a low tempering temperature of 230°C can be safely ruled out. The Mn-Mo steel used for the production of track shoes is susceptible to tempered martensite embrittlement (TME) when tempered at 220 to 400°C, whereas the specified Si-Cr-Mn-Mo steel is not susceptible at the specified tempering temperature. The usage of a steel prone to embrittlement at the tempering temperature resulted in P segregation at prior-austenite grain boundaries, causing grain boundary embrittlement and brittle failure of the track shoe in intergranular mode. The preexisting cracks acting as stress raisers aggravated the cracking tendency.

Conclusion

The failure of the track shoe was due to TME. The use of the wrong steel (different from the specified grade) and tempering it at a temperature that is in its TME range resulted in the failure of the track shoe. Cracks acting as stress raisers aggravated the cracking tendency.

1.4 Failure of a Draw Hook of a Railway Car

Key Words: 0.6% Plain carbon steel, Cleavage fracture, Lamellar pearlite, Notch toughness

Introduction

A high-speed, 21-car passenger train was involved in an accident, resulting in derailment of 17 cars and considerable casualties [8]. The accident was suspected to have been caused due to failure of a draw hook fitted at the front end of the third car, which was the first to derail. The draw hook was specified to be forged from 0.6% C steel and hardened and tempered to a minimum hardness of 255 HB and Izod impact strength of 27 J.

Experimental Results

A portion of the failed draw hook is shown in Figure C1.18. The fracture surface was rough and bore radial marks pointing to a triangular region at one corner of the fracture surface (Figure C1.19). A portion of the triangular region of the fracture surface near the surface of the hook was smooth and covered with dark oxide (Region A) followed by a relatively coarse fracture (Region B) (Figure C1.20). Thin shear lips were observed near all the edges of the fracture surface.

FIGURE C1.18
Portion of the failed draw hook.

FIGURE C1.19
Fracture surface showing radial marks pointing to a triangular region (arrow).

FIGURE C1.20
Fractograph showing the fracture origin. Region A is smooth and covered with dark oxide. Region B is rough.

Suitably sectioned fracture surface was de-rusted by ultrasonic cleaning and examined in SEM. Regions A and B showed equiaxed dimples (Figure C1.21). The rest of the fracture surface containing radial marks showed transcrystalline fracture features with cleavage facets (Figure C1.22). A section perpendicular to the fracture surface showed a number of elongated sulphide and globular oxide inclusions. On etching with 2% Nital the sam-

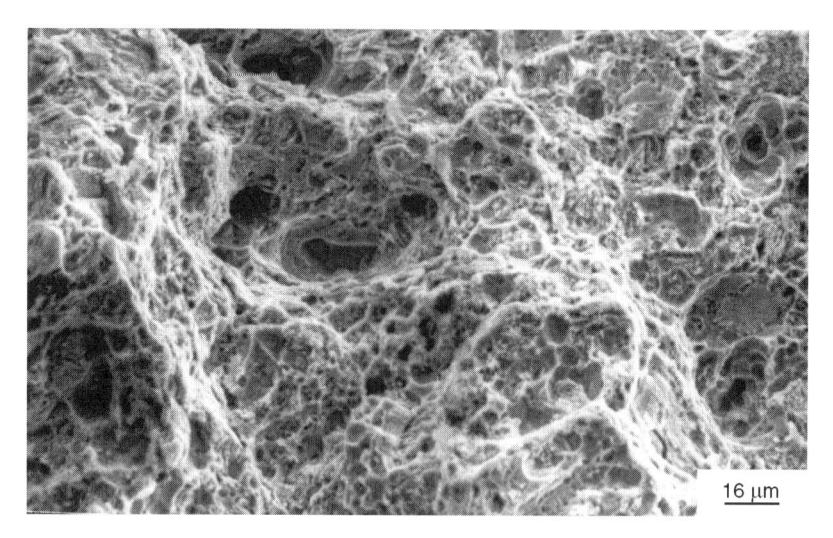

FIGURE C1.21
Fractograph showing equiaxed dimples in Regions A and B.

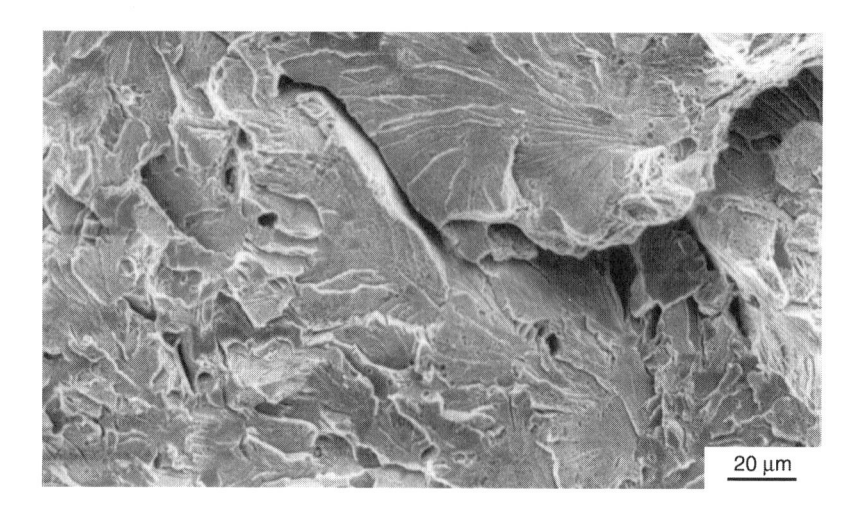

FIGURE C1.22
Fractograph showing transcrystalline cleavage fracture on the surface containing radial marks.

ples revealed a network of ferrite around lamellar pearlite (Figure C1.23). The grain size was in the range of ASTM No. 5 to 6.

The average hardness of the component was 220 HB. The chemical composition (wt%) of the failed component is given in Table C1.4. Izod impact tests were conducted on samples extracted from the failed component in as-received and quenched and tempered conditions. The results are given in Table C1.5.

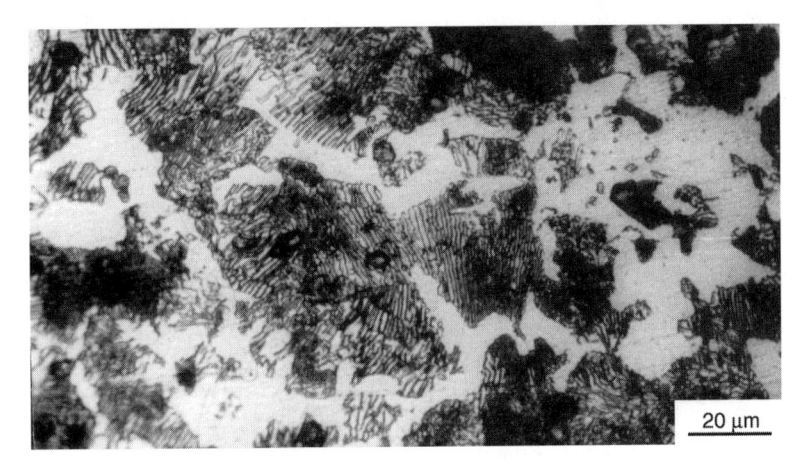

FIGURE C1.23
Optical micrograph showing ferrite network and lamellar pearlite.

TABLE C1.4
Chemical Composition of the Failed Draw Hook

	C	Si	Mn	S	P	Fe
Wt%	0.62	0.28	0.74	0.02	0.02	Bal

TABLE C1.5
Impact Test Results

Condition	Impact energy (J)
As-received	14
Quenched and tempered	27

Discussion

Visual and SEM examination indicated that the failure had originated from one corner of the rectangular section. The presence of a dark oxide layer and dimpled rupture at Region A suggested that this region represented a crack formed during forging. The presence of equiaxed dimples at Region B indicated that the preexisting crack had propagated to critical length under conditions of overload.

Metallographic examination indicated that the component was in pearlitic condition but not in the tempered martensitic condition, as specified for the component. Generally, at identical strength levels, a pearlitic steel will have lower notch toughness than the same steel in tempered martensitic condition. In fact, the impact test results showed that the steel had much lower toughness in pearlitic condition than in tempered martensitic condition. Lower impact toughness of the pearlitic high-carbon steel had made the draw hook notch brittle, which was confirmed by the presence of cleavage facets over most of the fracture surface.

Conclusion

Failure of the draw hook was due to its notch-sensitive microstructure and the stress-raising effect of the preexisting crack, which propagated to critical length during service, leading to catastrophic final failure.

1.5 Failure of a 30-mm Armor-Piercing Shell

Key words: Si-Mn-Cr steel, Heat treatment, Inclusion stringers

Introduction

A number of 30-mm armor-piercing shells were found to split into two pieces when the first explosive pellet was pushed in at a pressure of 320 MPa [9]. The shells were specified to be manufactured from steel grade 35 HGS of GOST B-10230-75. The machined shells were hardened by oil quenching from 840°C followed by tempering at 210°C for 3 h. In view of the difficulty experienced in obtaining the specified mechanical properties, the heat treatment schedule was altered to oil quenching from 890°C and tempering at 150°C for 3 h.

Experimental Results

A failed shell is shown in Figure C1.24. The fracture surface showed a chevron pattern pointing to the origin located at a region where the central disc portion is flush with the inner cylindrical surface of the component. The wall thickness on one side of the failed shell was distinctly lower than that on the other. Variation in wall thickness was observed on a number of other shells.

The fracture surface containing the origin showed mixed intergranular fracture with transgranular quasi-cleavage features and large stringers of inclusions (Figure C1.25). Longitudinal sections of the failed shell revealed very long stringers of silicate and sulphide inclusions (Figure C1.26). On etching with 2% Nital, the sample showed a banded structure consisting of acicular tempered martensite and bainite (Figure C1.27). The prior-austenite grain size was ASTM 6. The chemical composition (wt%) of the failed component is given in Table C1.6. The hardness of the component was in the range 593–613 HV.

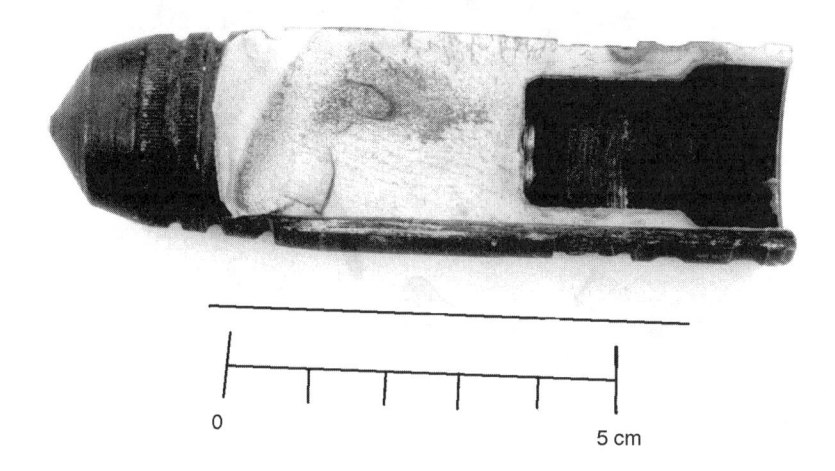

Failures Due to Improper Material Selection and Heat Treatment 145

FIGURE C1.24
Failed shell showing the fracture surface.

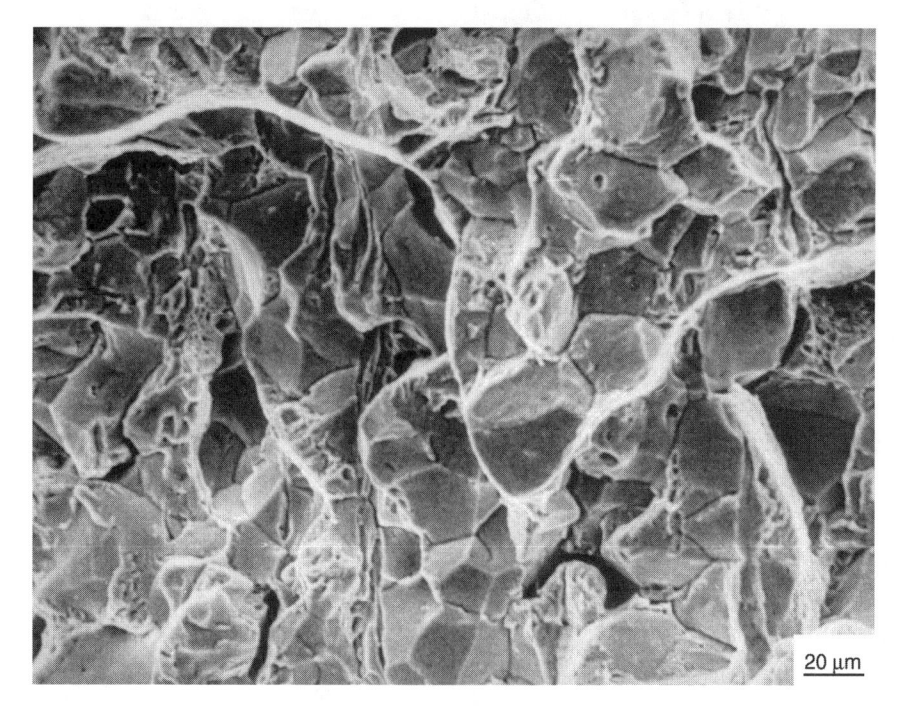

FIGURE C1.25
Fractograph showing mixed intergranular fracture with transgranular quasi-cleavage features and inclusion stringers.

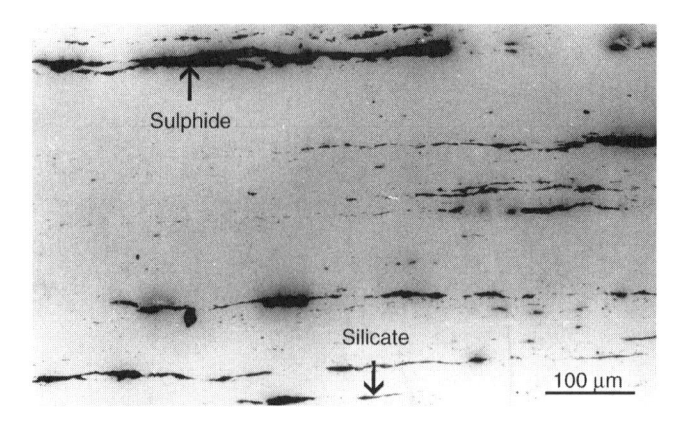

FIGURE C1.26
Photomicrograph showing long stringers of silicate and sulphide inclusions.

FIGURE C1.27
Microstructure of the shell showing banded structure consisting of tempered martensite and bainite.

TABLE C1.6

Chemical Composition of the Failed Shell

Element	Actual	Specified
C	0.54	0.32–0.39
Si	0.27	1.10–1.40
Mn	0.66	0.80–1.10
Cr	1.10	1.10–1.40
Ni	0.44	≤0.4
Mo	0.26	—
S	0.02	0.02 max
P	0.02	0.02 max
Fe	Bal	Bal

Failures Due to Improper Material Selection and Heat Treatment 147

Discussion

The chemical composition of the steel used did not conform to the specified grade. The first signal of changed steel composition was available to the producer when the components failed to meet the strength requirements after imparting the specified heat treatment. Instead of analyzing the problem, the heat treatment schedule was altered to achieve the desired properties. While with the revised heat treatment (increased austenitization temperature and decreased tempering temperature) the required strength could be achieved, the significant loss in ductility due to very low tempering temperature was not given due importance. The presence of a few pockets of bainite suggested that even the hardening was not properly done. The abnormally large population of stringers of silicate and sulphide inclusions, the large variation in wall thickness at diametrically opposite ends and the low-temperature–tempered martensite all contributed to the failure of the armor-piercing shell.

Conclusion

Failure of the shell was due to (a) wrong steel, (b) improper heat treatment and (c) high inclusion content.

Recommendations

1. Component machining to the correct drawing dimensions
2. Strict enforcement of quality assurance standards for material quality
3. Adherence to the process parameters stipulated in the technology documents

1.6 Failure of the Balancing Gear Rod of a Gun Carriage

Key words: *Ni-Cr-Mo steel, Cleavage fracture, Stress concentration, Impact loading*

Introduction

The balancing gear rod of a gun carriage of a medium-range gun failed while the gun assembly was hauled over rough terrain after a successful run of about 4000 km [10]. The gear rod was specified to be fabricated from BS 970 En24 steel, hardened and tempered to a strength of 855 MPa and an Izod impact strength of 55 J.

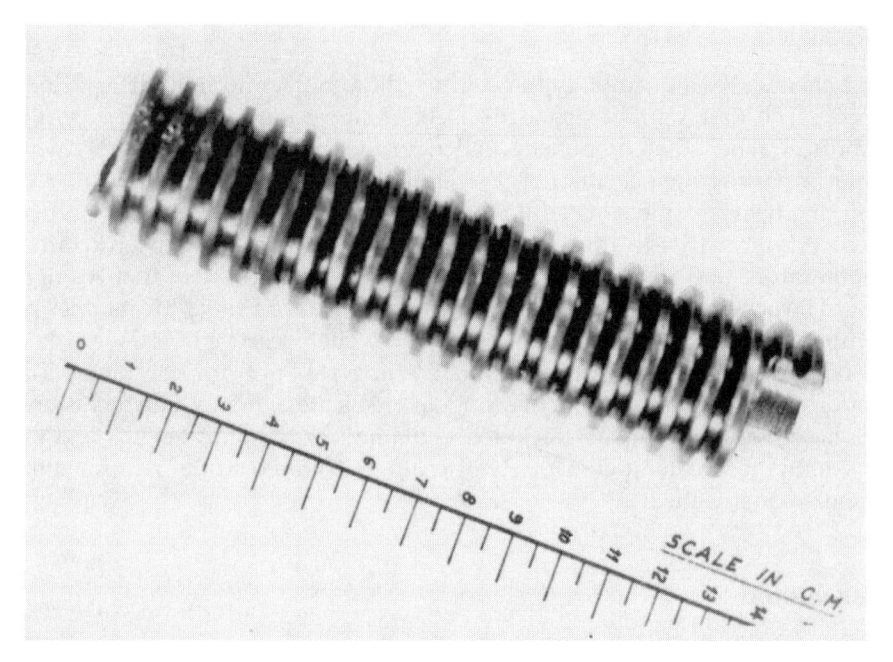

FIGURE C1.28
Photograph of the failed balancing gear rod.

Experimental Results

The balancing gear rod was a hollow component with the outer surface threaded with acme thread (Figure C1.28). Visual examination of the de-rusted fracture surface indicated that the failure had originated at the thread root. The fracture surface at the origin was bright and crystalline, covering about 10% of the fracture surface. The rest of the fracture surface had a dull appearance and showed fibrous fracture (Figure C1.29). Relatively deep tool marks were found on the machined surface, including the critical area near the thread root.

SEM examination of the region near the fracture origin revealed transgranular fracture with cleavage facets and a few secondary cracks (Figure C1.30). The region showing fibrous fracture revealed equiaxed dimples (Figure C1.31). The chemical composition of the failed component is given in Table C1.7. The average hardness of the component measured at a 10-kg load on the sample used for microexamination was 216 HV.

A longitudinal section, taken away from the failed end of the component, was prepared for metallographic examination. In the unetched condition, a number of sulphide and oxide inclusions were seen. On etching with 2% Nital, the sample revealed a normalized structure containing ferrite and pearlite (Figure C1.32).

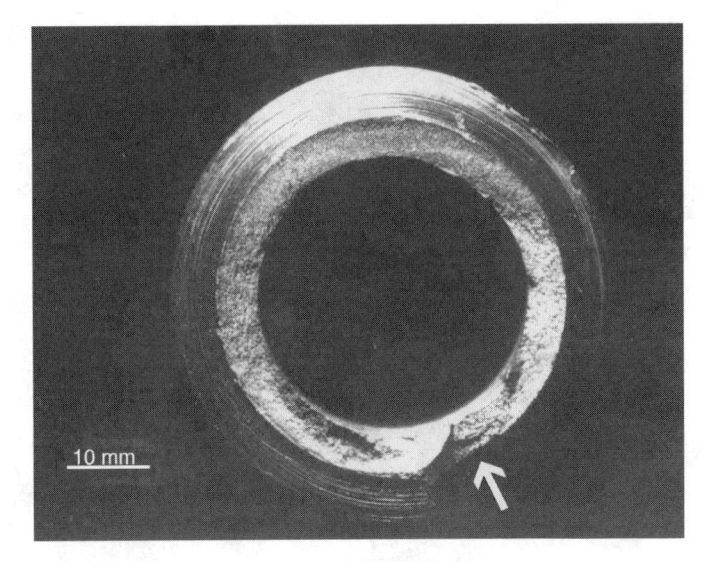

Failures Due to Improper Material Selection and Heat Treatment 149

FIGURE C1.29
Close-up view of the fracture surface. Arrow shows fracture origin.

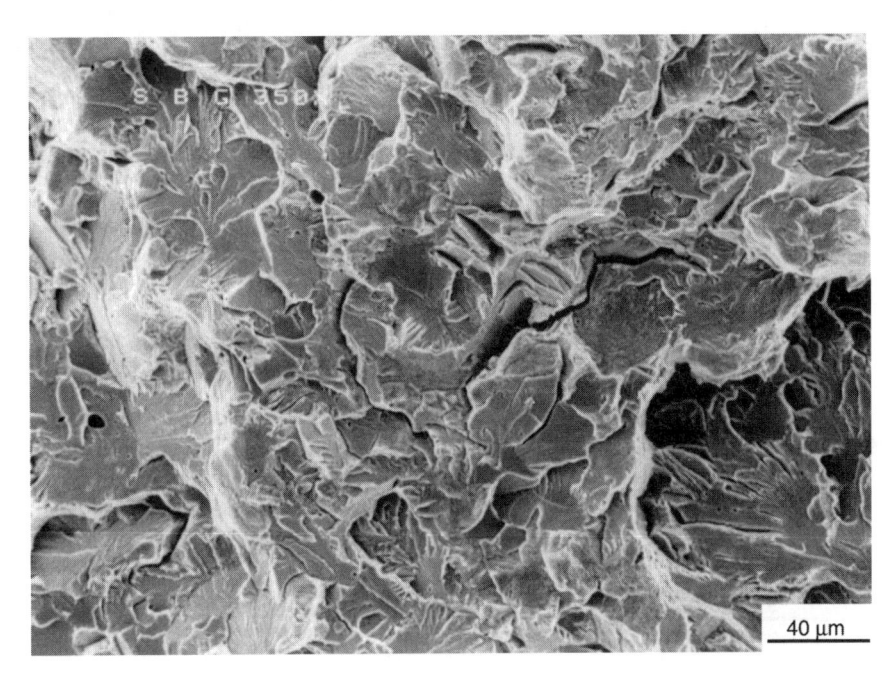

FIGURE C1.30
SEM fractograph showing transgranular fracture with cleavage facets and secondary cracks near the fracture origin.

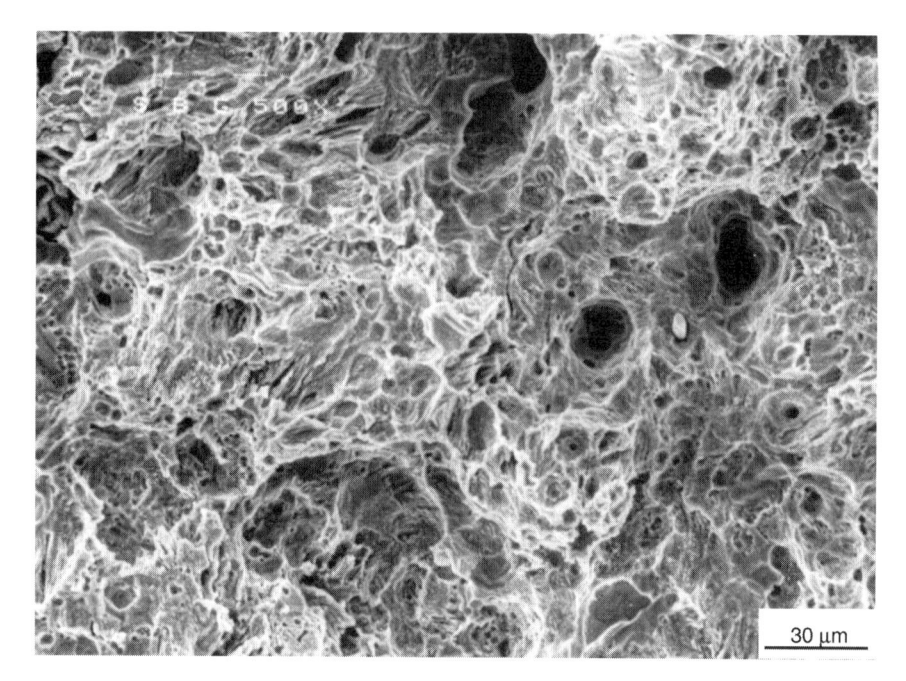

FIGURE C1.31
SEM fractograph showing dimpled rupture away from the fracture origin.

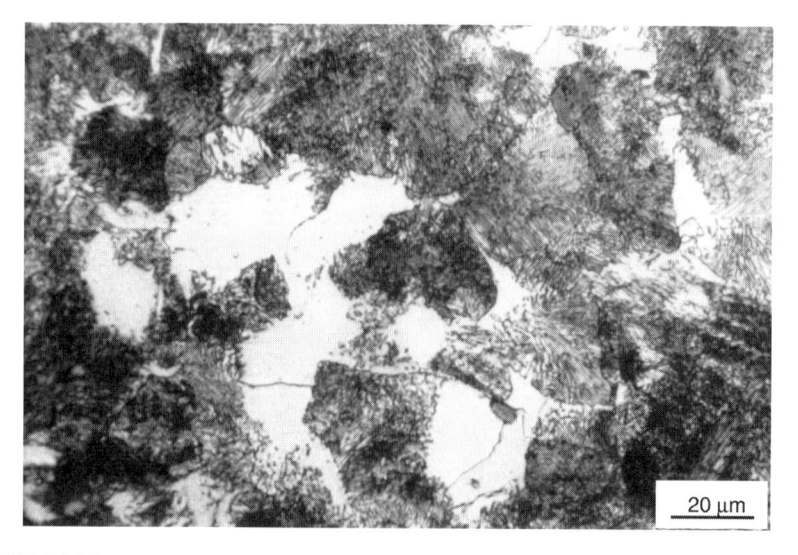

FIGURE C1.32
Photomicrograph showing lamellar pearlite and ferrite.

TABLE C1.7

Chemical Composition of the Failed Balancing Gear Rod

Element	Wt%
C	0.41
Si	0.26
Mn	0.64
Ni	1.64
Cr	0.95
Mo	0.26
S	0.025
P	0.03
Fe	Bal

Discussion

The chemical composition of the steel used for the balancing gear rod conformed to BS 970 En 24. Metallographic examination indicated that the component was in normalized condition containing ferrite and pearlite. If the steel of this grade was quenched and tempered to obtain the specified properties, it should have tempered martensitic structure with hardness in the range 260 to 320 HV. In the normalized condition, the hardness was found to be only 216 HV. It is well known that at similar strength levels, lamellar pearlitic structure results in a lower impact toughness than that of tempered martensite.

While the gun carriage was hauled over rough terrain, the balancing gear rod was subjected to severe shock loading in addition to the static load imposed on it. The transgranular fracture with cleavage facets at the fracture origin was due to the stress-raising effect of the deep tool marks at the thread roots. The change from cleavage mode to dimpled rupture mode might be due to changes in operational loading rates. The presence of ductile fracture features over the rest of the fracture surface indicated that the final failure was due to overload. The failure of the component was thus due to lower strength and toughness of the material in the normalized condition.

Conclusion

The failure of the component was due to gun haulage shock loading imposed on a steel whose toughness was lower than that required for the application due to wrong heat treatment.

1.7 Failure of a Gigli Saw

Key words: Austenitic stainless steel, Non-metallic inclusions, Fatigue

Introduction

A Gigli saw is a device used extensively for cutting hard tissue by neurosurgeons in craniotomy and by orthopedic surgeons in implant procedures. Some Gigli saws started failing after performing one craniotomy instead of the minimum specified life of four craniotomies [11]. The saw was made of vacuum-remelted AISI 321 grade austenitic stainless wire in spring-hard condition. The surgeon cuts the skull by sliding the saw against the hard tissue with force by holding at end loops. The rope experiences tension-tension type cyclic loading.

Experimental Results

The construction details of a Gigli saw are shown in Figure C1.33a/b. The ridges and grooves in the wire rope act like saw teeth and provide cutting action. The fracture surface of the rope that failed after one craniotomy was dull and fibrous and that of the one that failed after four craniotomies showed a smooth area followed by a small area of fibrous fracture.

The prematurely failed saw on fractographic examination revealed that the fracture had originated at a large inclusion near the surface of the wire and that the fracture surface contained elongated dimples (Figure C1.34a). On energy-dispersive spectroscopic analysis, the inclusion was found to contain Al, Mg, Si, and Ca in addition to the base metal elements (Figure C1.34b). The sample that failed after four craniotomies showed the presence of fatigue striations originating at the surface covering over half the area, and the rest of the fracture surface showed elongated dimples (Figures C1.35a and 35b). Microstructural examination of the wire revealed the presence of a large number of carbonitride and a few sulphide and oxide inclusions (Figure C1.36a). On etching, severely deformed grains of austenite were observed (Figure C1.36b). The chemical composition of the stainless steel wire is given in Table C1.8. The ultimate tensile strength of the wire was 1465 MPa.

Failures Due to Improper Material Selection and Heat Treatment 153

(a)

(b)

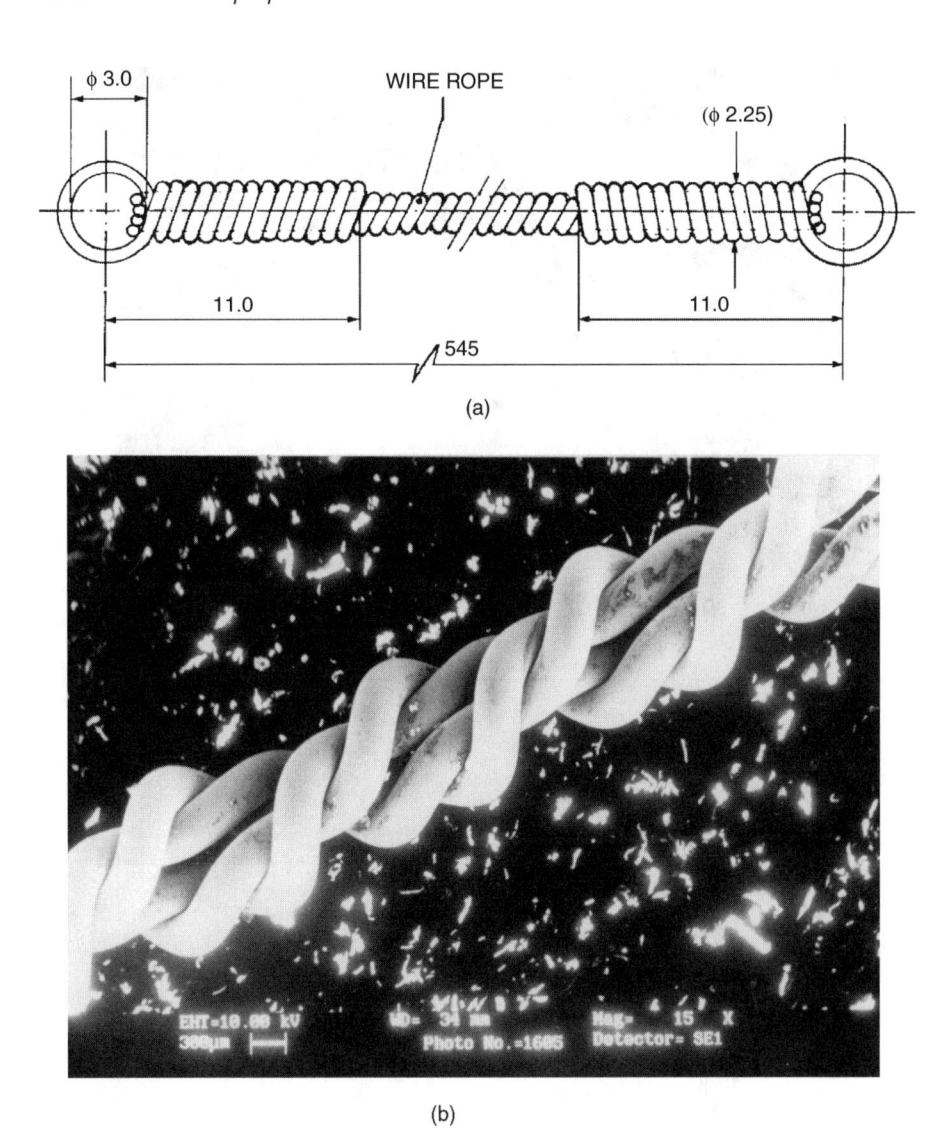

FIGURE C1.33
(a) Construction details of Gigli saw. (b) Ridges and grooves in the saw.

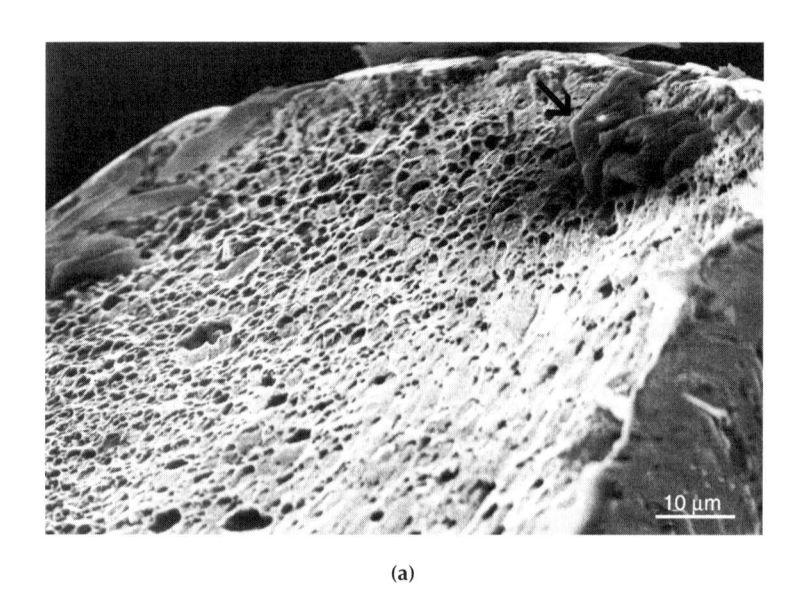

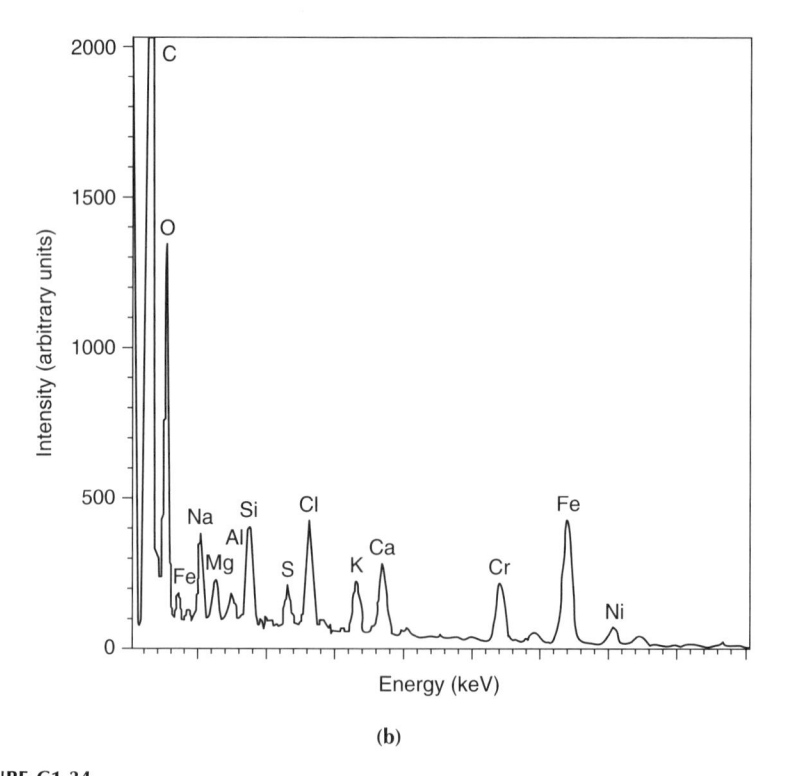

FIGURE C1.34
(a) Fracture surface of the prematurely failed saw showing elongated dimples and a large inclusion at the origin (arrow). (b) EDAX spectrum of the inclusion.

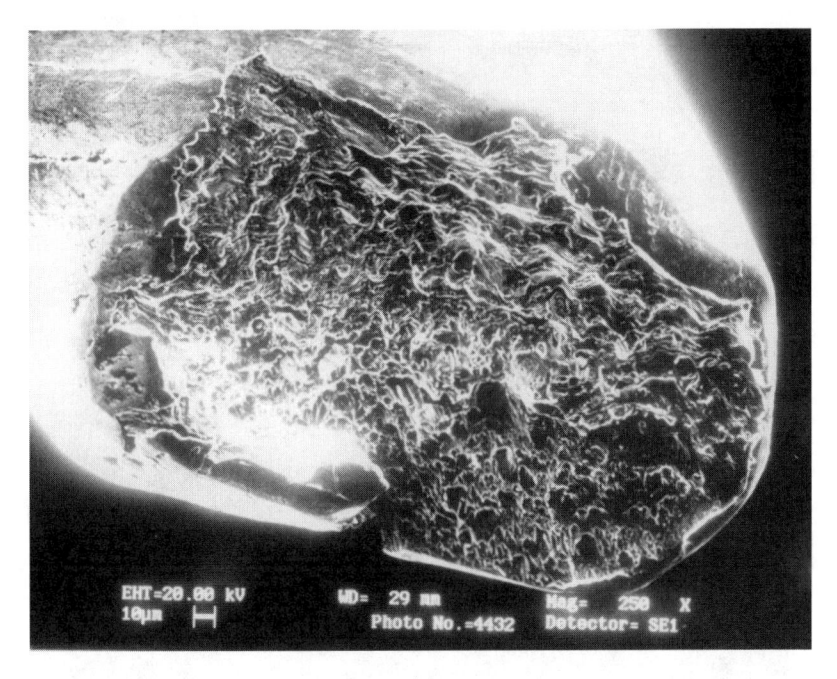

Failures Due to Improper Material Selection and Heat Treatment 155

(a)

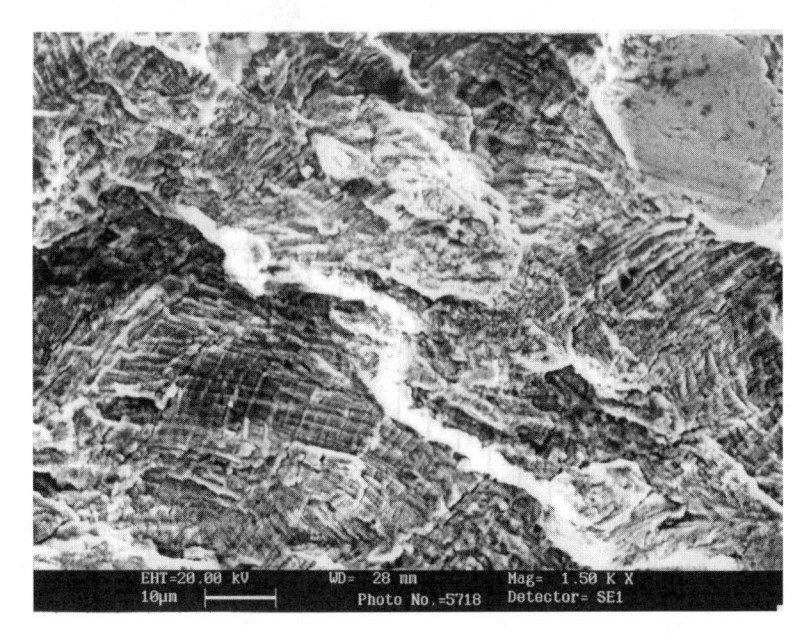

(b)

FIGURE C1.35
(a) Fracture surface of the saw that failed after meeting the specified life. (b) Fatigue striations on the fracture surface.

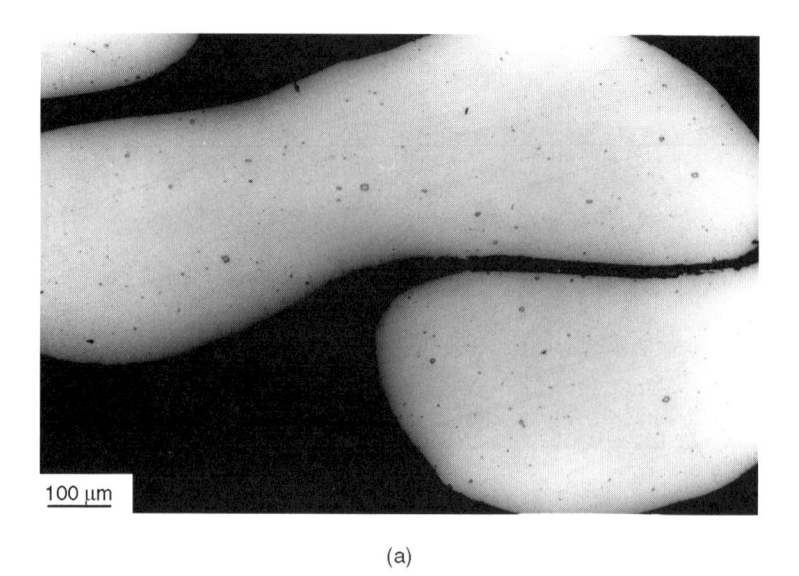

FIGURE C1.36
(a) Inclusion in the wire used for the saw. (b) Microstructure of the wire showing deformed grains of austenite.

TABLE C1.8
Chemical Composition of the Stainless Steel Wire

	C	Mn	Si	Cr	Ni	Ti	S	P	Fe
Wt%	0.08	1.52	0.58	18.3	10.5	0.6	0.009	0.012	Bal

Discussion

The chemical composition, microstructure and tensile properties of the wire conformed to the specified AISI 321 grade in spring-hard condition. The general cleanliness condition of the steel was in conformity with the requirements. An unusually large (28 µm) slag inclusion containing Ca, Mg, Al, and Si was observed at the surface of the wire in the sample that failed after one cut; such large inclusions have not been observed in other saws that have met the stipulated life. The incidence of such a large defect in 350-µm-diameter wire is obviously adequate to initiate failure of one of the four wires in the saw in shear mode under the applied load, as evidenced by the presence of elongated dimples. The presence of fatigue striations over about half the fracture surface and elongated dimples on the rest of the fracture surface in the saws that performed well indicated that the failure occurred under cyclic loading conditions in fatigue mode up to a certain depth and that final failure in shear overload mode was caused by the reduction in cross section of the wire by the fatigue crack. The absence of large inclusions appears to prolong life and favor failure in fatigue mode.

Conclusion

The premature failure of one of the Gigli saws was caused by the presence of a large slag pocket close to the surface of the wire.

1.8 Failure of the Camshaft of a Combat Vehicle Engine

Key words: Carbon steel, Induction hardening, Quench cracks, Intergranular fracture

Introduction

The camshafts of a heavy-duty engine fitted in an infantry combat vehicle failed during a trial run [12]. While camshafts supplied by one manufacturer (M1) failed prematurely, those supplied by another manufacturer (M2) performed satisfactorily. The camshafts were forged from plain carbon steel grade C45, normalized and induction-hardened and tempered at 170°C for

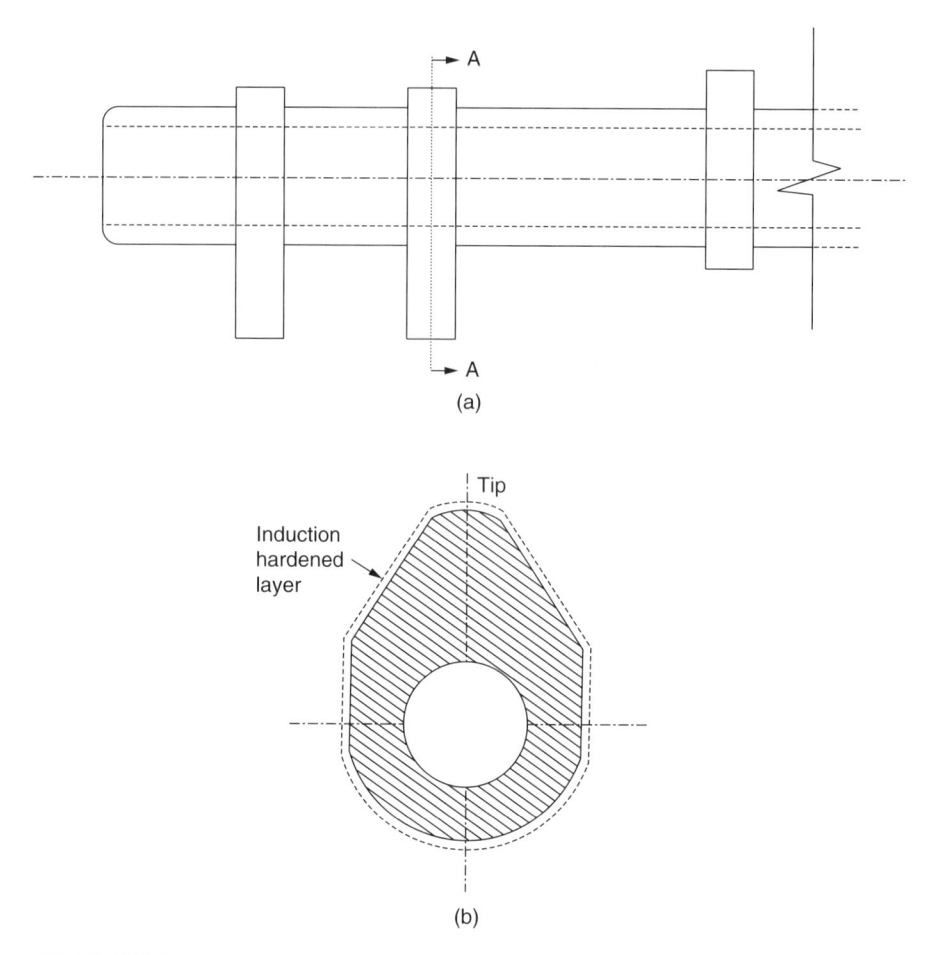

FIGURE C1.37
Schematic of the camshaft. (a) Segment of the camshaft. (b) Section AA showing the induction-hardened zone.

3 h at cam zone. The camshafts were ground to finish dimensions, stress-relieved and protected by oxidation treatment. The specified case depth at the cam region was 2.25 to 3.6 mm at the sides and 5.8 mm maximum at the tip with a case hardness of R_C 54 to 60. The camshafts of M1 were inspected after each process step, and it was found that the cracks were noticed only after induction hardening treatment.

Experimental Results

A segment of the camshaft and the cut section at the induction hardened zone are shown schematically in Figure C1.37. A few typical cracks on the cams were opened. The fracture surfaces in general were covered with black oxide similar to that observed on the surface. The fracture features were

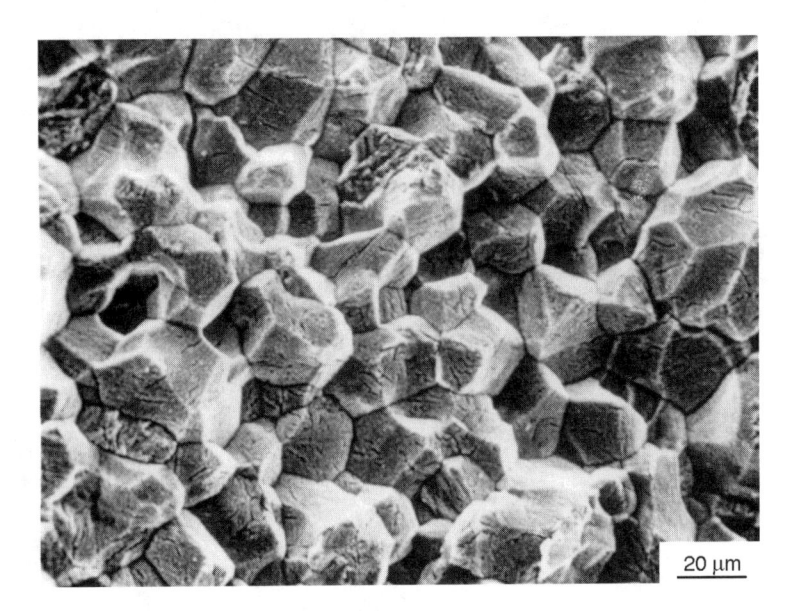

FIGURE C1.38
Fractograph showing intergranular fracture features.

coarse and crystalline in appearance. The entire fracture surface representing the crack showed intergranular fracture (Figure C1.38).

Transverse sections of the cams from M1 and M2 were subjected to microstructural examination. The depths of the induction-hardened layer in M1 and M2 are shown in Figure C1.39. A number of intergranular cracks originating at the surface and running along the case consisting of tempered martensite structure were observed in M_1, whose core consisted of a ferrite network enclosing colonies of coarse lamellar pearlite (Figure C1.40). The cam M2 showed ferrite and tempered martensite in the case and ferrite and colonies of fine lamellar pearlite in the core (Figure C1.41). The details of case depth, hardness, and microstructure of M1 and M2 are given in Table C1.9. The chemical composition of the cams M1 and M2 is given in Table C1.10.

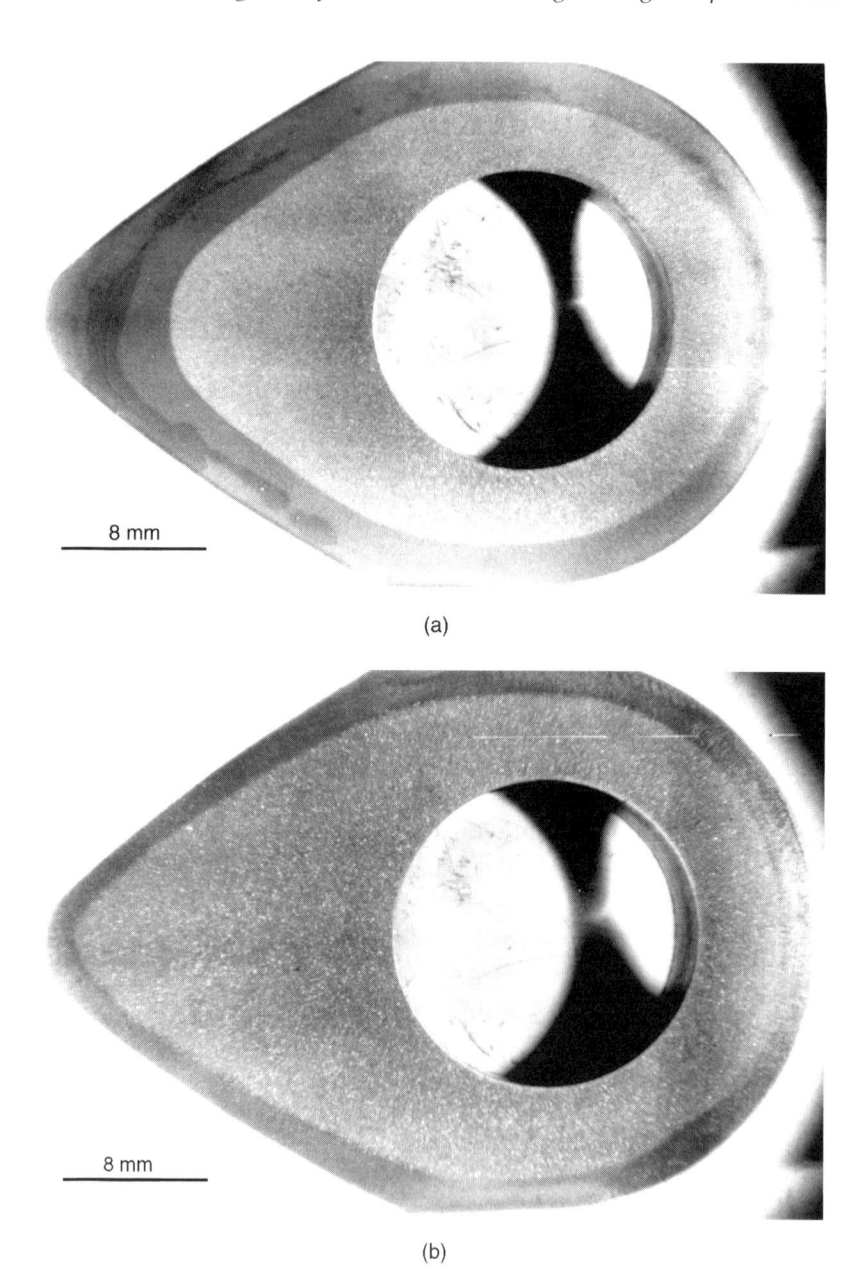

FIGURE C1.39
Induction-hardened layers in the camshafts. (a) M1. (b) M2.

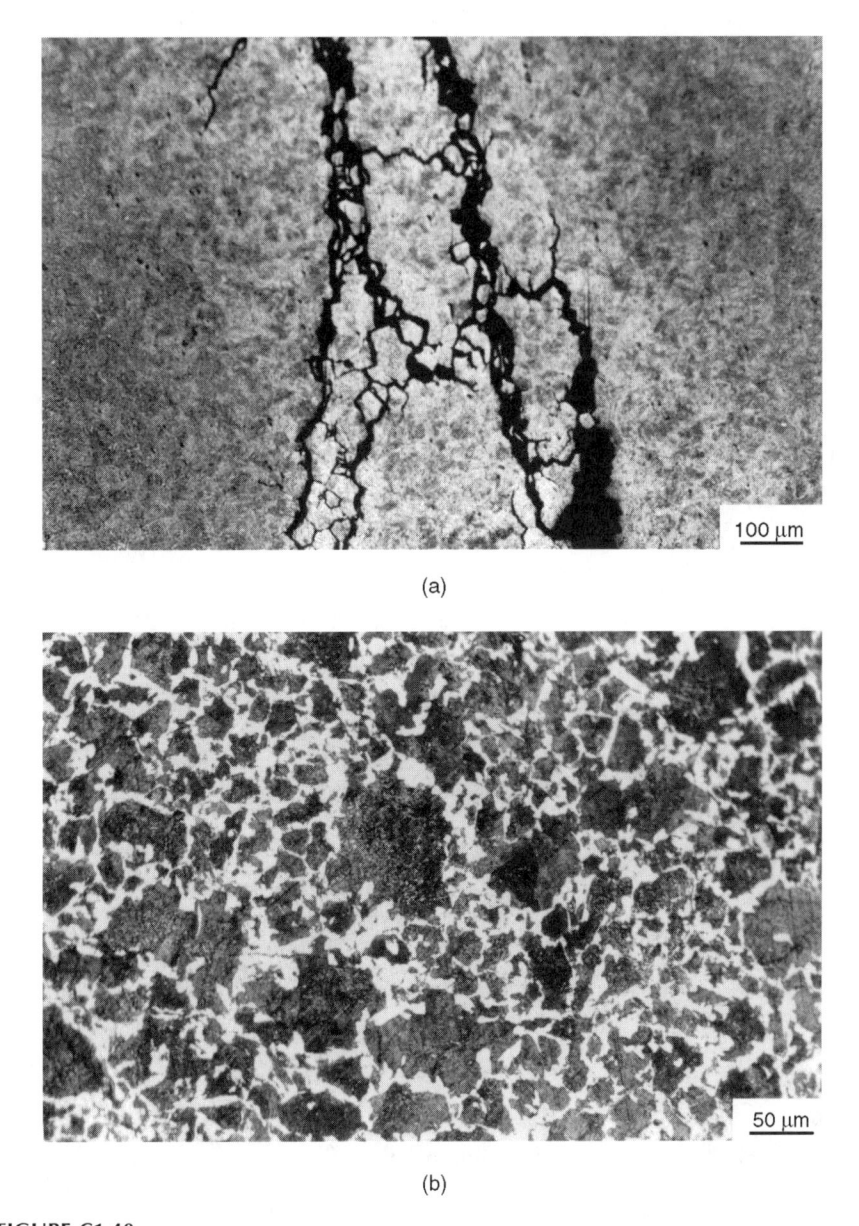

FIGURE C1.40
Microstructure of the camshaft M1. (a) Intergranular cracks in the case. (b) Ferrite network enclosing colonies of coarse lamellar pearlite in the core.

(a)

(b)

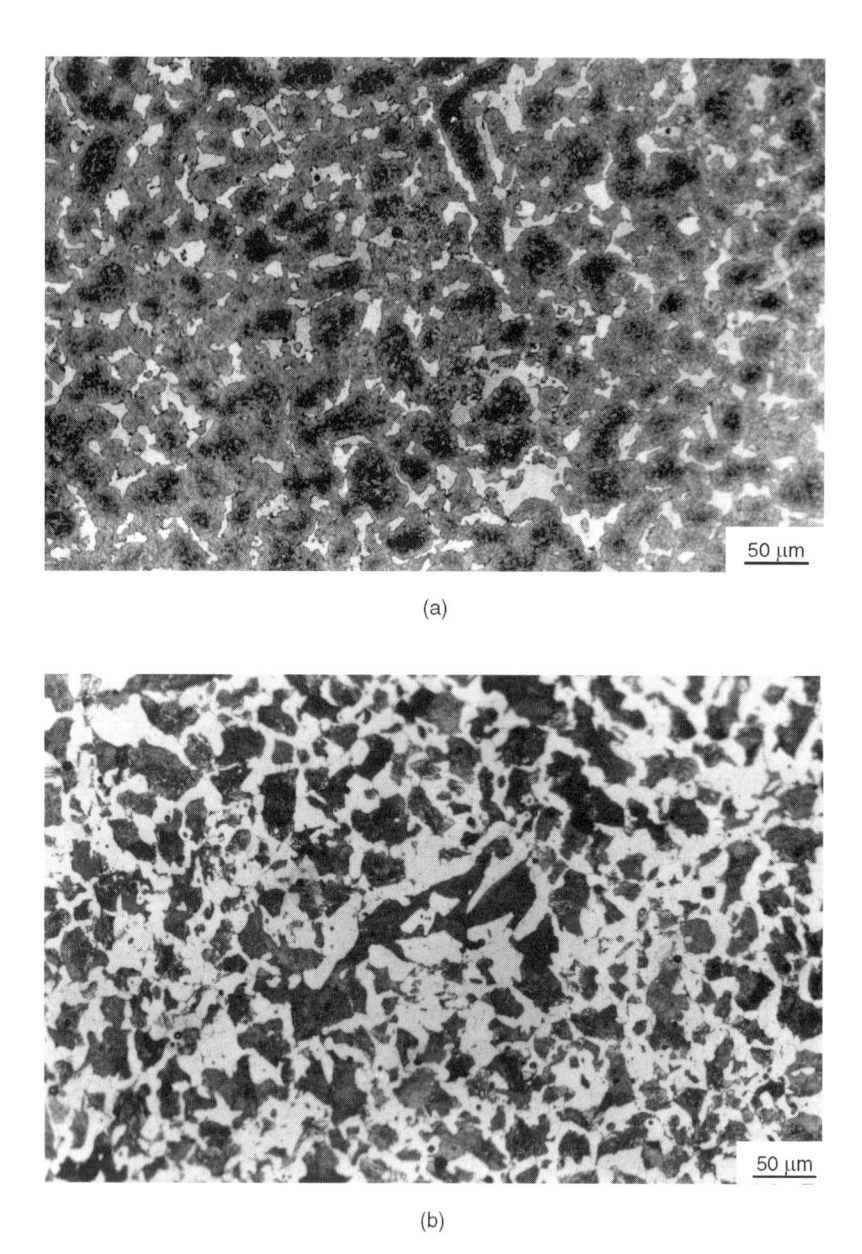

FIGURE C1.41
Microstructure of camshaft M2. (a) Case. (b) Core.

TABLE C1.9

Case Depth, Microstructure and Hardness Details of M1 and M2 Camshafts

Property	Location	M1	M2
Depth of case	Case, tip	9 mm	5 mm
	Case, sides	5 mm	3.5 mm
Structure	Case	Tempered Martensite	Tempered Martensite + 15% Ferrite
	Core	30% ferrite 70% pearlite	50% ferrite 50% pearlite
Hardness	Case	655 HV	584 HV
	Core	199 HV	210 HV

TABLE C1.10

Chemical Composition (wt%) of the M1 and M2 Camshafts

	Source		
Element	M1	M2	Specified
C	0.49	0.44	0.42–0.47
Si	0.17	0.27	0.05–0.35
Mn	0.91	0.54	0.50–0.80
S	0.02	0.011	0.03 max
P	0.023	0.014	0.03 max
Fe		Bal	

Discussion

The chemical composition of the steel used by the manufacturer of the failed camshafts did not conform to the specified grade. Both C and Mn levels were higher. The higher case depth and absence of ferrite in the hardened layer of the failed camshafts suggested that the induction hardening temperature was far above AC_3 temperature, leading to the formation of martensite on quenching. It is well known that high austenitization temperature and case depth give rise to high tensile residual stresses. Higher C and Mn increase the hardenability and decrease the M_S temperature. As a result, the transformation to martensite occurs at a lower temperature in the steel containing higher C and Mn when austenitized at higher temperature. The susceptibility to quench cracking increases because of the high hardness and low ductility of high-carbon martensite formed at lower M_S temperature. It is thus clear that the cracks in the induction-hardened layer of the failed camshafts were formed due to the combined effect of high tensile residual stresses generated due to poor control on austenitization temperature, case depth and composition of the steel. The intergranular mode of failure and the presence of an oxide layer similar to the one formed during surface treatment confirmed that the cracks formed during quenching after induction hardening.

The lower hardness of the case, presence of free ferrite in the case (austenitization temperature below AC_3) and lower C and Mn levels in the steel used by M2, whose camshafts performed well in service, were indicative of judicious induction hardening treatment employed by the manufacturer.

Conclusion

The cracks in the induction-hardened layer of the camshafts were quench cracks formed due to faulty induction hardening technique and nonconformity of the steel to the specified grade.

Part 2

Fatigue Failures

2.1 Failure of First-Stage Compressor Blades of an Aeroengine

Key Words: *Martensitic stainless steel, Beach marks, Fatigue, Non-metallic inclusions*

Introduction

Many aircraft fitted with a particular type of aeroengine were involved in serious accidents. The accidents were investigated and failure of the first-stage compressor blades was identified to be the main cause for each of the accidents [13]. The compressor blades were forged from electroslag remelted martensitic stainless steel conforming to Russian grade AE961W. The blades were hardened and tempered to achieve yield and tensile strengths of 900 MPa and 1100 MPa, respectively. All the blades were sand-blasted and were coated with 22-μm-thick Ag at the root region and with 25-μm-thick Ni-Cd on the airfoils by electroplating. Many failed compressor blades were examined to identify the cause for the incidence of premature failures. The compressor blade failures can be categorized into two classes: (1) failures that have the cracks originating from the groove at the root/aerofoil junction and (2) aerofoil failures. While a majority of the failures were of the first category, only a few were of the second category.

Category 1

Experimental Results

A failed blade containing a crack at the root portion is shown in Figure C2.1. The crack was located along the aerofoil–root junction on the convex side (Figure C2.2). The fracture surface, after opening the crack, showed characteristic beach marks originating at the groovelike feature above the root resulting from the radius provided to flush the root with the aerofoil (Figure C2.3).

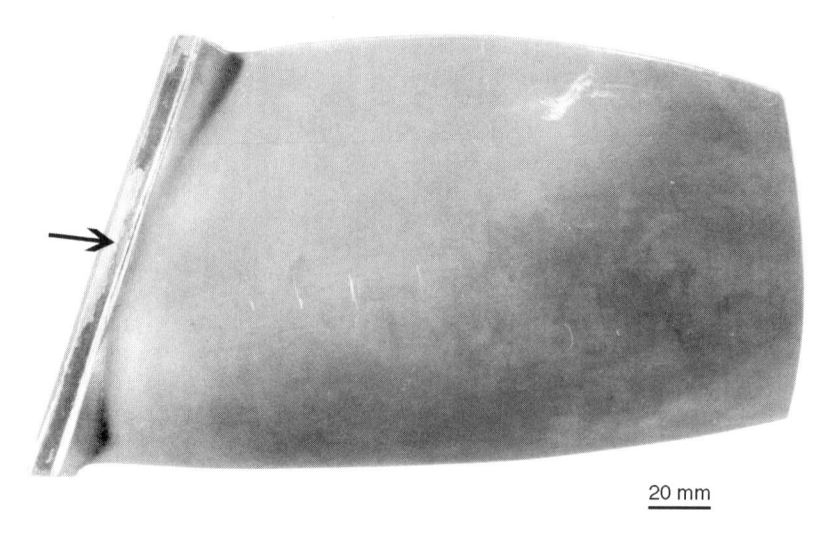

FIGURE C2.1
Photograph of the failed compressor rotor blade in as-received condition.

FIGURE C2.2
Photograph showing a crack at the root region (arrow).

In some cases, multiple fatigue crack origins were observed. Embedment of fine, angular, nonmetallic particles to a depth of about 10 µm was observed at each of the fatigue crack origins (Figure C2.4). The particles, upon EDAX analysis, were found to contain Si and traces of Al in addition to the elements

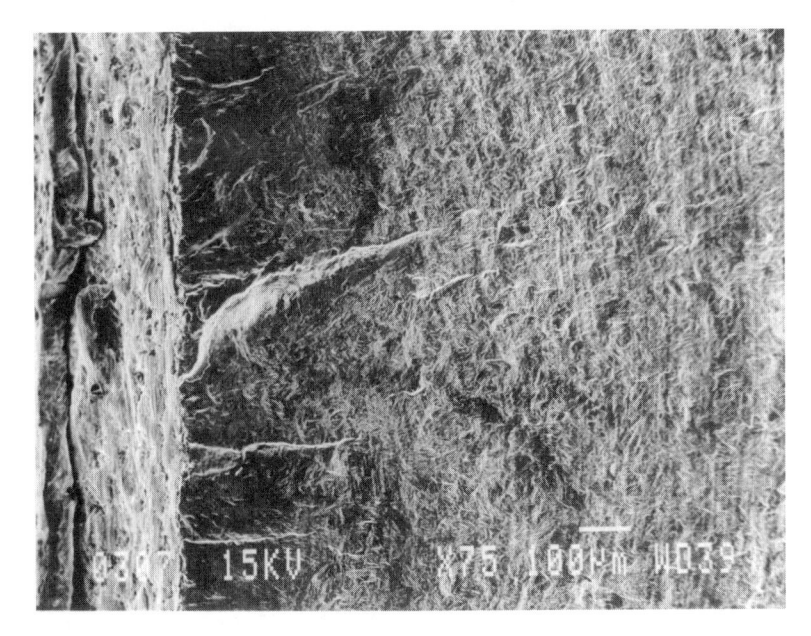

FIGURE C2.3
Beach marks originating at the groove.

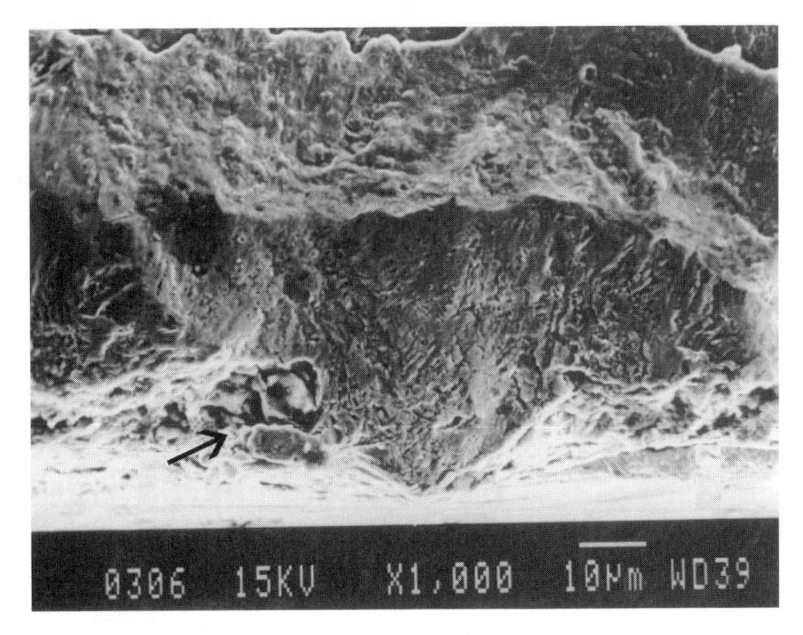

FIGURE C2.4
SEM fractograph showing fine, angular, nonmetallic particles at the fatigue crack origin (arrow).

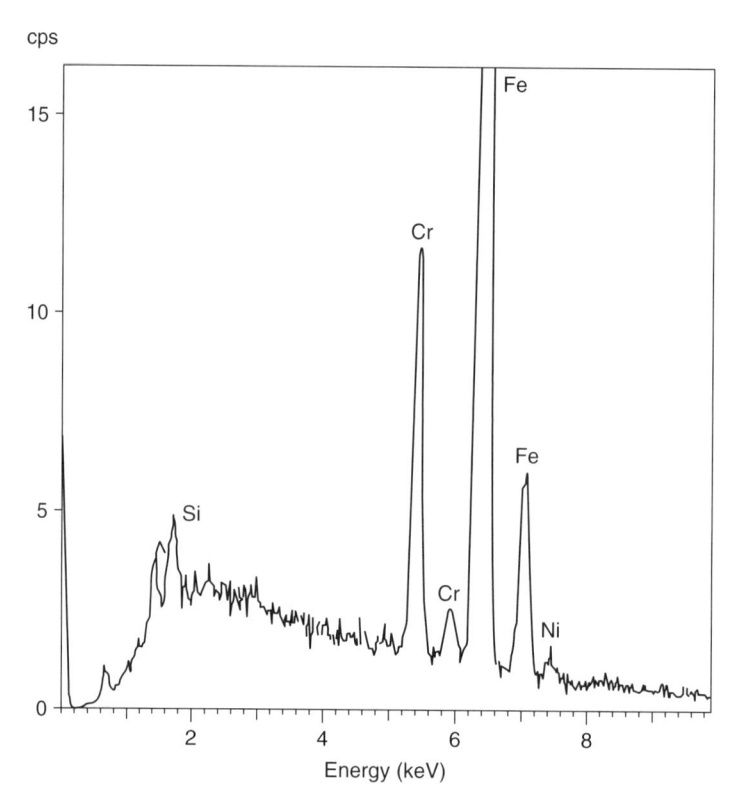

FIGURE C2.5
EDAX spectrum obtained on the embedded particles.

in the steel (Figure C2.5). EPMA analysis confirmed the presence of Si and O in embedded particles. The fracture surface showed fine fatigue striations superimposed on the fracture steps (Figure C2.6). Ni-Cd coating on the aerofoil and Ag coating on the root were observed. Sections at the root and aerofoil regions showed the presence of angular particles embedded in the steel surface at the coating-steel substrate interface (Figure C2.7). The microstructure consisted of tempered lath martensite free from δ ferrite (Figure C2.8). The chemical composition of the failed component is given in Table C2.1. The hardness of the component was 352 HV.

Fatigue Failures

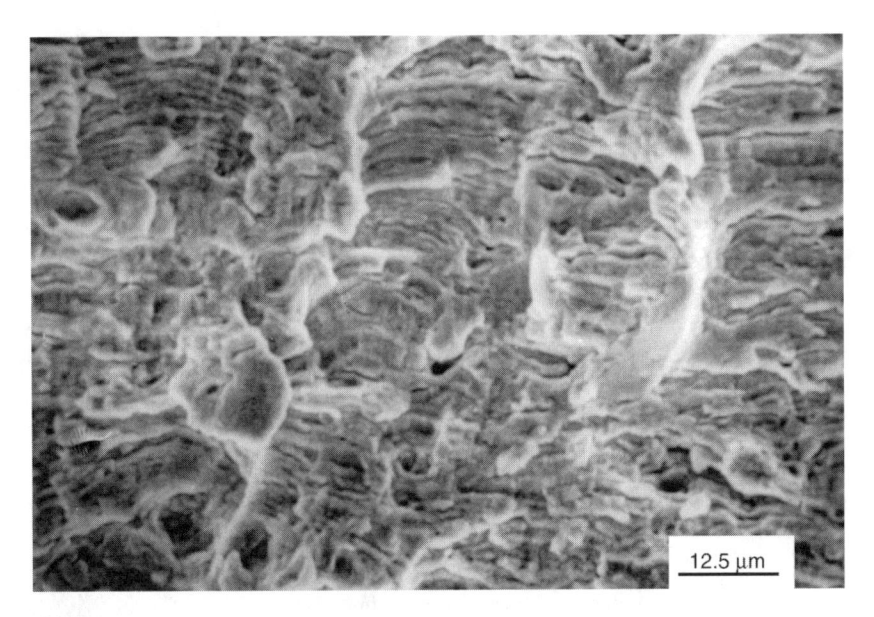

FIGURE C2.6
SEM fractograph showing fatigue striations on the fracture surface.

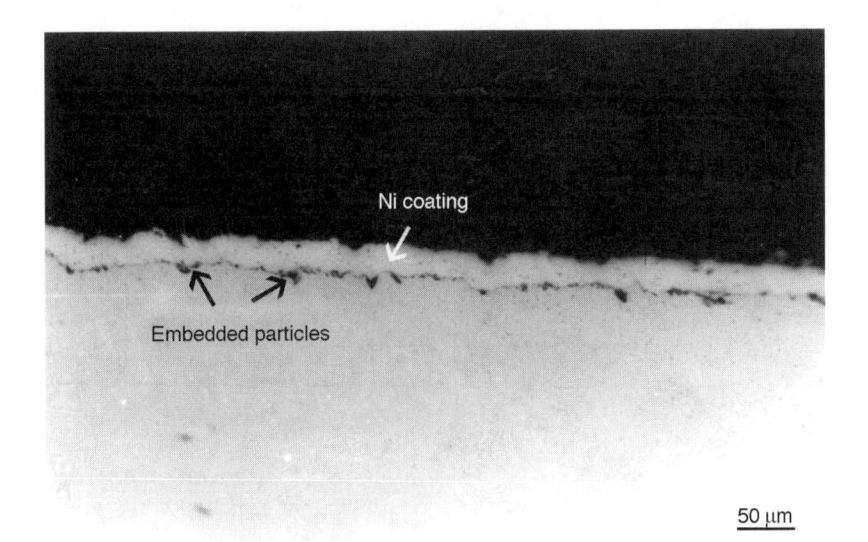

FIGURE C2.7
Micrograph showing embedded particles below the coating.

FIGURE C2.8
Photomicrograph showing tempered martensite structure.

TABLE C2.1

Chemical Composition of the Failed Compressor Blade

Element	Wt%		Element	Wt%
C	0.13		W	1.94
Si	0.06		Mo	0.38
Mn	0.43		V	0.24
Cr	12.1		S	0.005
Ni	1.90		P	0.018
Fe		Bal		

Discussion

The chemical composition, hardness and microstructure were in conformity with the relevant material standard. The component was protected with the specified Ni-Cd and Ag coatings at the aerofoil and root regions, respectively. A large number of embedded angular particles were observed on the steel surface. From the analysis of the particles, it was clear that the particles contained essentially Si and O, the constituents of the sand used for sand blasting. It is well known that (a) angular, friable particles fragment and become embedded on impact in the metallic surface if the sand-blasting variables such as flow rate and velocity and particle characteristics are not precisely controlled and (b) the composite layer thus formed due to particle

Fatigue Failures 171

embedment adversely influences the corrosion characteristics and fatigue properties. It is thus evident that cracks had originated at the particle-steel composite layer and propagated in fatigue mode as evidenced by beach marks and striations over most of the cracked area. The groovelike defect at the aerofoil–root interface could have acted as an additional stress raiser and aided fatigue crack initiation.

Conclusion

The failure of the first-stage compressor blades was due to fatigue caused by the embedment of sand particles during sand blasting.

Category 2

Experimental Results

There were a few incidents of failure of first- and second-stage compressor blades made of AE-961W, and all these failures were located in the aerofoil region. A typical first-stage blade containing a crack is shown in Figure C2.9. The crack running parallel to the root was 75 mm long and was located at a distance of 125 mm from the root. The crack was wide at its midlength and the tips were located about 30 mm from the edges of the aerofoil. The fracture surface showed beach marks originating at the aerofoil surface (Figure C2.10). In another case, the crack was located about 10 mm above the root at the leading edge (Figure C2.11), which on opening also showed fracture with beach marks originating at a nonmetallic pocket (Figure C2.12). At the origin, a defect with spongy features was noticed (Figure C2.13). EDAX analysis on the spongy debris indicated that it contained Si, Mg, Al, S, Cl, K, and Ca in addition to the elements in the steel (Figure C2.14). The areas containing beach marks showed fatigue striations (Figure C2.15).

On microstructural examination a few fine sulphide and globular oxide inclusions were observed in the unetched condition. On etching, the sample revealed tempered lath martensitic structure (Figure C2.16). The chemical composition of the failed blade is given in Table C2.2. The average hardness of the blade was 341 HV.

In order to confirm the presence of such defects, a number of first- and second-stage compressor blades were subjected to ultrasonic examination at the manufacturer's premises. Samples showing defect indications were carefully sectioned and fractured at the defect location and examined. All the defect indications were associated with slag pockets, some of which were 0.7 mm in size. A typical slag pocket and the EDAX spectrum of the slag are shown in Figures C2.17 and C2.18, respectively.

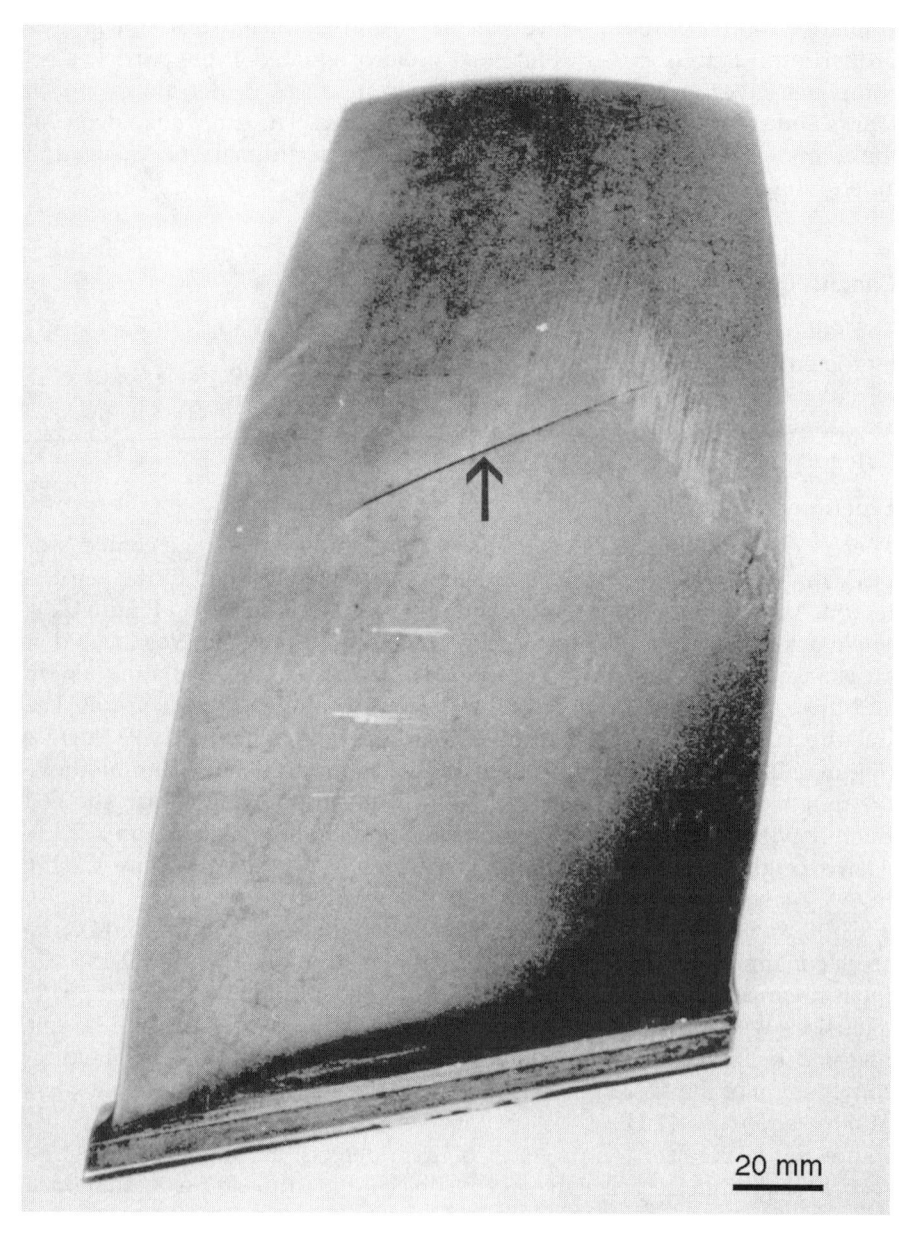

FIGURE C2.9
Photograph of a first-stage compressor rotor blade containing a crack (arrow).

Fatigue Failures

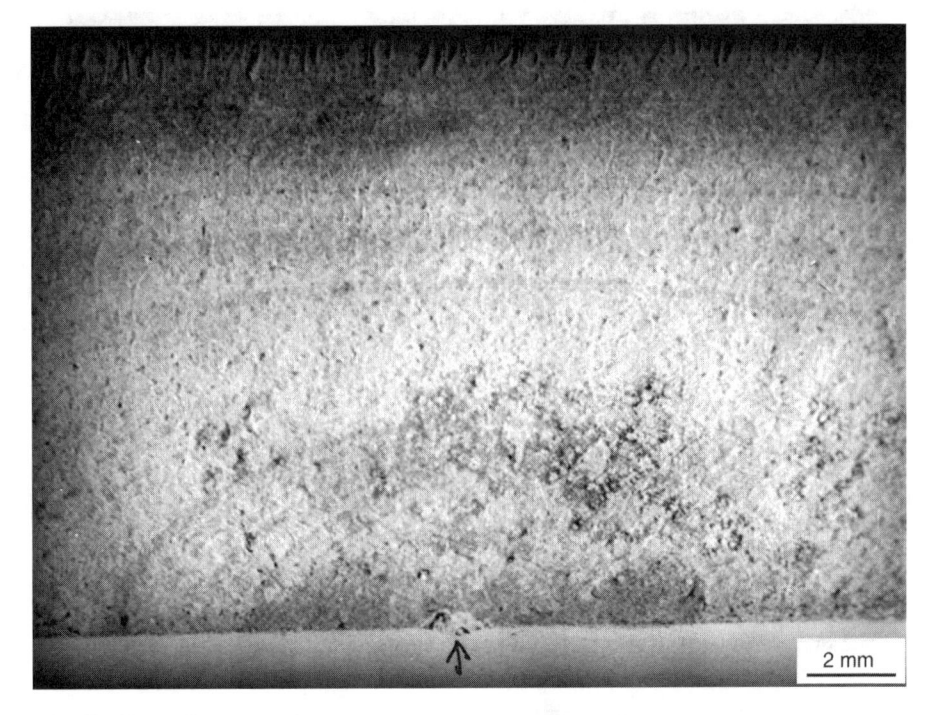

FIGURE C2.10
SEM fractograph showing beach marks originating at the aerofoil surface.

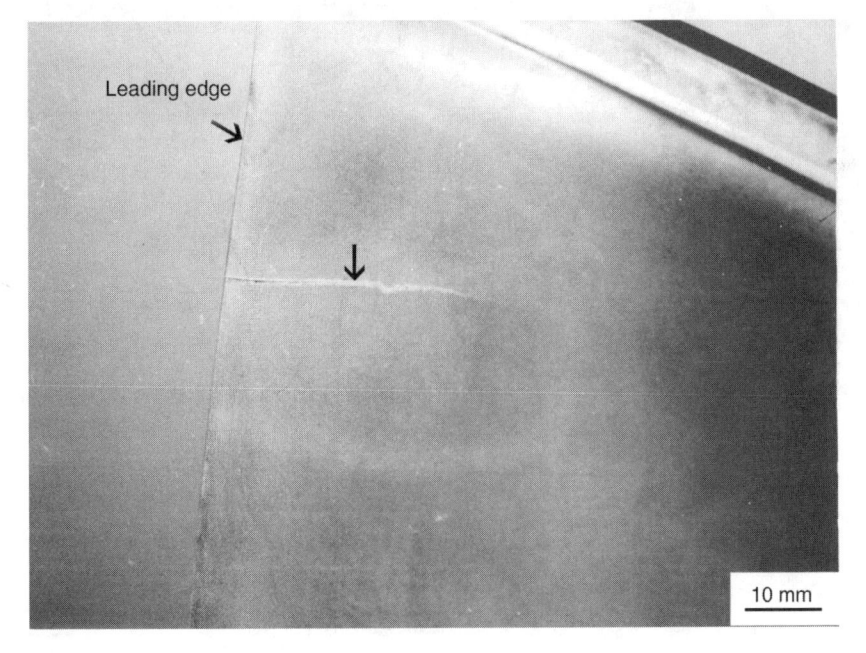

FIGURE C2.11
Crack at the leading edge of another compressor blade.

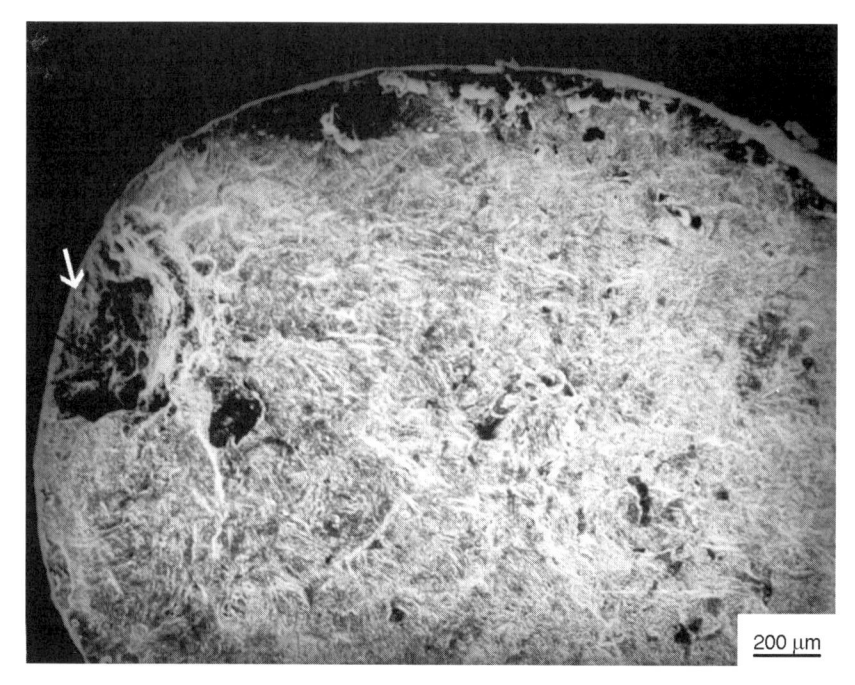

FIGURE C2.12
Fractograph showing beach marks originated from the surface at a defect near the leading edge of the blade (arrow).

FIGURE C2.13
Fractograph showing a surface defect with spongy features.

Fatigue Failures

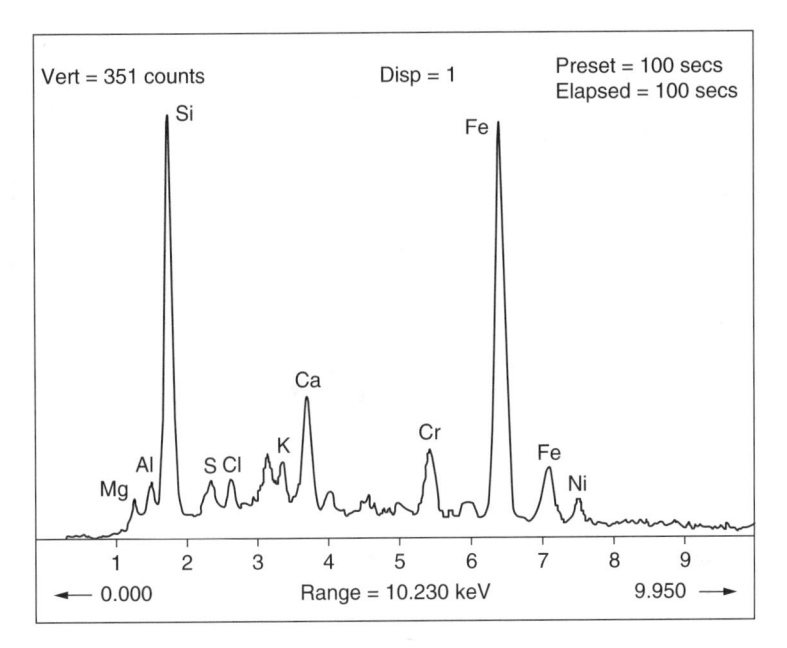

FIGURE C2.14
EDAX spectrum obtained on the slag pocket.

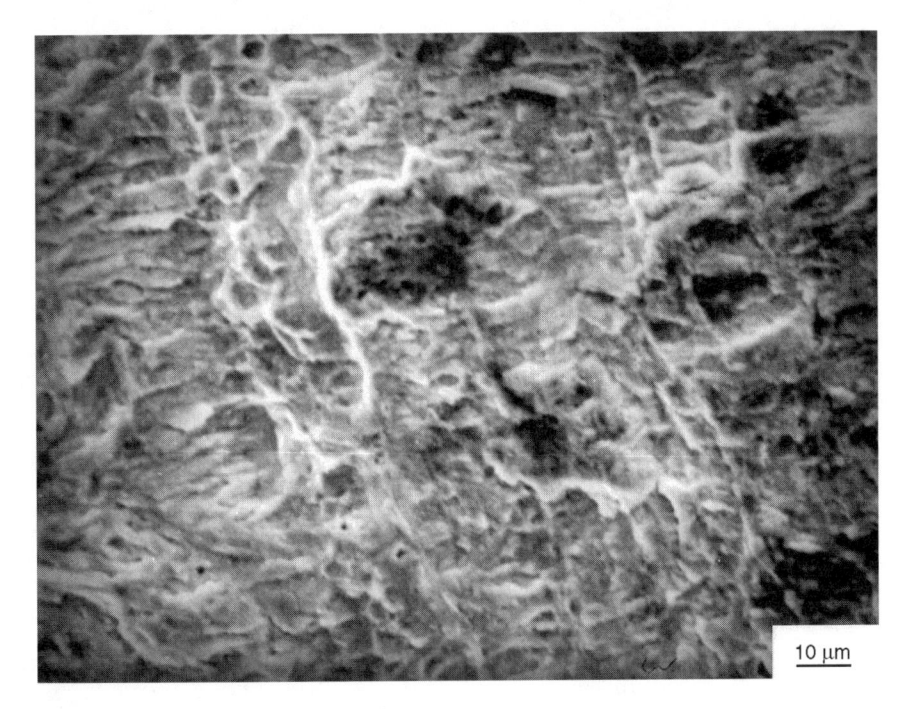

FIGURE C2.15
SEM fractograph showing fatigue striations.

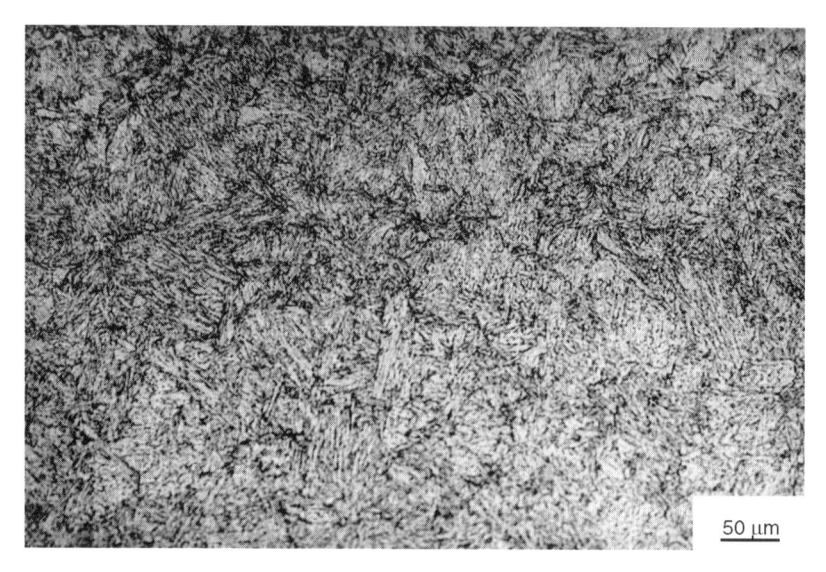

FIGURE C2.16
Photomicrograph showing tempered lath martensite.

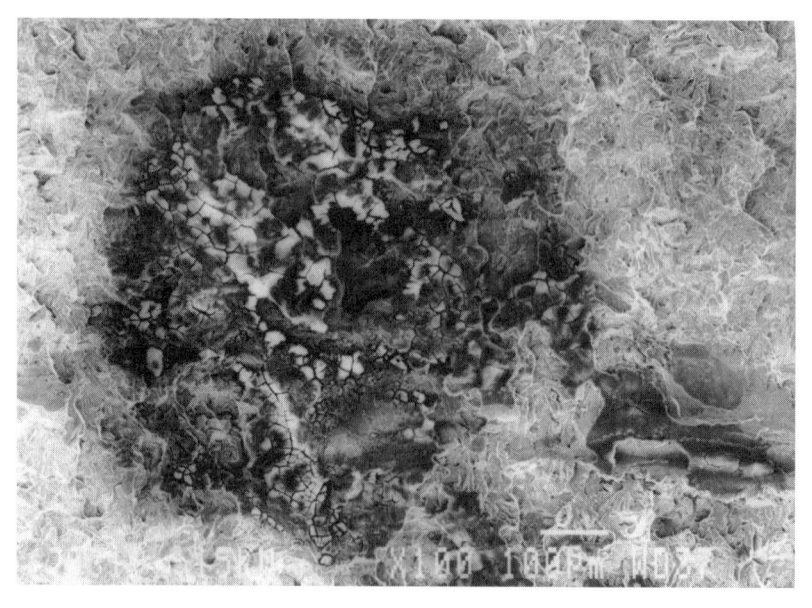

FIGURE C2.17
SEM picture showing a large slag pocket in the blade.

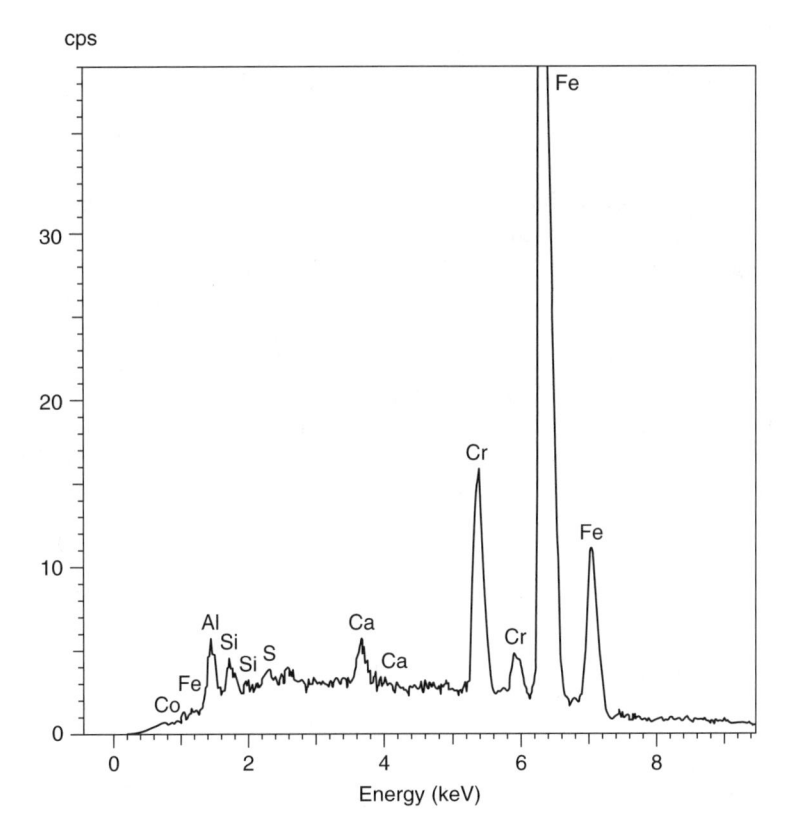

FIGURE C2.18
EDAX spectrum of the slag.

TABLE C2.2

Chemical Composition of the Failed Component

Element	Wt%	Element	Wt%
C	0.13	Mo	3.5
Si	0.31	W	2.14
Mn	0.30	V	0.27
Ni	1.77	S	0.012
Cr	11.0	P	0.032
Fe	Bal		

Discussion

The chemical composition conformed to the specified grade. The microstructure and hardness were in agreement with the heat treatment imparted to the blade. The fracture origin contained a spongy debris that on analysis was found to contain Si, Al, Mg, and Ca, elements characteristic of slag. The presence of Cl is considered incidental due to the operation of the aeroengine in a coastal area. The incidence of slag pockets was also confirmed in the new blades by ultrasonic testing. The influence of slag pockets on the fatigue life of the blades depends on their location. If the slag pocket is located in the aerofoil volume, where the cross section is small, fatigue cracks could easily initiate. Their role in crack initiation will not be dramatic if located inside the root volume. The problem becomes acute if the slag pocket is located either at the surface or at the leading and trailing edges of the aerofoil, as was evidenced by the failure of the two blades; one had a slag pocket at the surface of the aerofoil and the other had the defect at the leading edge.

Conclusion

The crack had initiated at a slag pocket located at the blade surface and propagated in fatigue mode.

2.2 Failure of Third-Stage Compressor Blades of an Aeroengine

Key words: Al-Cu-Mn alloy, Beach marks, Corrosion debris, Fatigue striations

Introduction

The third-stage compressor rotor blades of an aeroengine failed, which was observed on inspection after a flight [14]. The blades were forged from an Al-Cu-Mn alloy, solution-treated, aged and anodized. The maximum working temperature was stated to be about 200°C.

Experimental Results

Two failed blades in as-received condition are shown in Figure C2.19. Both the failures are similar and are at identical locations, that is, at the platform on the concave side below the leading edge. The fracture pattern suggested that the failure had originated at multiple locations on the bottom face of the platform.

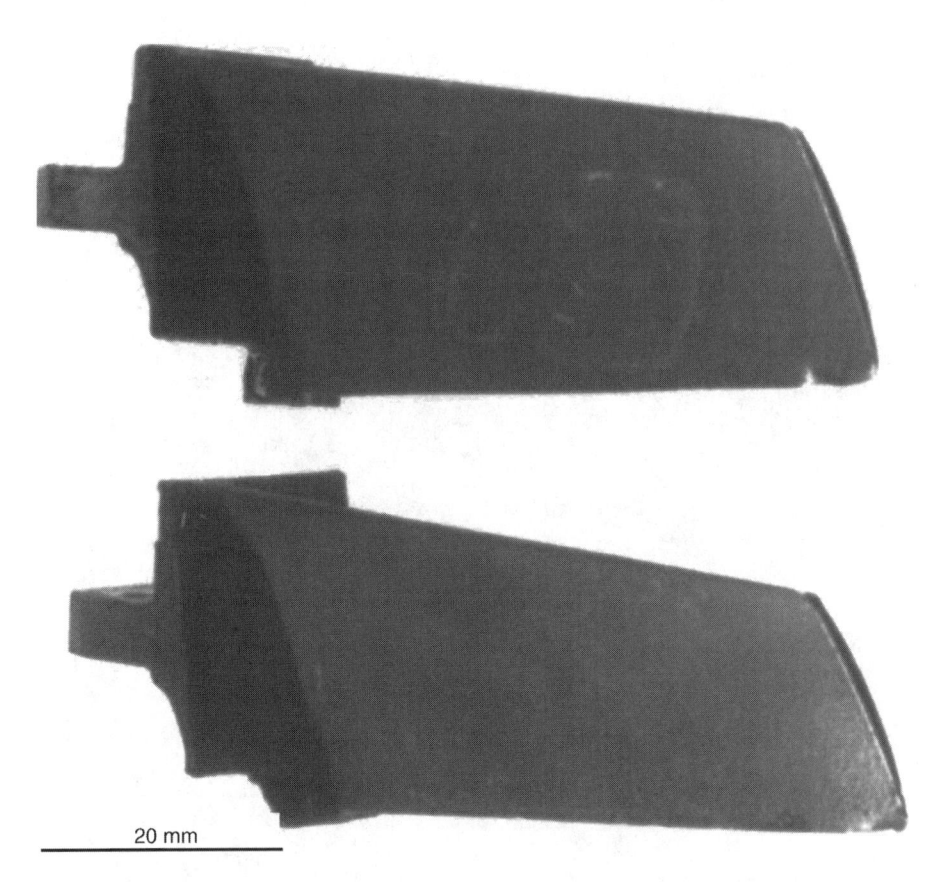

FIGURE C2.19
Photomicrograph of the compressor rotor blades in as-received condition.

Fractographic examination revealed beach marks radiating from multiple crack origins superimposed on fracture steps (Figure C2.20). Striations were observed in the area containing beach marks (Figure C2.21). All the fracture origins contained corrosion debris (Figure C2.22), which on EDAX analysis (Figure C2.23) revealed the presence of Cl, K, Ca, and Si. The final failure regions showed dimpled rupture. The longitudinal section close to the fracture revealed large grains of aluminum solid solution with a few undissolved particles (Figure C2.24). The chemical composition of the failed blade is given in Table C2.3. The hardness of the blade was 108 HV.

FIGURE C2.20 SEM fractograph showing beach marks radiating from multiple crack origins.

Fatigue Failures

FIGURE C2.21 Fractograph showing striations between beach marks.

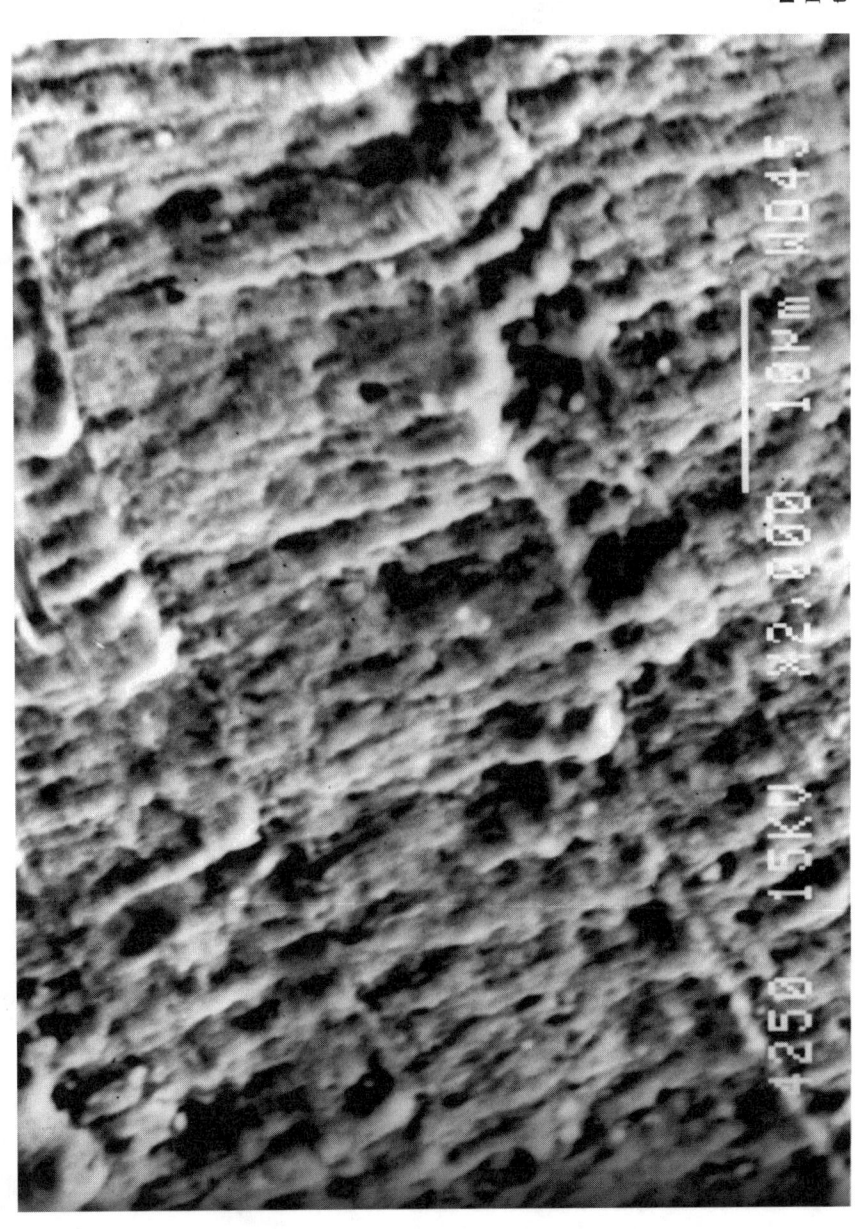

FIGURE C2.22 SEM fractograph showing corrosion debris near the fracture origin.

Fatigue Failures

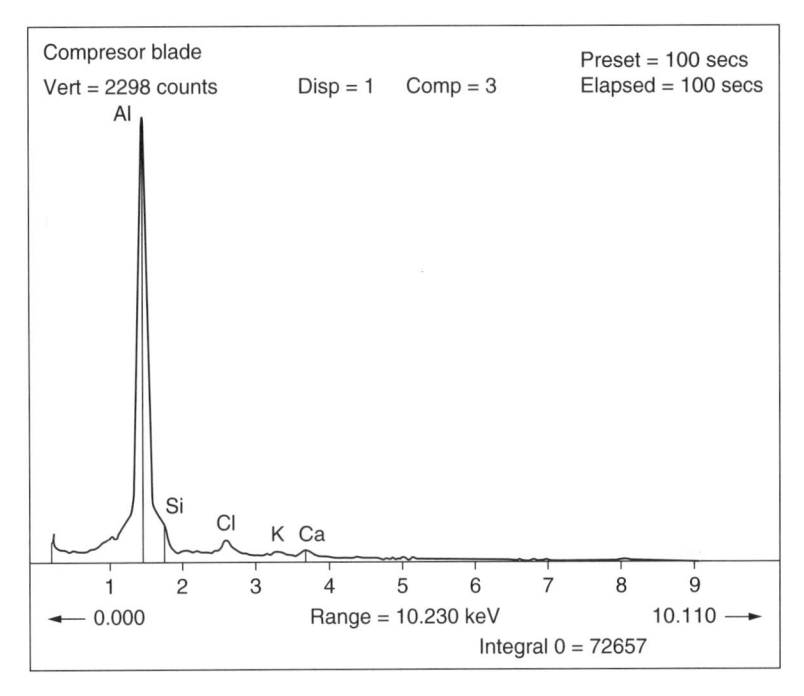

FIGURE C2.23
EDAX spectrum of the corrosion product showing the presence of Ca and Cl.

FIGURE C2.24
Microstructure of the blade showing unrecrystallized grains of aluminum solid solution with slip bands.

TABLE C2.3
Chemical Composition of the Failed Compressor Blade

Element	Wt%
Cu	5.73
Mn	0.22
Fe	0.10
Si	0.15
Al	Bal

Discussion

The chemical composition conforms to the specification DTD 5004. It is not clear how Al-Cu-Mn alloy was used for stage III compressor blades when stage I and II compressor blades of the same engine were made of Al-Cu-Mg-Ni-Si-Fe (DTD-5084) alloy, which has better temperature capability. The grain size was coarser and the hardness was lower than those expected of the alloy for the compressor blade application in solution-treated and aged condition. Both these metallurgical conditions are detrimental to fatigue properties. The presence of corrosion debris at each of the crack origins indicated that the primary damage was due to corrosion in a chloride-containing environment, as a result of local damages to the anodized layer. The presence of beach marks and striations suggested that the corrosion damage triggered initiation of fatigue cracks at many locations. As the fatigue crack propagated the remaining section thickness was inadequate to withstand the applied load, leading to the final failure that occurred in overload.

Conclusion

The blades failed due to fatigue initiated by corrosion.

2.3 Failure of the High-Pressure Turbine Blades of an Aeroengine

Key words: *Ni-base superalloy, Beach marks, Stress raisers, Hot corrosion, Fatigue*

Introduction

Many high-pressure turbine (HPT) blades of an aeroengine failed prematurely in service [15]. The blades were cast from a Russian Ni-base superalloy

Fatigue Failures

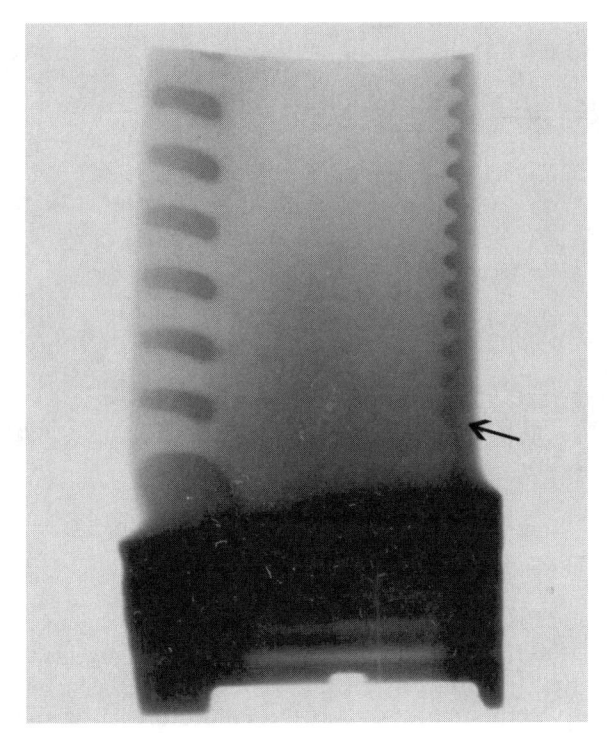

FIGURE C2.25
Radiograph of the turbine blade of an aeroengine showing the location of the first cooling ridge.

grade BZL-12UVD, suitably heat-treated and coated with 0.02- to 0.06-mm-thick aluminide. The blades were cooled with the help of a perforated air deflector inserted into the hollow blade. The total length of the blades was about 84 mm and the aerofoil was 48 mm wide and 62 mm long (Figure C2.25). The operating temperature of the blade was about 1150°C. Failures were observed between 180 to 370 h, compared to the specified overhaul life of 1650 flying hours.

Most of these HPT blade failures fall broadly into two categories. In one category, the failure occurs at about two thirds of the aerofoil length from the platform of the root (Case 1). In the second category, failure occurs at about one sixth of the aerofoil length from the platform (Case 2). The analysis of the failures in each category is dealt with separately in the succeeding sections.

Case 1

Experimental Results

The failed blade along with the cooling air deflector is shown in Figure C2.26. The blade failed at approximately 40 mm from the platform and the deflector

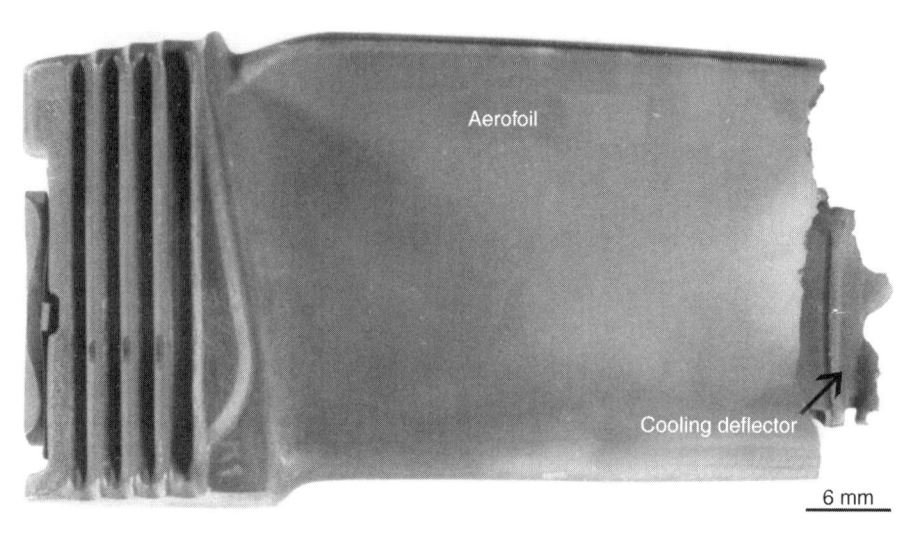

FIGURE C2.26
Root side of the failed turbine blade.

also failed about 10 mm above the blade failure. The fracture surface was smooth at the leading edge, followed by an area having beach marks and crystalline fracture features toward the trailing edge.

The fracture pattern suggested that the failure origin was located at the outer surface of the leading edge and the fracture origin was covered with debris, which was different from the general oxidation of the fracture surface (Figure C2.27). EPMA analysis on the debris showed that it contained Na, Cl, S, Ca, and K. The BSE and elemental X-ray images of the debris are shown in Figure C2.28.

Distinct beach marks were observed on the fracture surface at regions away from the leading edge (Figure C2.29). SEM examination at higher magnification of the fracture surface covered with corrosion debris (region A in Figure C2.29) showed fatigue striations and corrosion pits (Figure C2.30a). The area on the fracture surface exhibiting distinct beach marks (region B in Figure C2.29) also showed fatigue striations (Figure C2.30b). The trailing edge region showed failure along dendrite boundaries. Microstructural features of the sections near the failed end and at the root were examined. The samples revealed cast structure with coarse dendrites, eutectic γ-γ' nodules and carbides with cuboidal γ' precipitates in austenitic matrix (Figure C2.31). The chemical composition of the failed blade is given in Table C2.4. The hardness at the failed region was 328 HB and at the root was 313 HB.

FIGURE C2.27 Fracture surface near the leading edge of the blade showing corrosion products and a cleavelike pattern.

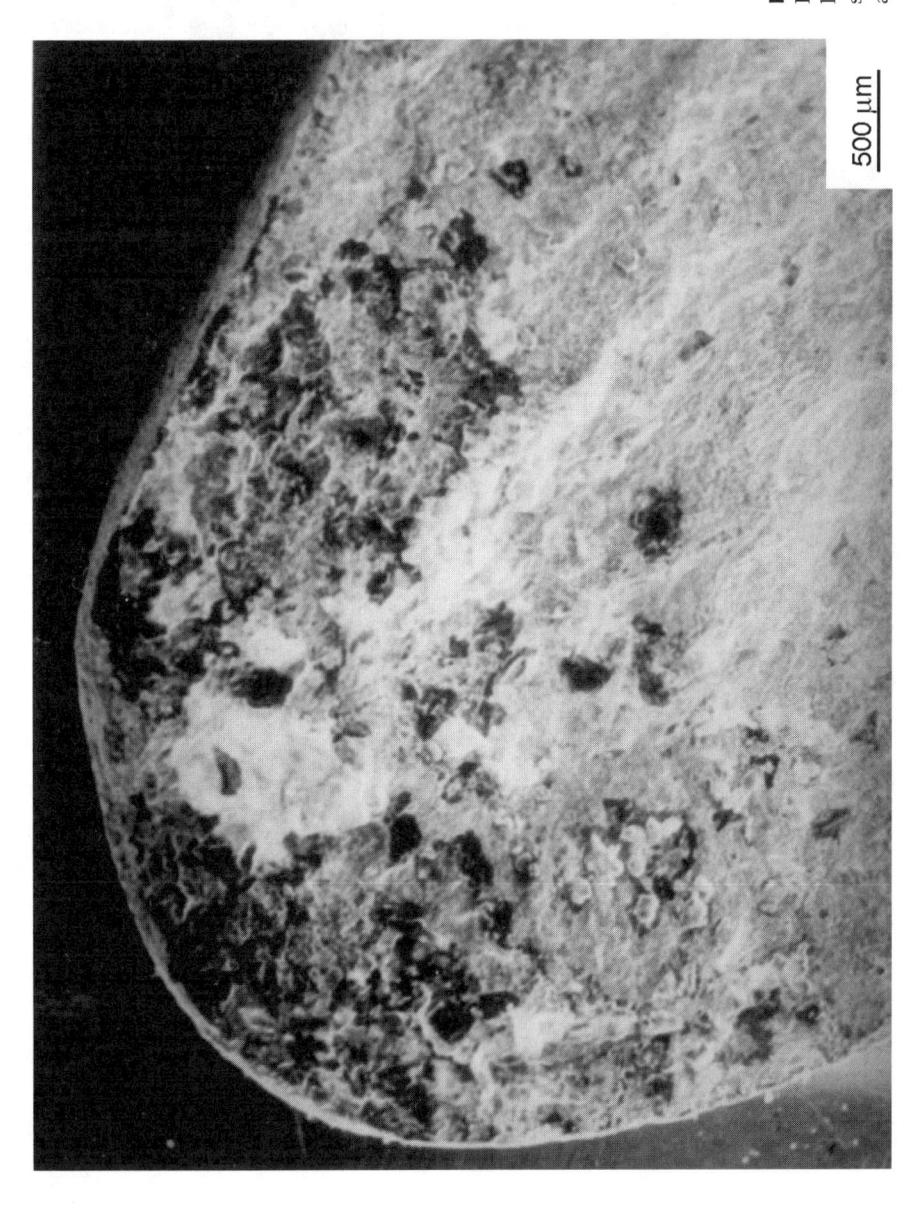

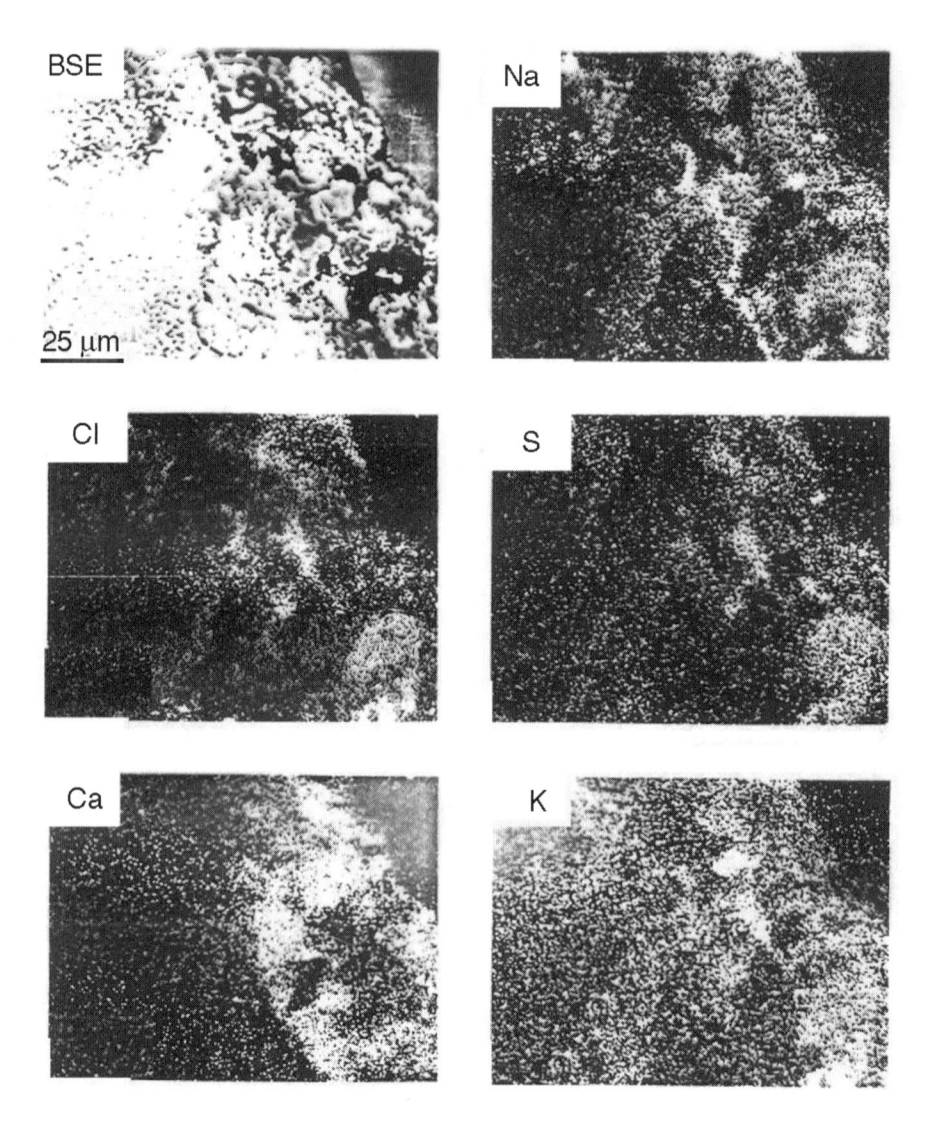

FIGURE C2.28
BSE image of the corrosion debris and the corresponding elemental x-ray images.

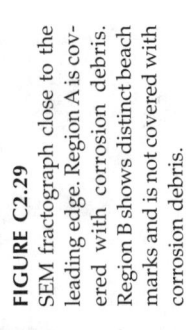

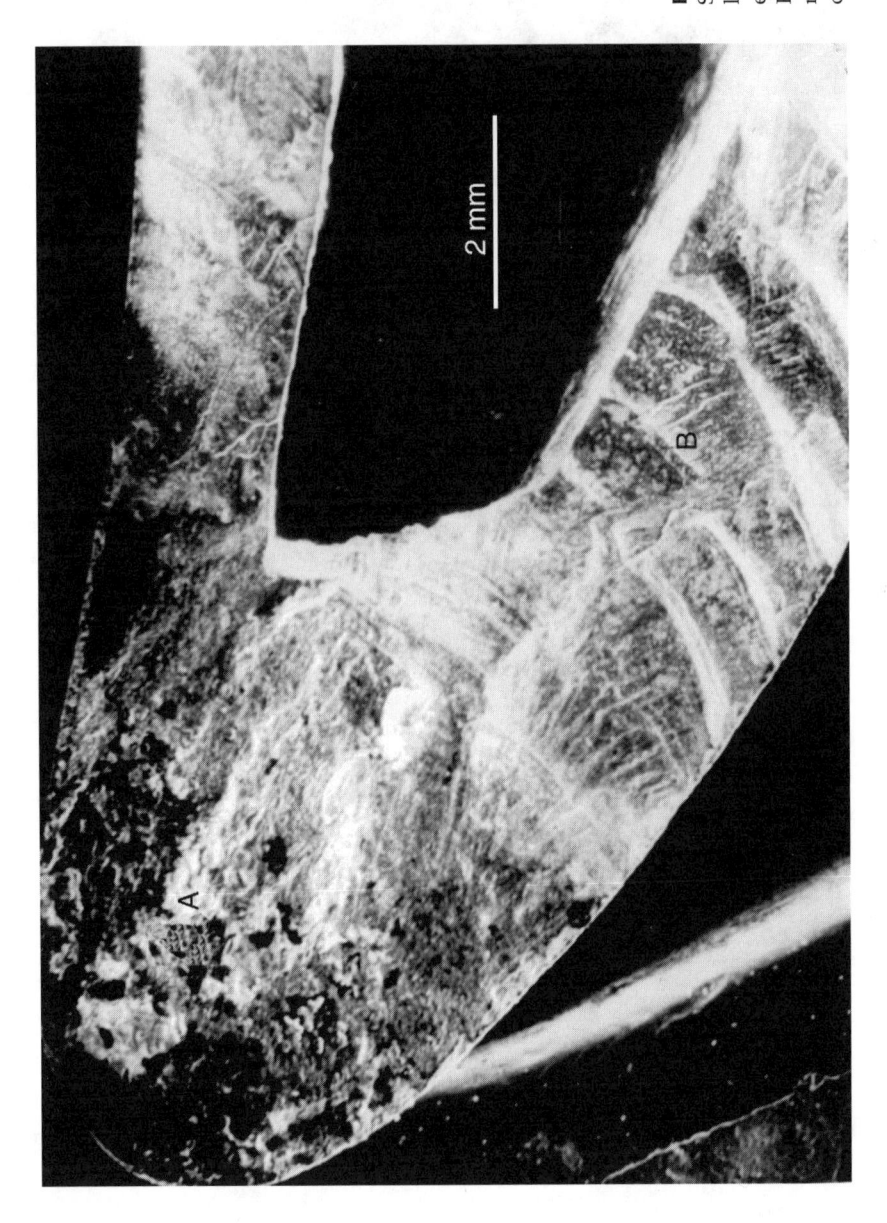

FIGURE C2.29
SEM fractograph close to the leading edge. Region A is covered with corrosion debris. Region B shows distinct beach marks and is not covered with corrosion debris.

(a)

(b)

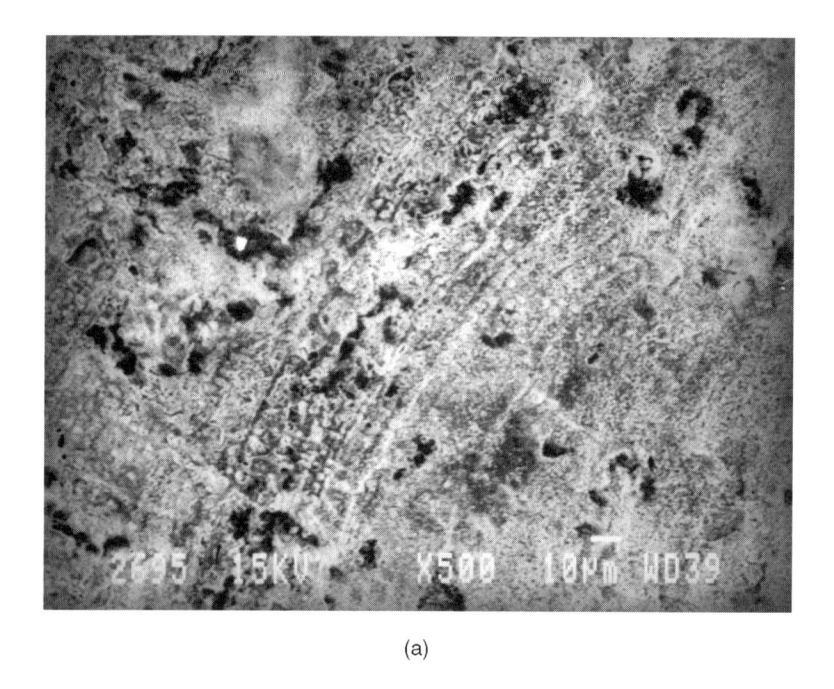

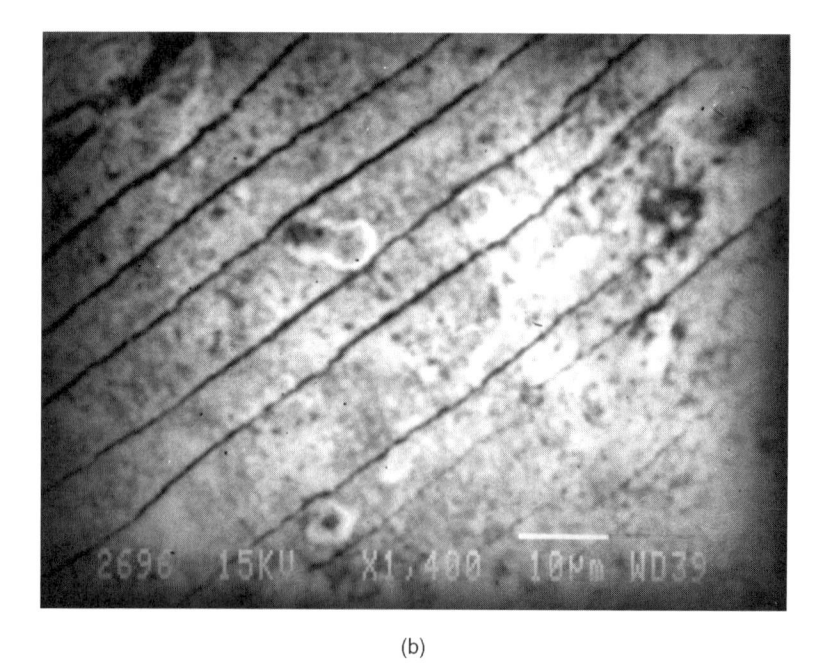

FIGURE C2.30
SEM fractographs (a) near the leading edge showing fatigue striations and corrosion pits and (b) away from the leading edge showing widely spaced striations.

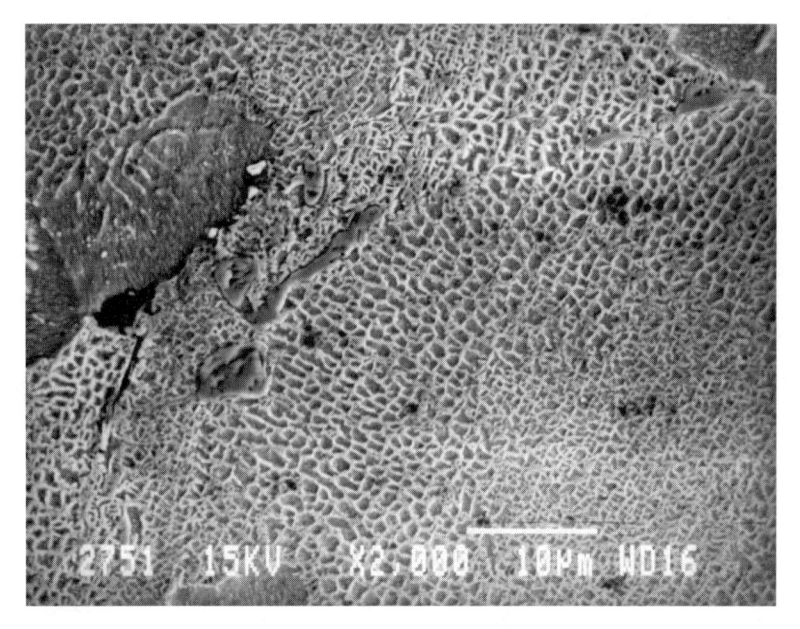

FIGURE C2.31
Photomicrograph of the turbine blade near the leading edge showing a cast structure with eutectic γ-γ′ nodules and carbides in a matrix of γ with precipitates.

TABLE C2.4

Chemical Composition of the Failed Turbine Blade

Element	Wt%	Element	Wt%
C	0.18	Al	5.4
Si	0.2	B	0.09
Mn	0.2	Cr	8.5
S	0.003	Co	14.0
P	0.012	Mo	4.5
Ti	4.3	Ni	Bal

Discussion

The material microstructure and hardness conform to the specified standard. The fracture pattern suggested that the fracture originated near the outer surface of the leading edge of the blade. The corrosion debris on the fracture surface close to the leading edge showed high concentrations of sodium, calcium, sulfur and some chlorine in addition to the other alloying elements present in the material. Porous pockets on the fracture surface near the leading edge and the presence of sulfur and chlorine in the debris indicated that the blade had suffered hot corrosion in an atmosphere containing sulphates and chlorides of sodium and calcium. It is reasonable to assume that local mechanical damage to the nickel aluminide coating could have triggered corrosion of the base metal.

The presence of beach marks and fatigue striations ahead of the corroded region indicated that the crack initiated due to corrosion damage and propagated under fatigue loading. The final failure towards the trailing edge was influenced by the reduced cross section and operating temperature.

Conclusion

The turbine blade failed due to fatigue initiated by hot corrosion in a sulfur- and chlorine-rich environment.

Case 2

Experimental Results

The failed blade and the cooling air deflector insert are shown in Figure C2.32a/b. The failure of the blade occurred about 10 mm from the platform. The fracture surface was relatively smooth at the leading edge and bore a crystalline appearance near the trailing edge.

The fracture pattern suggested that the crack had originated at the first root of the cooling ridge on the leading edge inside the hollow space of the blade. Beach marks, originating from the cooling ridge at the inner surface of the leading edge, superimposed on the fracture steps were observed (Figure C2.33). A few cavities and a linear casting defect were observed at the fracture origin (Figure C2.34). The fracture surface in the region containing

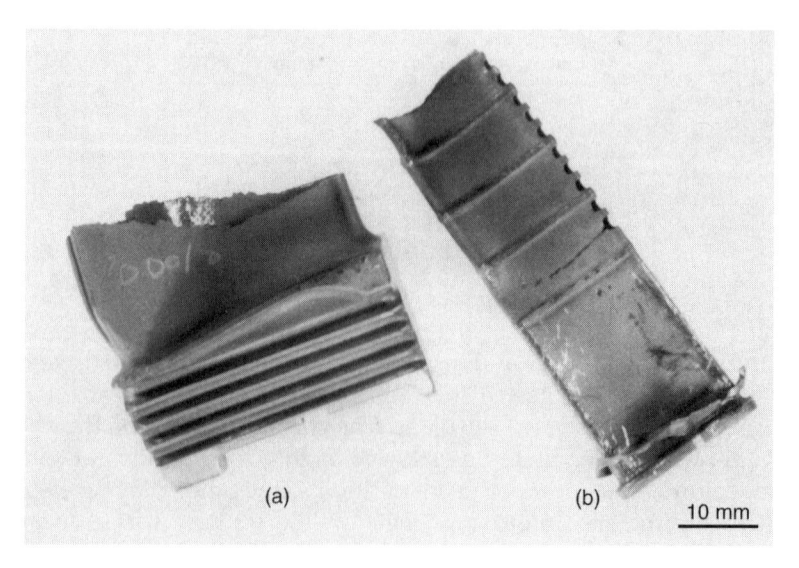

FIGURE C2.32
(a) Root side of the failed turbine blade. (b) Cooling air deflector.

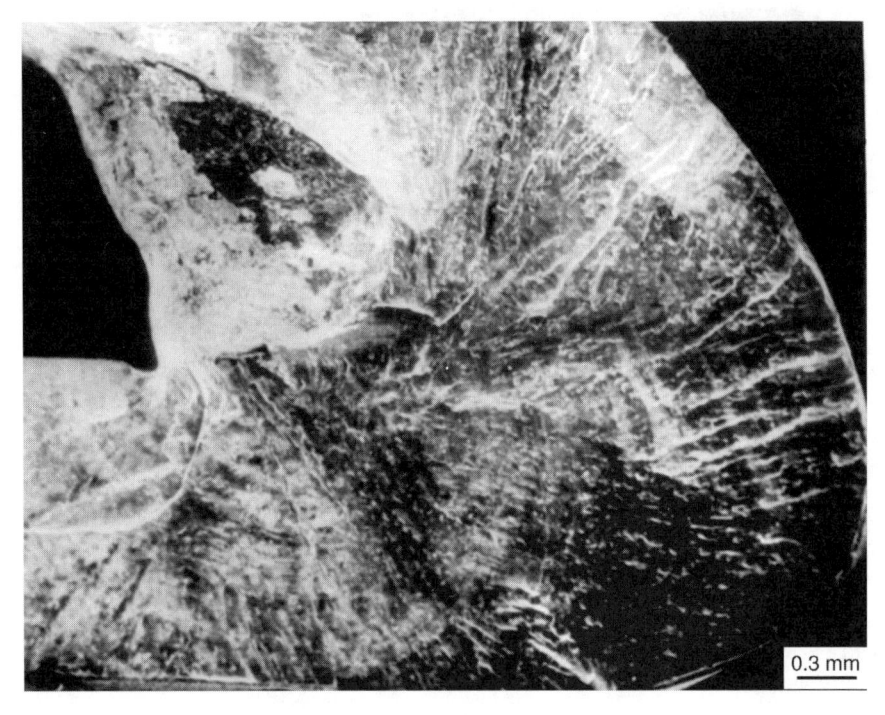

FIGURE C2.33
SEM fractograph of the fracture surface near the leading edge showing beach marks superimposed on the fracture steps.

beach marks showed fatigue striations (Figure C2.35). The trailing edge portion showed failure in interdendritic mode.

Sections from the leading edge close to the fracture were examined. A few interdendritic cavities were observed (Figure C2.36). The microstructure consisted of cuboidal γ' precipitates in austenitic matrix along with eutectic γ-γ' nodules and carbides (Figure C2.37). The blade's surface contained an aluminide coating 10 μm thick. The structure was similar at the root region. The chemical composition of the failed blade is given in Table C2.5. The hardness of the blade on the leading edge was 351 HB and that at the root was 339 HB.

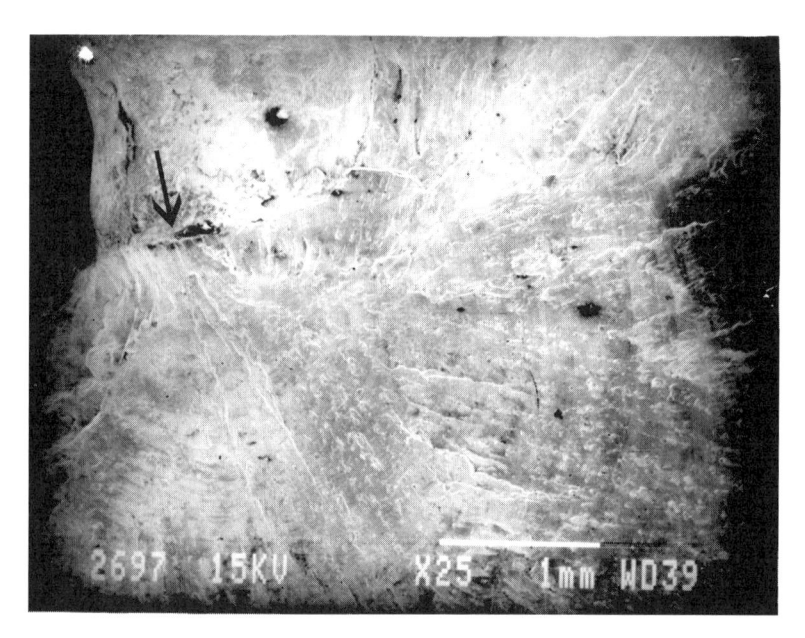

FIGURE C2.34
SEM fractograph of the fracture surface showing a few cavities and a linear casting defect (arrow) at the fracture origin.

FIGURE C2.35
SEM fractograph near the leading edge showing beach marks and fatigue striations.

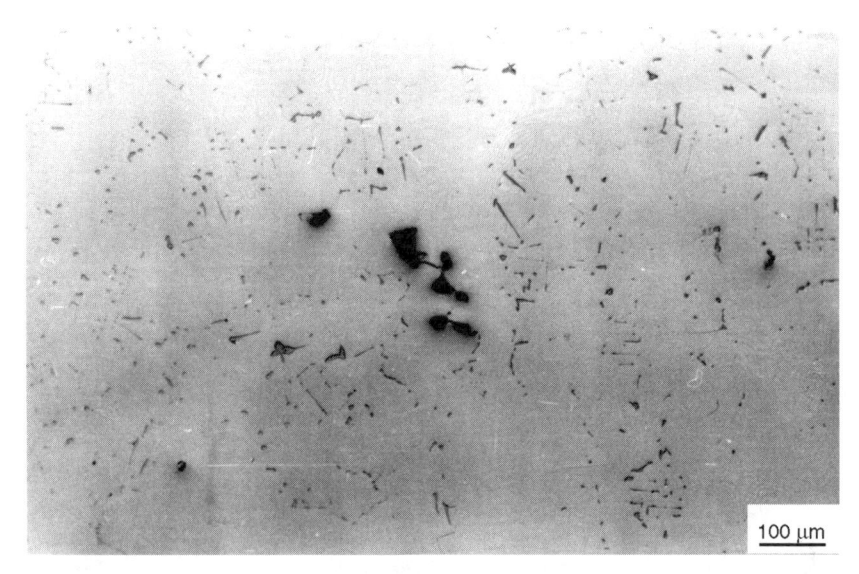

FIGURE C2.36
Typical micrograph of a section near the leading edge showing interdendritic cavities.

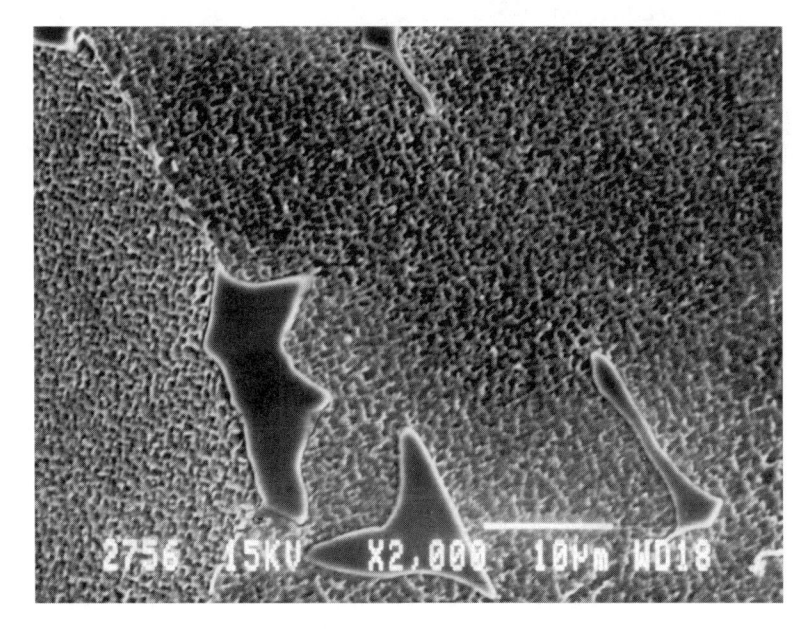

FIGURE C2.37
SEM micrograph of a section near the leading edge showing cuboidal γ' precipitates and carbides.

TABLE C2.5

Chemical Composition of the Failed Component

Element	Wt%	Element	Wt%
C	0.18	Al	5.5
Si	0.2	B	0.08
Mn	0.2	Cr	8.74
S	0.009	Co	14.5
P	0.011	Mo	4.6
Ti	4.25	Ni	Bal

Discussion

The microstructure and hardness are in conformity with those expected from the Ni-based alloy of the specified composition and heat treatment. The fracture pattern suggested that the fracture had originated at the root of the first cooling ridge on the inner surface of the leading edge. The origin contained sharp corners due to improper flushing of the cooling ridge with the inner wall of the blade and concentration of casting defects. The presence of fatigue striations clearly confirmed that the failure occurred due to fatigue. The fatigue originating at the inner surface of the leading edge of a cooled hollow blade is uncommon because of the relatively lower stresses and temperatures experienced by this region of the blade compared to the outer surface of the leading edge. The casting defect, probably a hot tear due to section changes, that existed at the fracture origin might have acted as stress raiser, boosting the applied stress to levels more than those experienced by the outer surface of the leading edge.

Conclusion

The turbine blades failed due to fatigue initiated by stress raisers at the inner surface of the leading edge.

2.4 Failure of the Fuel Pipes of a Jet Engine

Key Words: *Austenitic stainless steel, Brazed joint, Stress concentration, Fatigue*

Introduction

Frequent failures of fuel supply pipes were observed in a particular type of aeroengine [16]. The fuel is supplied at a pressure of 150 kg/cm^2 through the pipes to a variable area jet nozzle. The pipes are made of austenitic

Fatigue Failures

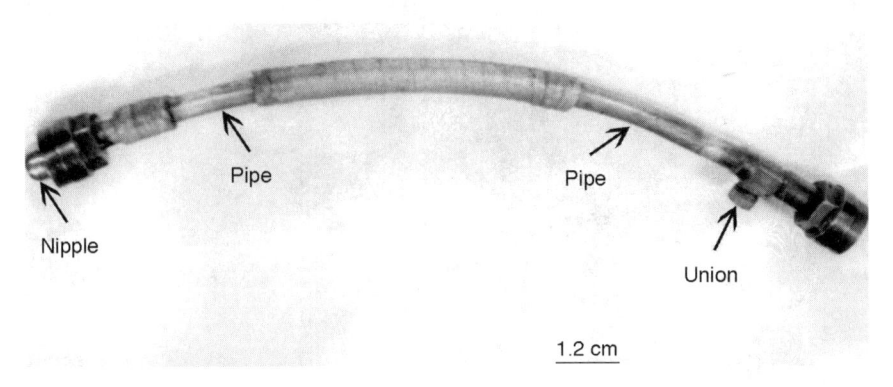

FIGURE C2.38
Photograph of the failed pipe in as-received condition.

stainless steel conforming to Russian grade 08X18H10T, equivalent to AISI 321. The 7.5-mm-diameter pipe with a wall thickness of 1.0 mm is connected to various types of unions and T-joints by brazing. The maximum operating temperature is 110°C. All the failures were located generally at the brazed pipe–union joint or pipe–nipple junctions.

Experimental Results

A typical segment of the fuel pipe line with the union joint and nipple is shown in Figure C2.38. The circumferential crack, located at the brazed joint between the pipe and the union, extending to about one third of the circumference, is shown in Figure C2.39. The crack was found to originate at and propagate along the groovelike feature between the brazed joint and the pipe where the braze material is not properly flushed with the outer surface of the pipe. The fracture surface of the crack, forced open, showed a lens-shaped region followed by fibrous fracture (Figure C2.40). The lens-shaped region showed fatigue striations (Figure C2.41). The remaining fracture surface showed equiaxed dimples (Figure C2.42).

Longitudinal sections containing the union, pipe and brazed joint were examined. It was observed that the pipe was inserted into the union and brazed. The outer surface of the pipe contained uniformly spaced grooves, probably to improve the braze joint. Braze material was observed between the pipe and inner wall of the union and also covering the fillet between the pipe and union (Figure C2.43). Such machine marks were not observed on the pipe away from the joint. Many fine cracks initiated at the bottom of the

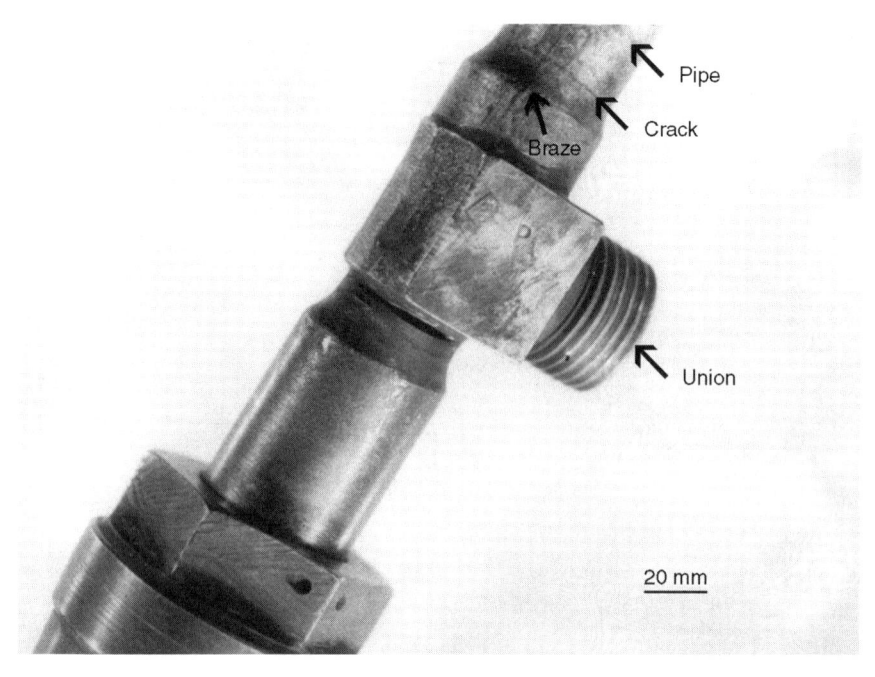

FIGURE C2.39
Photograph of the failed pipe showing the crack near the union.

grooves and propagating into the pipe wall were observed at the end of the braze joint on the surface of the pipe (Figure C2.44). The pipe showed equiaxed grains of austenite. The chemical composition of the pipe is given in Table C2.6. The hardness of the pipe was 155 HV.

FIGURE C2.40
Fractograph showing a darker, lens-shaped area near the outer surface (arrow) and the remaining fibrous region.

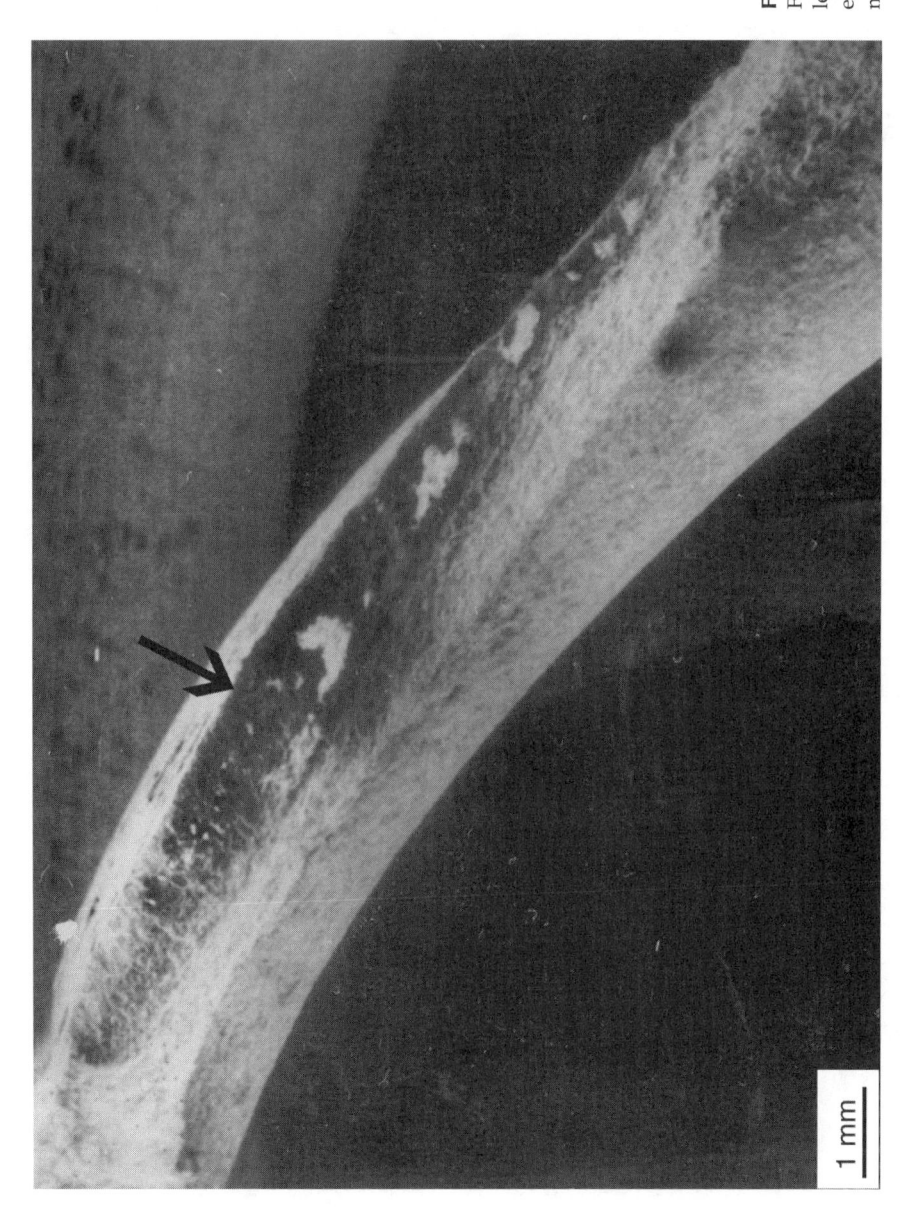

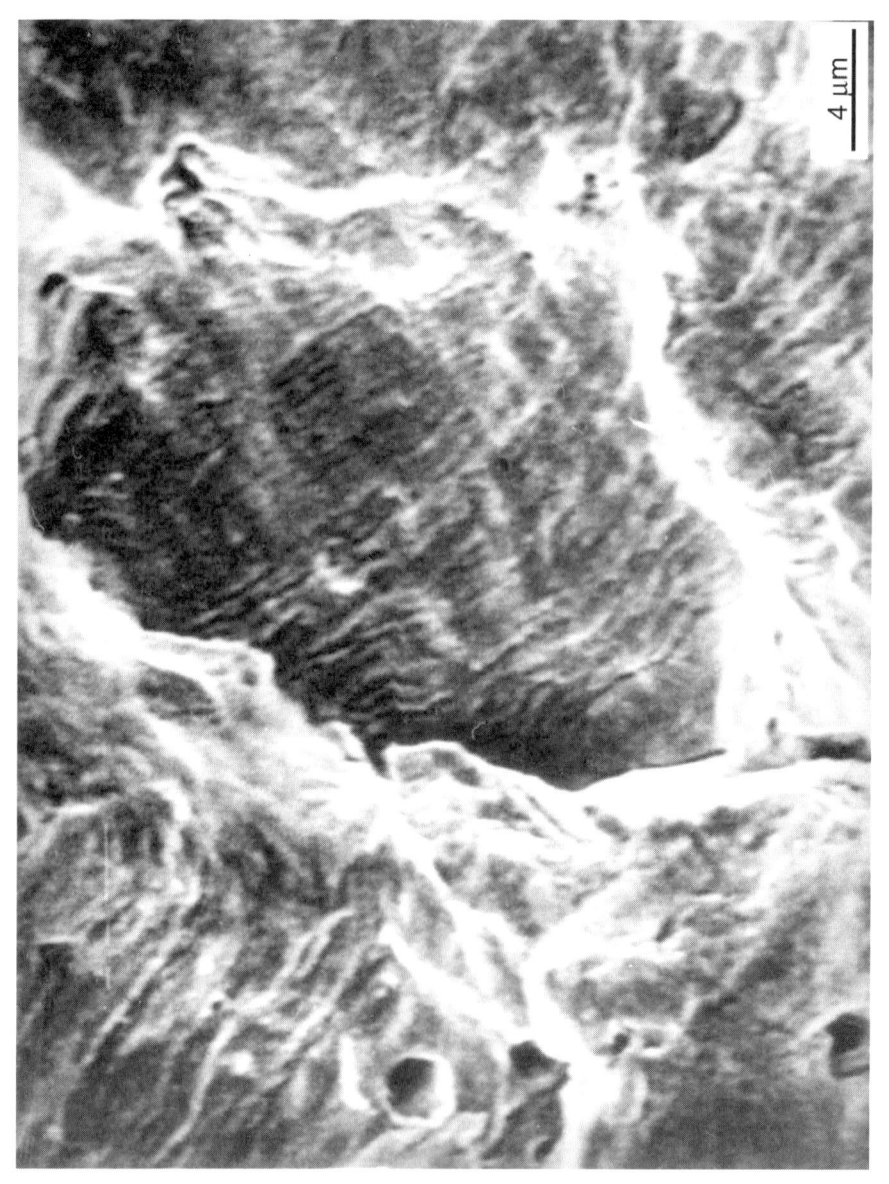

FIGURE C2.41
SEM fractograph of the dark, lens-shaped area of the fracture surface showing the characteristic fatigue striations.

Fatigue Failures

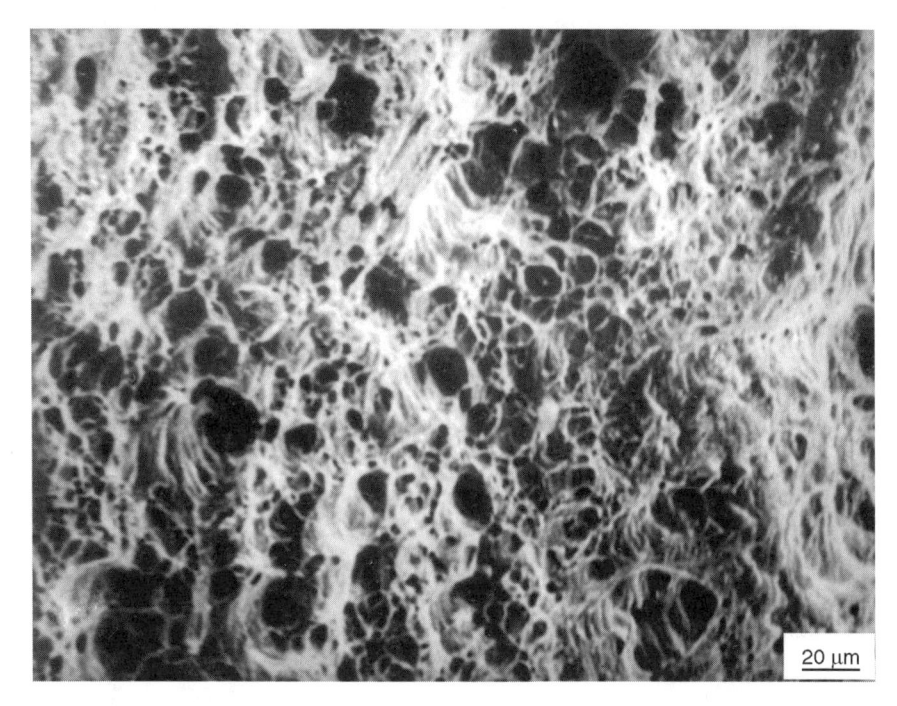

FIGURE C2.42
SEM fractograph of the remaining one third of the fracture surface showing equiaxed dimples.

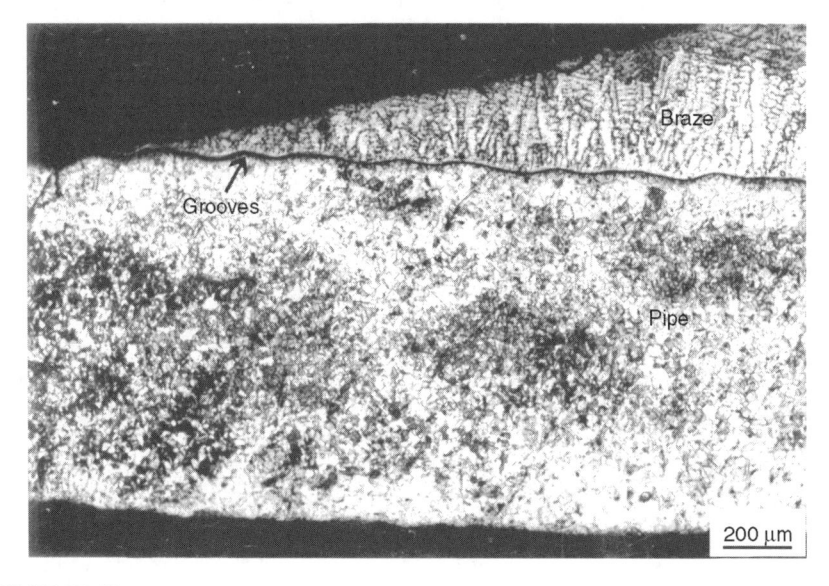

FIGURE C2.43
Photomicrograph of the pipe showing machined surface grooves extending outside the brazed joint region.

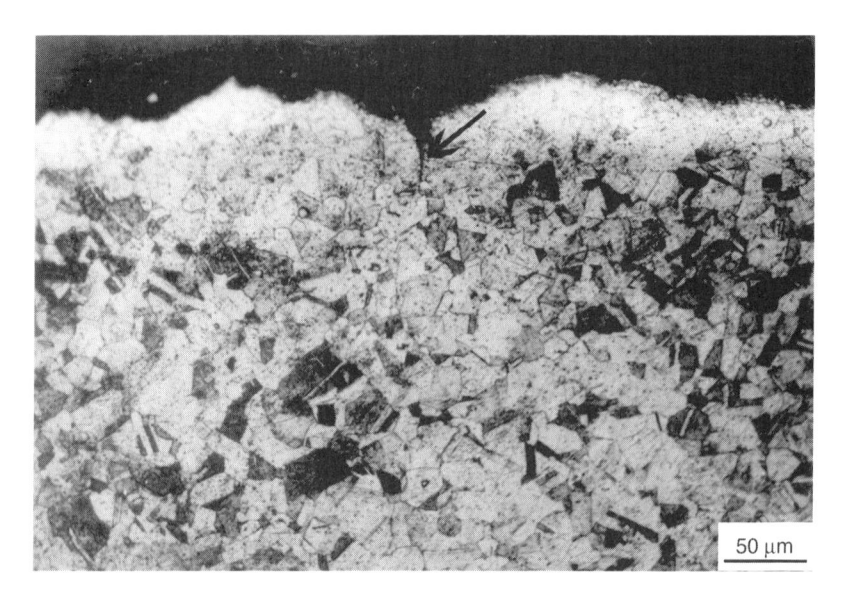

FIGURE C2.44
Micrograph of the inner pipe showing cracks (arrow) at the surface grooves.

TABLE C2.6
Chemical Composition of the Pipe

Element	Wt%	Element	Wt%
C	0.06	Cr	18.95
Si	0.45	Ti	0.55
Mn	1.33	S	0.012
Ni	9.12	P	0.01
Fe	Bal		

Discussion

The chemical composition, microstructure and hardness conformed to the material standard. The experimental results indicated that (a) the crack had originated at the edge of the braze fillet, which was not properly flushed with the pipe, leaving a groovelike circumferential defect; (b) the pipe was made rough by machining at the portion to be brazed to improve the soundness of the joint; (c) the machining marks existed slightly beyond the braze joint; (d) cracks were found to originate at the root of the machine marks and (e) the initial fracture was in fatigue mode. The circumferential defect at the edge of the fillet in combination with cracks at the root of the machine marks acted as a stress raiser for fatigue crack initiation that had propagated under the influence of cyclic stresses developed as a consequence of repeated pressure application during the service of the aeroengine. The dimpled rupture features on the fracture surface ahead of the fracture origin indicated that the final fracture in overload mode was due to a reduction in cross-sectional area.

Conclusion

The fuel pipe failed by fatigue initiating at the grooves on the outer surface.

2.5 Failure of the Impeller of a Turbo Starter

Key Words: Al-Cu-Mg-Ni alloy, Quasi-cleavage fracture, Fatigue

Introduction

Frequent failure of the impellers of a turbo starter of an aeroengine was recorded. The failures occurred after completion of as few as 170 starts [17]. The component was specified to be examined only after a mandatory 1500 starts. The impellers, made from an Al-Cu-Mg-Ni alloy, were solution-treated and aged and anodized. The maximum operating temperature of the component was 50°C in varied environments.

Experimental Results

A damaged impeller with a broken blade and many damaged blades are shown in Figure C2.45. The fracture features on one of the blades, indicated by an arrow on the figure, are different from the rest. The damage pattern indicated that one blade failed first and caused damage to the successive blades in the direction of rotation of the impeller. The fracture on the first-to-fail blade occurred close to the hub at the root. The fracture surface of the blade that consisted of an oval-shaped region whose origin was at the surface of the blade (Figure C2.46). The oval-shaped region was rough and showed some debris at the origin (Figure C2.47). Distinct fatigue striations superimposed on fracture steps were observed in the oval-shaped region (Figure C2.48). The fracture surface away from the oval-shaped region showed quasi-cleavage fracture features (Figure C2.49).

Sections taken close to the fracture region and away from it showed coarse grains of aluminum solid solution with undissolved particles. The chemical composition of the failed impeller is given in Table C2.7. The hardness of the impeller blade was 136 HV.

204 Investigation of Aeronautical and Engineering Component Failures

FIGURE C2.45
Photograph of the damaged impeller. The arrow shows the first-to-fail blade.

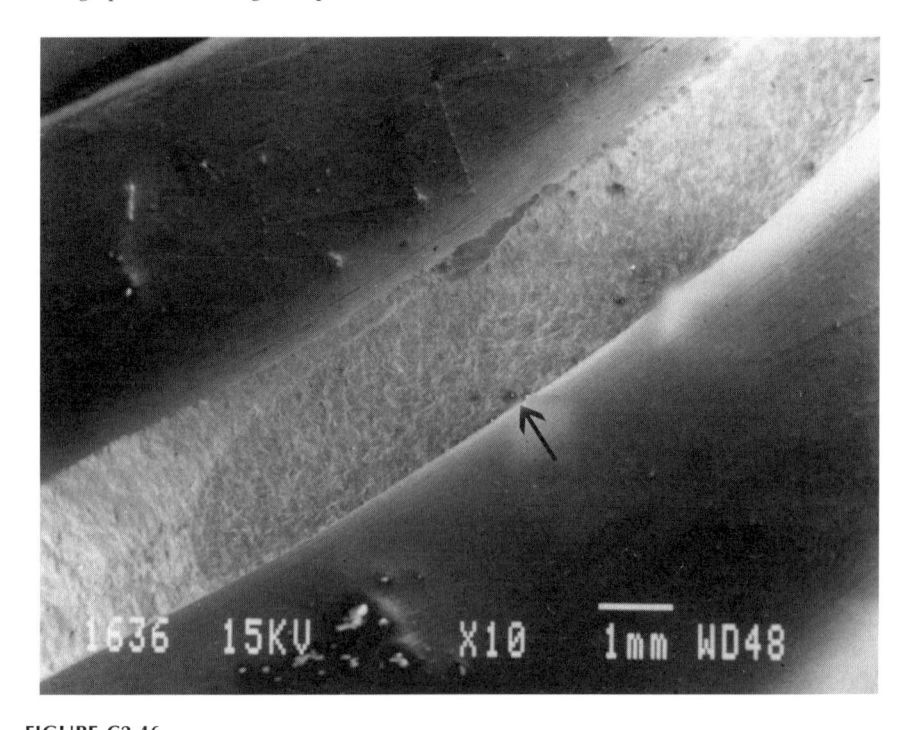

FIGURE C2.46
Fractograph showing the oval-shaped region on the fracture surface. The arrow shows the fracture origin at a dark spot.

FIGURE C2.47 Fractograph showing the crack origin and corrosion debris.

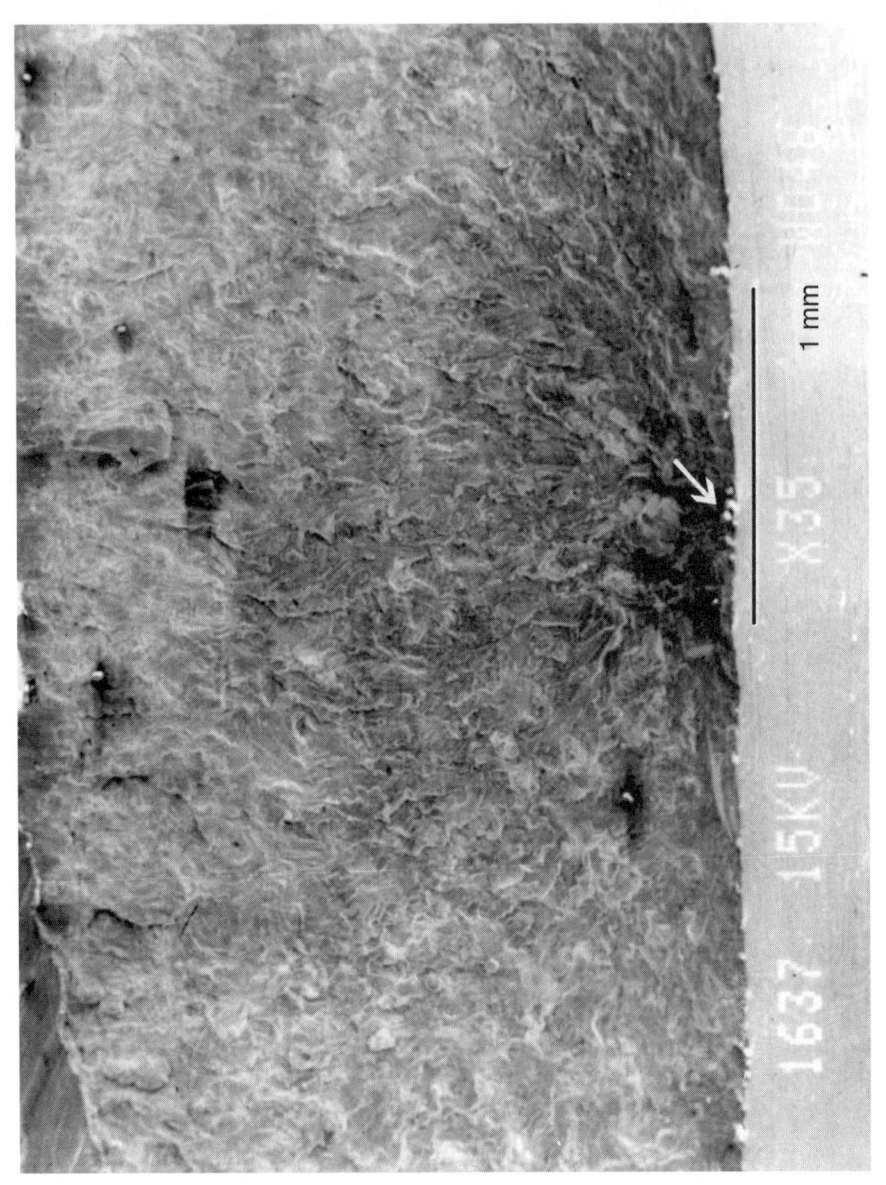

FIGURE C2.48
Fractograph of the impeller blade showing striations on the fracture surface.

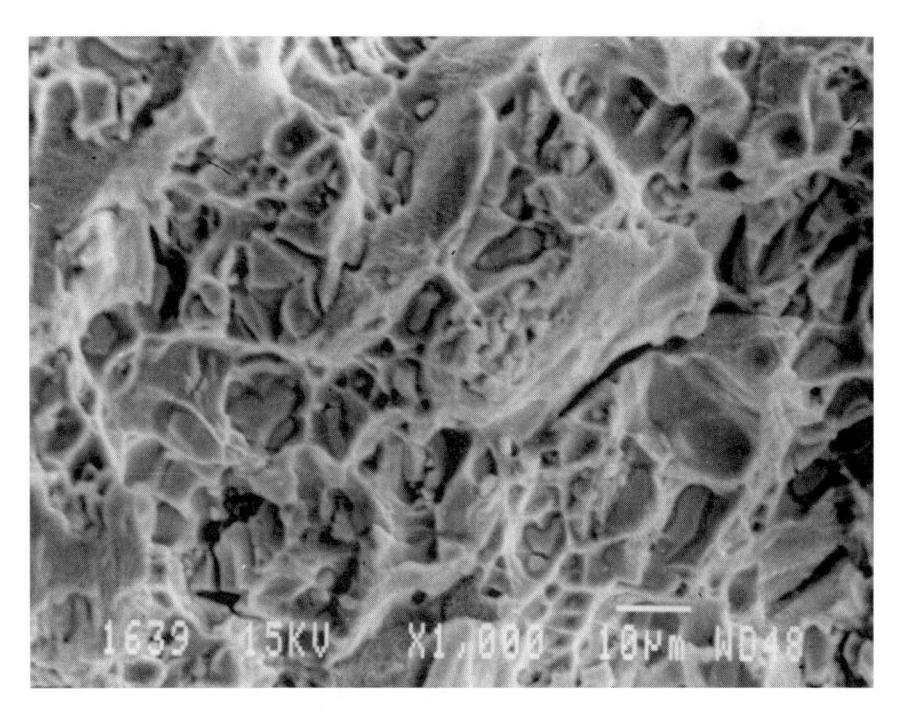

FIGURE C2.49
Fractograph of the fracture surface away from the origin showing quasi-cleavage features.

TABLE C2.7

Chemical Composition of the Failed Impeller

Element	Wt%	Element	Wt%
Cu	2.10	Ni	0.96
Mg	1.30	Si	0.20
Fe	0.79	Ti	0.20
Al	Bal		

Discussion

The chemical composition, microstructure and hardness indicated that the failed impeller was made from an Al-Cu-Mg-Ni alloy and was solution-treated and aged to a hardness level of 140 HV. The fracture features revealed that (a) the crack origin was on the surface of the blade close to the hub, (b) the initial fracture corresponding to the oval-shaped region was due to fatigue as evidenced by the presence of striations and (c) the final failure was in quasi-cleavage mode, caused by overload conditions at high strain rates. The debris at the origin is probably due to corrosion. There were no other obvious abnormalities at the fracture origin to which the initiation of fatigue crack can be attributed.

Conclusion

The failure of the impeller was due to fatigue of one of the blades and consequential damage to the rest of the blades.

2.6 Failure of the Landing Gear Component of an Aircraft

Key words: Al-Zn-Mg alloy, Corrosion, Intergranular cracks, Fatigue striations

Introduction

The aluminum alloy forging of the main leg assembly of a fighter aircraft failed during landing [18]. The forging was stated to have a calendar life of more than 14 years. According to the manufacturer of the aircraft, the undercarriage main fittings are likely to fail by stress corrosion cracking when they get older. The undercarriage fittings, however, did not reveal the presence of any incipient crack during periodic inspections before the failure event.

The component was found to have failed between the compression switch and picketing lug, resulting in circumferential fracture. The forging was found to be broken into four pieces, two of them much smaller than the other two.

Experimental Results

The failed forging is shown in Figure C2.50 and various fracture surfaces of the failed component are shown in Figure C2.51 along with the two smaller broken pieces.

A close look at the circumferential fracture revealed a bright, lens-shaped region. The rest of the fracture surface had a dull appearance. The entire fracture surface including the bright region had a fairly rough, woody texture. A close-up view of the fracture surface including the bright, lens-shaped area is shown in the photograph in Figure C2.52. The outer surface of the component was found to be painted. On removal of the paint layer, it was found that the entire component was anodized except a small area near the lug region. Deep surface scratches were seen at this region, as shown in Figure C2.53. The chemical analysis of the aluminum alloy forging is given in Table C2.8. The average hardness of the component was found to be 178 HV.

A sample containing the bright fractured area was cut and the curved surface near the fracture end was prepared for metallographic examination. Slightly elongated, nonuniform grains and a few intergranular cracks were seen (Figure C2.54). The longitudinal section in the sample showed a fibrous

Fatigue Failures

FIGURE C2.50
Photograph of the component after bringing the broken pieces closer.

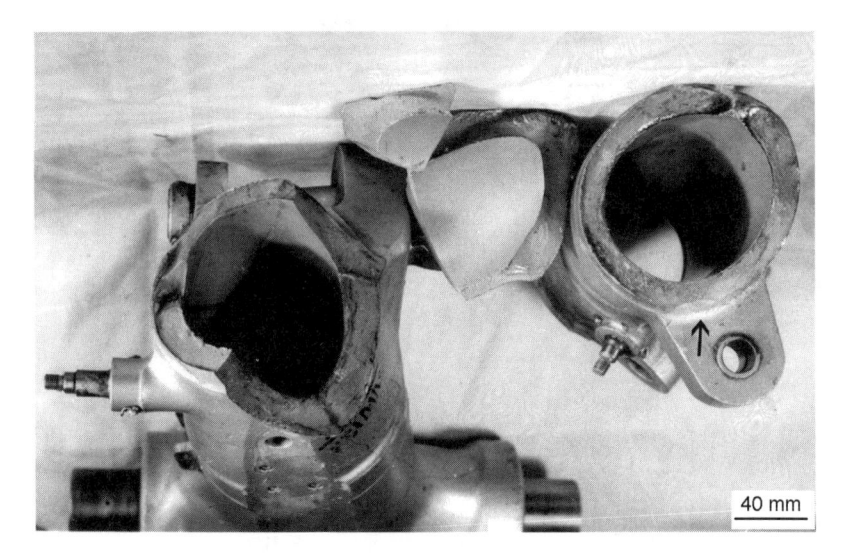

FIGURE C2.51
Photograph of broken pieces of the component. The arrow shows the fracture origin.

structure with fine grains and some undissolved particles aligned at about 45° to the longitudinal axis of the forging (Figure C2.55). The bright, lens-shaped region of the fracture surface showed fatigue striations (Figure C2.56). The dull fracture surface away from the boundary of the lens-shaped region showed dimple rupture (Figure C2.57).

FIGURE C2.52
Photograph showing a close-up view of the fracture surface showing the bright, lens-shaped area at the origin.

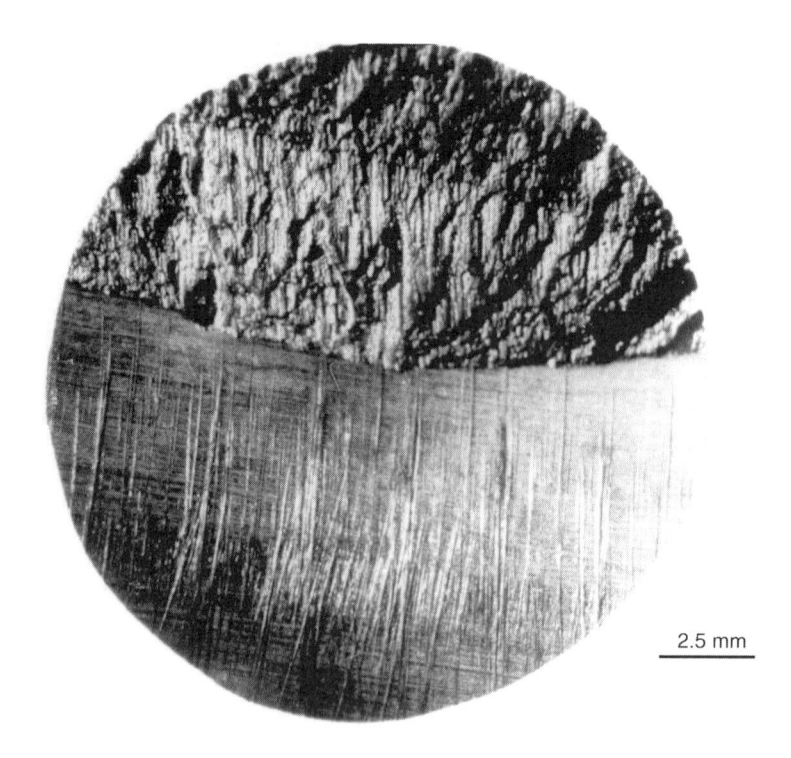

FIGURE C2.53
Macro photograph showing the bright fracture area and a number of deep scratches on the outer surface of the component where the anodized layer was absent.

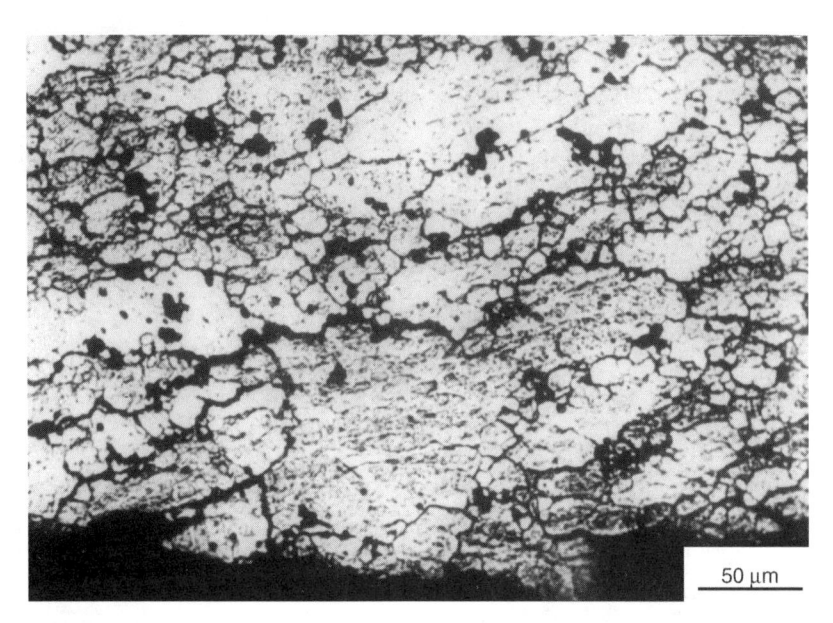

FIGURE C2.54
Photomicrograph of the lightly polished outer surface of the component near the fracture end showing intergranular cracks.

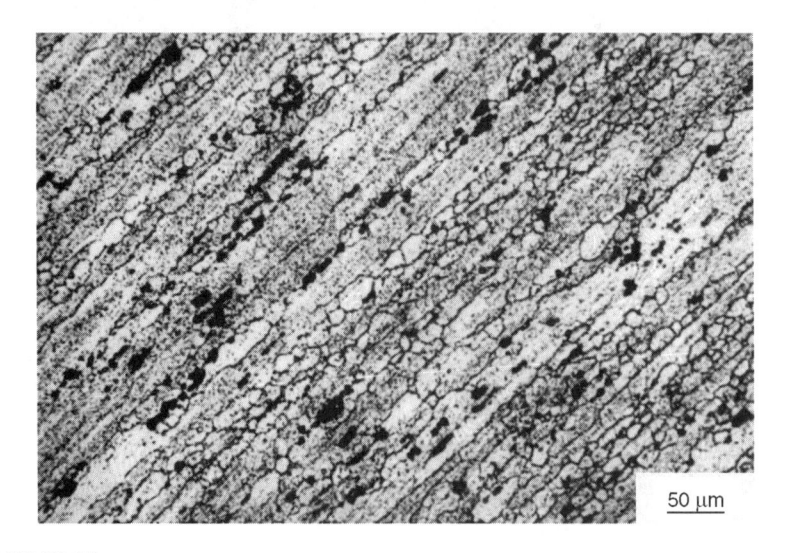

FIGURE C2.55
Microstructure near the failed end showing aligned fine-grained structure with some undissolved particles.

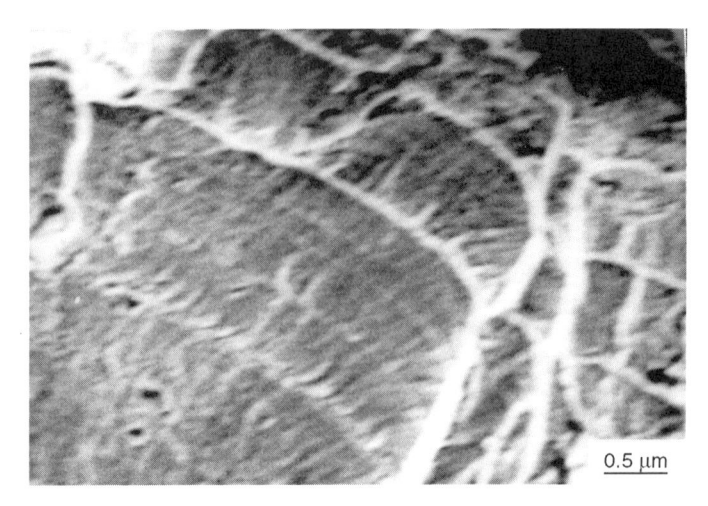

FIGURE C2.56
SEM photograph of the lens-shaped area showing fatigue striations.

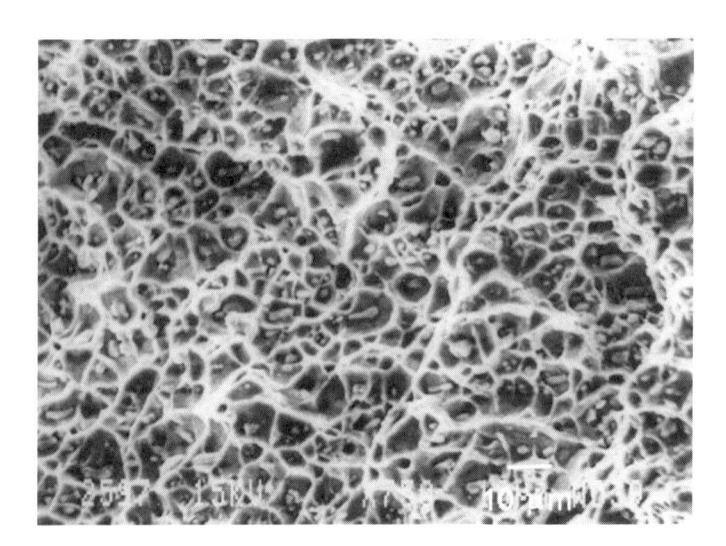

FIGURE C2.57
Fibrous fracture region showing equiaxed dimples.

TABLE C2.8

Chemical Composition of the Al Alloy Forging

Element	Wt%	Element	Wt%
Zn	6.6	Si	0.21
Mg	3.2	Cr	0.02
Cu	0.43	Ti	0.03
Mn	0.5	Al	Bal
Fe	0.25		

Discussion

Visual examination of the fracture surfaces indicated that the failure of the aluminum alloy forging had started from the outer surface at the bright, lens-shaped region. The material of the forging is an Al-Zn-Mg alloy, which is used in the solution-treated and aged condition. The hardness of the component indicates that the material was artificially aged to the peak hardness level. It is well known that fully aged Al-Zn-Mg alloys are susceptible to intergranular corrosion. Metallographic examination of the outer surface of the forging near the fractured end containing the bright, lens-shaped area revealed the presence of intergranular cracks. The outer surface of the forging was found to be devoid of the protective anodized layer at this location. The presence of deep surface scratches in this region is an indication that abrasive material was used, probably at the time of mandatory inspection during overhaul, resulting in removal of the anodized layer. The removal of the anodized layer on the fully aged material appears to have caused intergranular corrosion, resulting in intergranular cracking.

The presence of fatigue striations in the lens-shaped region suggests that the crack had propagated under cyclic loading conditions. Fatigue in this region is considered to have been initiated by some of the intergranular cracks observed on the outer surface near the fracture origin. The progress of the fatigue crack caused a reduction in the cross-sectional area of the component, resulting in final overload failure.

Conclusion

Due to improper handling, the protective anodized layer near the picketing lug was removed, which resulted in intergranular cracking due to corrosion, which in turn initiated fatigue failure of the landing gear.

2.7 Spalling of the Work Roll of a Cold Rolling Mill

Key Words: *Cr-Mo-V Steel, Spalling, Beach marks, Fatigue, Inclusions*

Introduction

The work roll (175 mm in diameter and 560-mm barrel length) of a four-high, reversing, cold rolling mill failed due to spalling after rolling 330 tons of steel [19]. The expected life of the roll was about 3000 tons of rolled steel. The roll was reportedly forged from a high-quality C-Cr-Mo-V steel and heat-treated to a surface hardness of 90 Shore.

Experimental Results

A few spalled fragments from the failed roll are shown in Figure C2.58a. Visual examination of the fracture surface (Figure C2.58b) revealed prominent beach marks on the fracture surfaces, suggesting failure due to subsurface fatigue. Fractographic examination revealed that fatigue had originated at two points (Figure C2.59). One of the origins contained a cluster of inclusions (Figure C2.60). EDAX analysis indicated that the inclusions contained Mg, Al, Si, S, and Ca (Figure C2.61).

Samples from the barrel surface and transverse section were prepared for metallographic examination. In the unetched condition both the samples showed large (10–20 µm) globular inclusions and a few elongated inclusions (Figure C2.62). On etching with 2% Nital, the samples revealed uniformly distributed fine carbide particles in a matrix of tempered martensite (Figure C2.63). The average hardness of the sample was 765 HV at 30-kg load. The chemical composition of the roll material is given in Table C2.9. The oxygen content of the steel was 140 ppm.

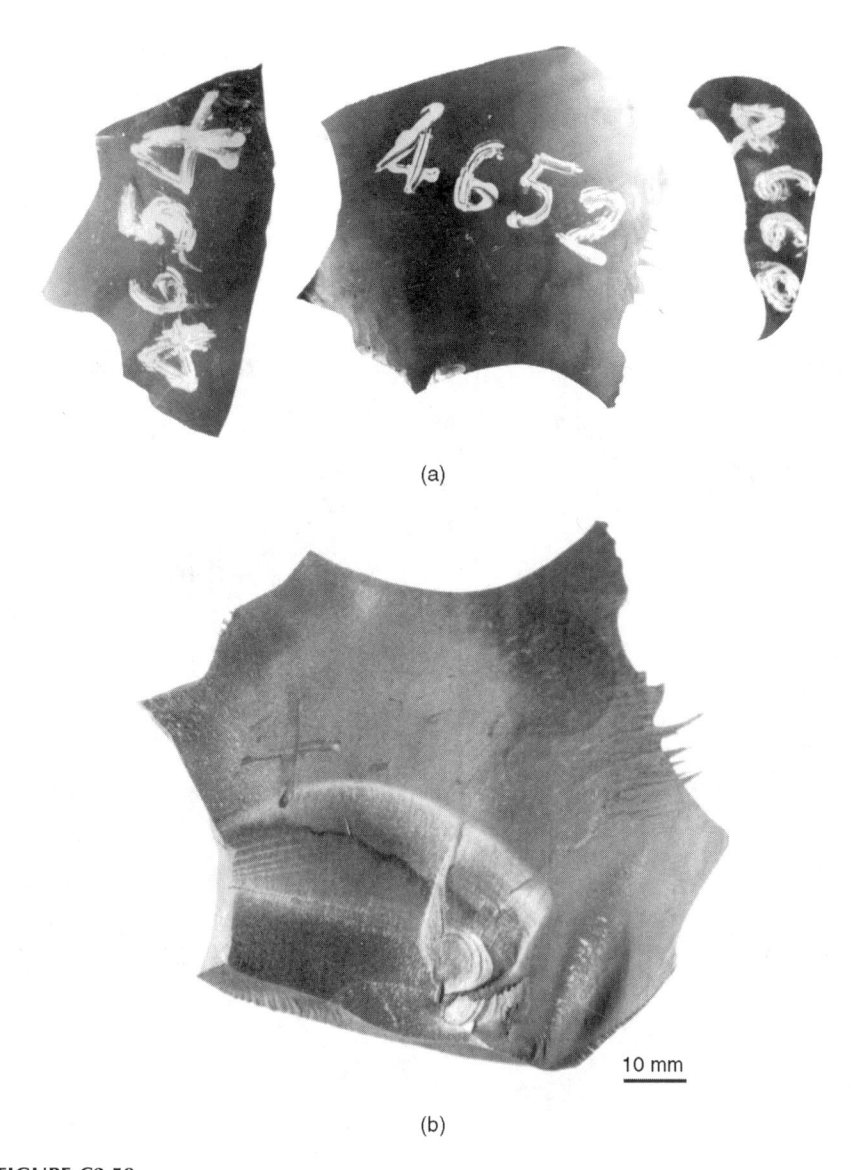

FIGURE C2.58
(a) Spalled fragments from the roll. (b) Fracture surface of the roll spall.

216 *Investigation of Aeronautical and Engineering Component Failures*

FIGURE C2.59 SEM fractograph showing two fracture origins (arrows) and beach marks.

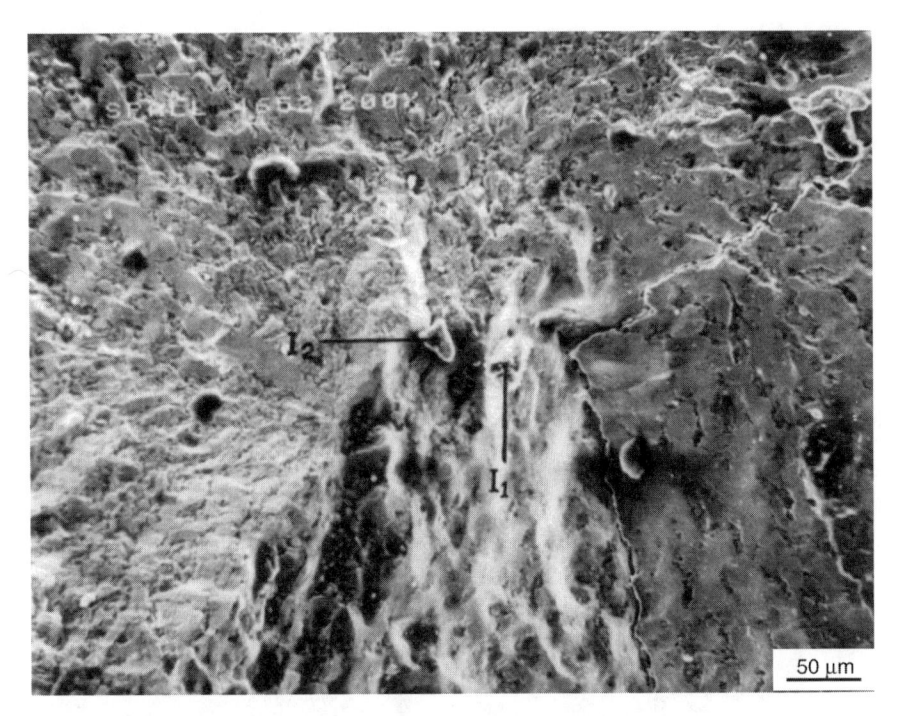

FIGURE C2.60
Fractograph showing inclusions (arrows) at one of the fracture origins.

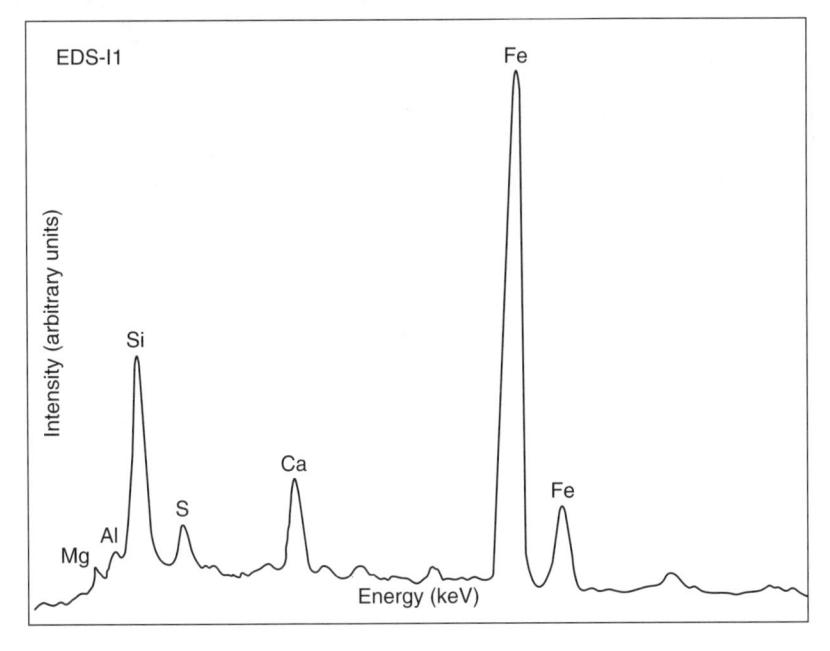

FIGURE C2.61
EDAX spectrum of the inclusions.

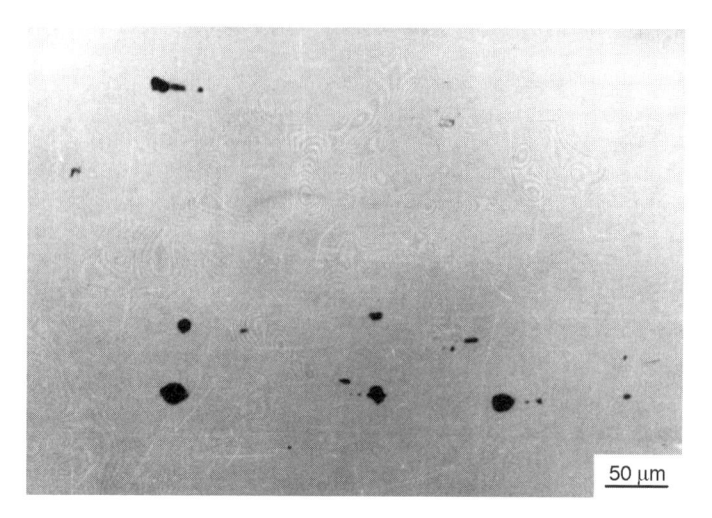

FIGURE C2.62
Globular and elongated inclusions in unetched roll sample.

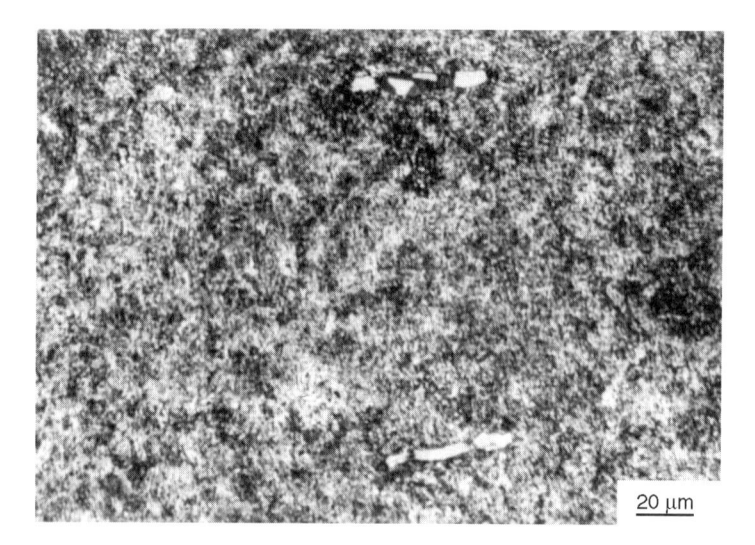

FIGURE C2.63
Microstructure of the roll sample showing fine carbides in a matrix of tempered martensite.

TABLE C2.9
Chemical Composition of the Roll

Element	Wt%	Element	Wt%
C	0.8	Mo	0.18
Si	0.24	V	0.14
Mn	0.4	S	0.015
Cr	1.94	P	0.01
Ni	0.25	Fe	Bal

Discussion

The examination of the fracture surface indicated that the spalling of the roll was due to subsurface fatigue. The inclusions present at the crack initiation zone appeared to be calcium-alumino-silicates that were nondeformable in nature. The steel used for the roll contained a very high level of oxygen, favoring formation of alumino-silicate inclusions.

It is well known that nonmetallic inclusions significantly influence fatigue life. Nondeformable inclusions such as calcium aluminate, alumina and spinnel result in tensile stresses in the surrounding matrix and are thus more harmful to fatigue life than deformable sulphide inclusions. Further, clusters of nonmetallic inclusions are more dangerous than randomly distributed ones and large inclusions can be more harmful than small ones.

The stress pattern experienced by the work rolls in a cold rolling mill is similar to that of ball and roller bearings. It is well known that the fatigue life of ball and roller bearings is seriously affected by nondeformable inclusions. In the same manner, the clusters of large, nondeformable calcium-alumino-silicate inclusions in the roll material are expected to affect the fatigue life of the rolls. It would therefore be reasonable to conclude that the spalling of the roll was due to the presence of a large number of nondeformable inclusions in the steel.

Conclusion

The premature failure of the cold rolling mill work roll by spalling was due to subsurface fatigue initiated at nondeformable inclusions.

2.8 Failure of the Main Gear Box Pinion of a Helicopter

Key words: *Ni-Cr-Mo steel, Case carburizing, Fatigue, Wear damage*

Introduction

A military helicopter was involved in a major accident. Decoding of the voice recorder and examination of evidence indicated malfunction of the main gear box (MGB) prior to the accident [20]. On strip examination, a large number of fragments of various components were noticed. Each of the fragments was carefully examined to identify the component that failed first and isolate consequential damages. Fracture surfaces of most of the fragments showed overload failure symptoms with dents and impact marks. One fragment of the second-stage pinion of the MGB was relatively undamaged and contained fracture features distinctly different from those of other fragments. Pinion failure leading to loss of power, resulting in the crash, fitted perfectly with the preliminary assessment. Examination of the process documents indicated that the component was forged from 16NCD13 grade steel. The pinion was stated to be carburized by liquid carburizing to a case depth of 0.8 to 1.0 mm and heat-treated to obtain case hardness of 550 HV min.

Experimental Results

The gear and pinion assembly of the MGB, which transmits the engine power to the main rotor, is shown schematically in Figure C2.64. The fracture surfaces of the pinion consisted of a smooth, semielliptical region containing beach marks to a depth of about 5 mm from the inner surface of the web followed by fibrous fracture. A few secondary cracks and severe rub marks were observed on the cylindrical surface leading to material removal near the fracture origin (Figures C2.65a and C2.65b).

The smooth region of the fracture surface showed beach marks (Figure C2.66). Distinct fatigue striations were not observed due to the interference of microstructural features with fracture features. The fracture surface away from the region containing beach marks showed dimpled rupture (Figure C2.67). Metallographic examination of a transverse section at an undamaged location revealed a case-hardened structure with a case depth of 0.14 mm (Figure C2.68). The core consisted of tempered martensite and the cast structure was found to be persistent (Figure C2.69). The hardness of the case and core was 620 HV and 485 HV, respectively. The chemical composition of the gear box is given in Table C2.10.

FIGURE C2.64
Schematic of the MGB.

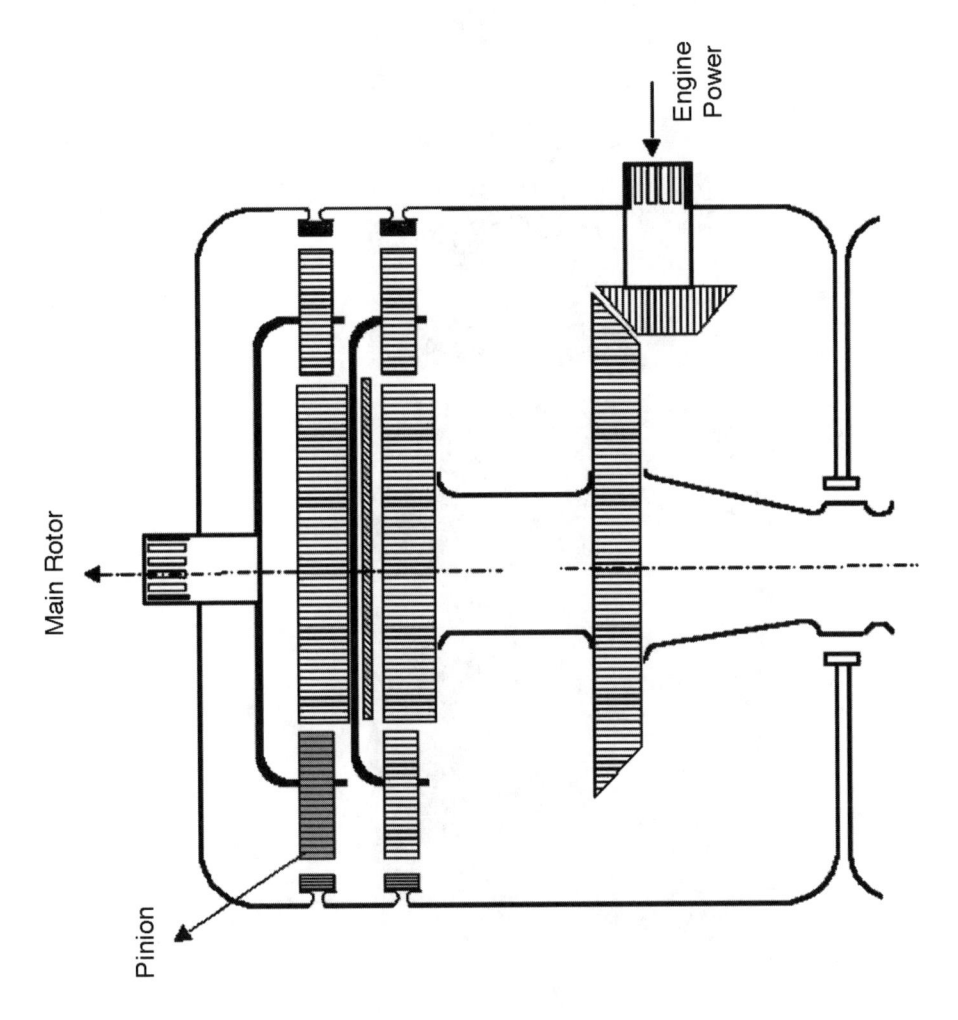

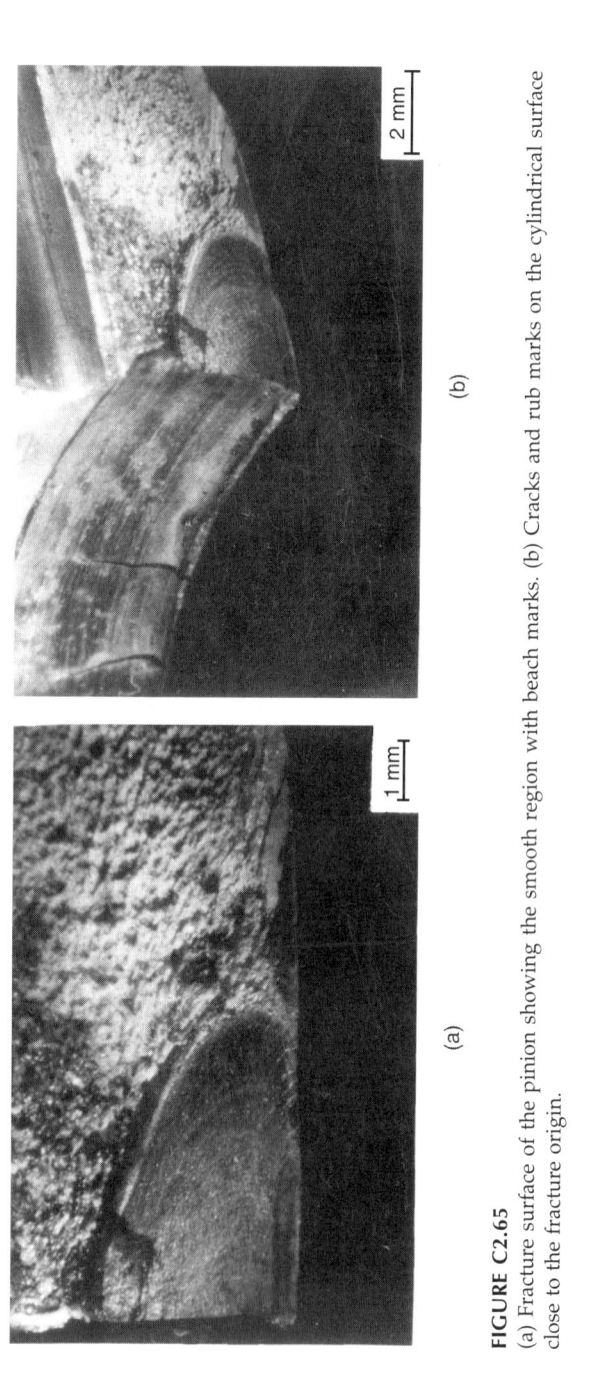

FIGURE C2.65
(a) Fracture surface of the pinion showing the smooth region with beach marks. (b) Cracks and rub marks on the cylindrical surface close to the fracture origin.

Fatigue Failures

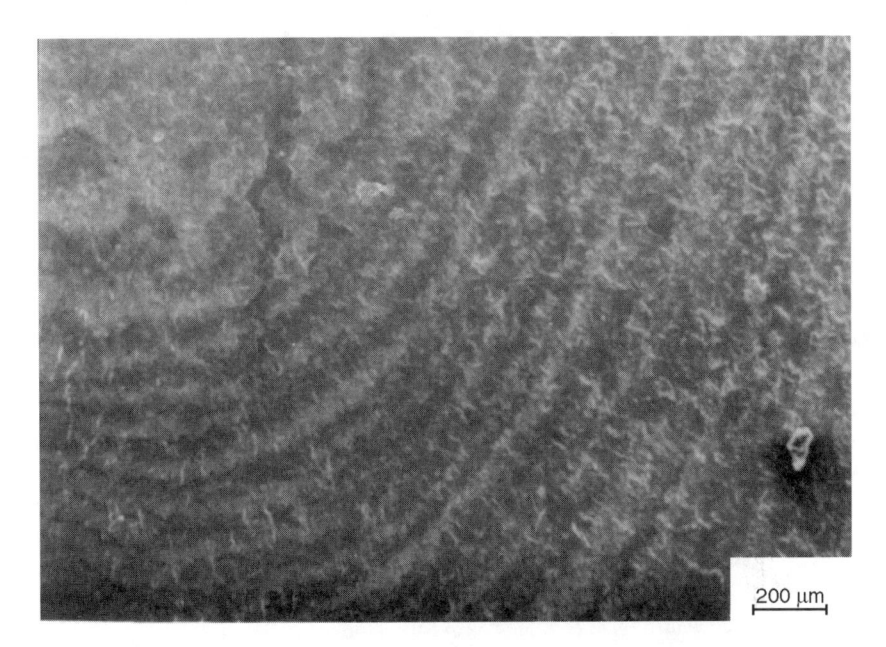

FIGURE C2.66
Beach marks in a semielliptical region of the fracture surface.

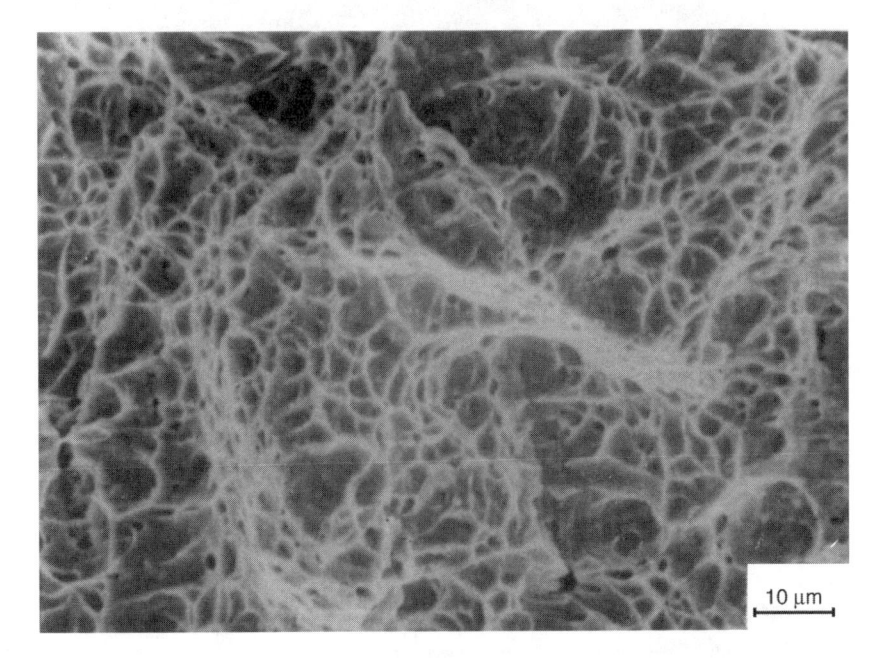

FIGURE C2.67
Dimpled rupture in the fibrous fracture region.

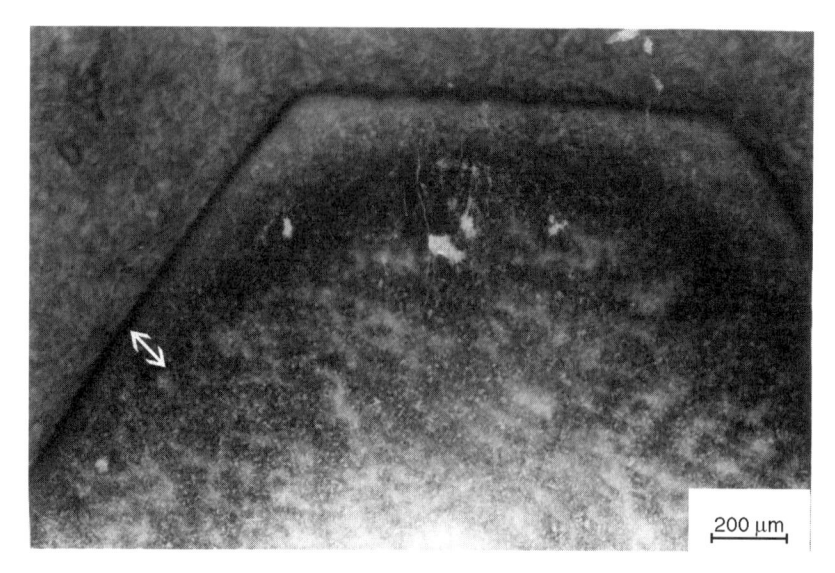

FIGURE C2.68
Case-carburized zone (arrow) at an undamaged region.

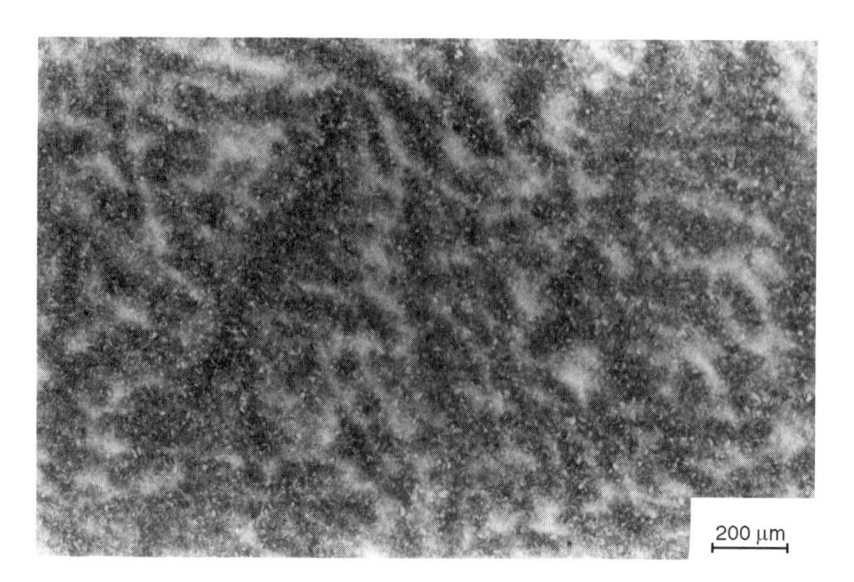

FIGURE C2.69
Tempered martensite and persistent dendritic structure at the case.

TABLE C2.10

Chemical Composition of the Main Gear Box

	C	Si	Mn	Ni	Cr	Mo	S	P	Fe
Wt%	0.16	0.31	0.42	3.21	0.97	0.23	0.01	0.015	Bal

Discussion

The chemical composition of the steel used conformed to the material standard. Fractographic examination suggested that the failure of the pinion initiated at the inner cylindrical surface and propagated to a depth of about 5 mm in fatigue mode, as evidenced by the presence of beach marks. Final failure was caused by overload conditions, as indicated by dimpled rupture. The case depth was only 0.14 mm instead of the specified minimum of 0.8 mm and the case hardness also was marginally lower than the specified minimum, indicative of improper case carburizing treatment and subsequent heat treatment. The extensive rubbing and formation of cracks at the cylindrical surface was due to friction caused by unintended relative motion between the inner surface of the pinion and the contact surface. The removal of material at the fracture origin due to rubbing resulted in loss of fatigue origin.

Conclusion

The failure of the pinion was due to fatigue and the exact cause of fatigue origin could not be identified because of loss of material at the origin due to wear.

2.9 Failure of the Bevel Gear of an Aeroengine Gear Box

Key words: Ni-Cr-Mo steel, Inclusions, Beach marks, Striations, Fatigue

Introduction

A bevel gear connecting the shaft to the main gear box of an aeroengine failed after the engine was run for about 40 h [21]. The gear was manufactured by a reputed gear manufacturer from a forged bar of 170-mm-diameter 3%Ni-Cr-Mo steel conforming to BS S 157-1976. The gears were case-carburized all over, except the bores and internal splines, to attain a case depth of 0.85 mm and heat-treated to a hardness level of 690 HV at the case and 400 HV at the core.

Experimental Results

One segment of the gear with five teeth intact had separated from the rest of the gear. All the teeth of the remaining portion of the gear were worn out. The exact location of the fracture and the fracture origin are shown on a schematic of the bevel gear (Figure C2.70a). The fracture close to the origin was smooth and contained beach marks, while the rest of the fracture was

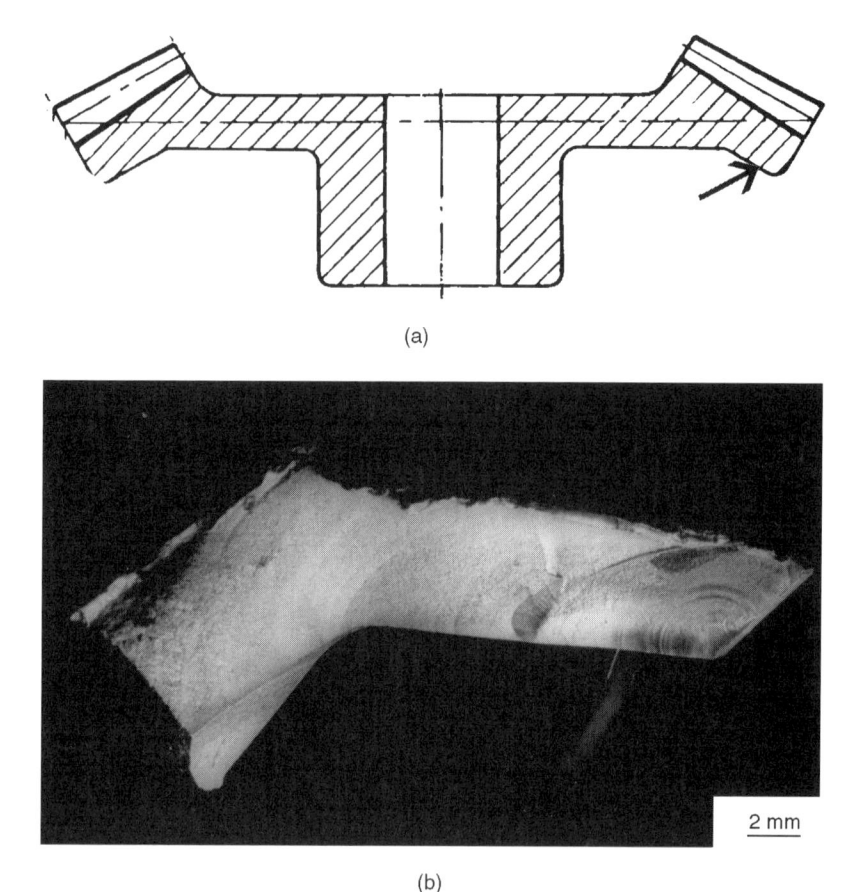

FIGURE C2.70
(a) Schematic of the bevel gear showing the location of fracture (arrow). (b) Fracture surface showing a smooth region containing beach marks at the origin.

fibrous (Figure C2.70b). The fracture surface revealed radial steps and beach marks originating at a corner from a small, stringerlike defect (Figure C2.71a). The stringerlike defect contained a large number of globular particles (Figure C2.71b). Fatigue striations were observed in the region containing beach marks (Figure C2.71c). The region showing fibrous fracture revealed dimples.

The particles in the stringer at the fracture origin were qualitatively analyzed using EPMA. The back-scattered electron image and the X-ray images of Mg, Al, O, Ca, Mn, and S are shown in Figure C2.72. A section parallel to the fracture surface was prepared for metallographic examination. In the unetched condition, a number of sulphides and globular oxides were noticed (Figure C2.73a). The inclusions were rated as per ASTM E45 and the results are given in Table C2.11. On etching with 2% Nital, the sample revealed a banded structure with a carburized case; the case showed a tempered

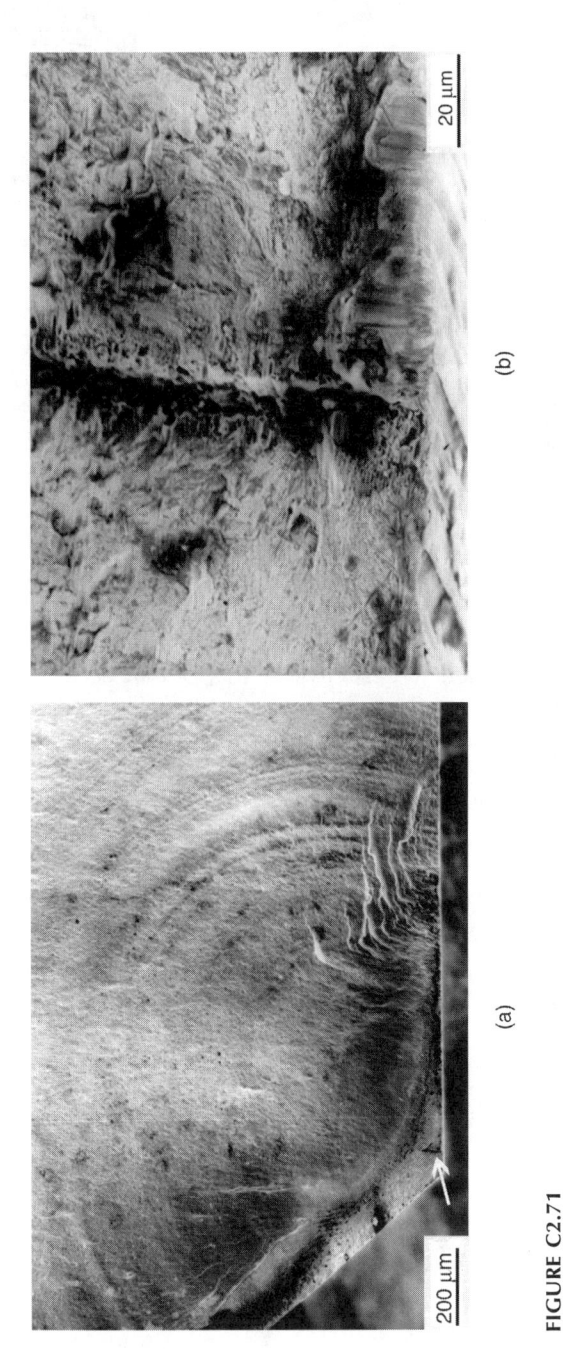

FIGURE C2.71
(a) Fracture surface revealing radial steps and beach marks originating at a small, stringerlike defect (arrow). (b) Stringerlike defect at the fracture origin containing globular particles. (c) Fatigue striations at the fracture origin (arrow).

(c)

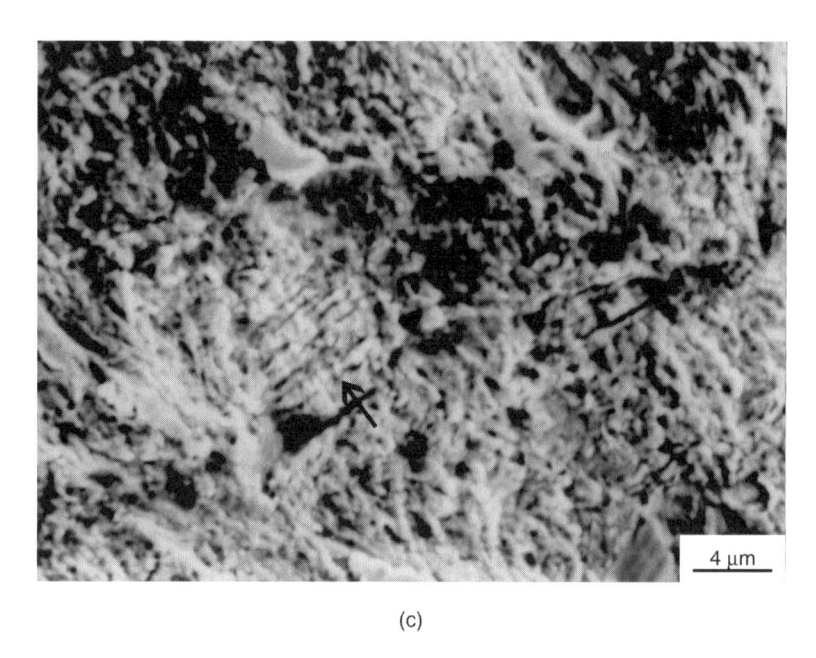

FIGURE C2.71 (continued)

martensite structure with prominent acicularity (Figure C2.73b) and the core showed a tempered martensite structure with globular carbides in a ferrite matrix. The case depth varied from 0.6 mm to 1.0 mm on the section examined. The average prior-austenite grain size was ASTM 7. The average microhardness of the case and the core measured at a 500-g load were 620 HV and 405 HV, respectively.

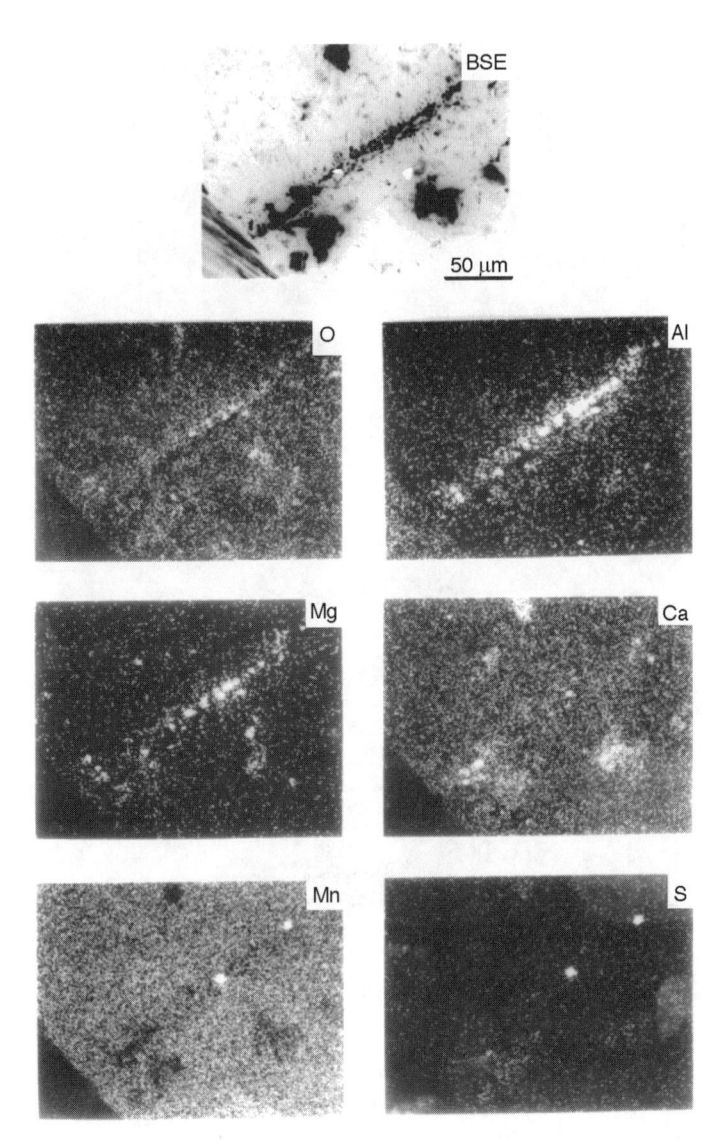

FIGURE C2.72
BSE and X-ray images on the stringerlike defect.

(a)

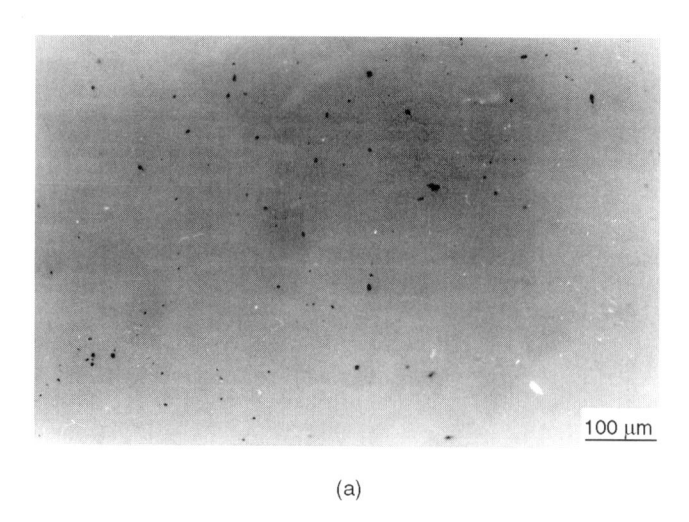

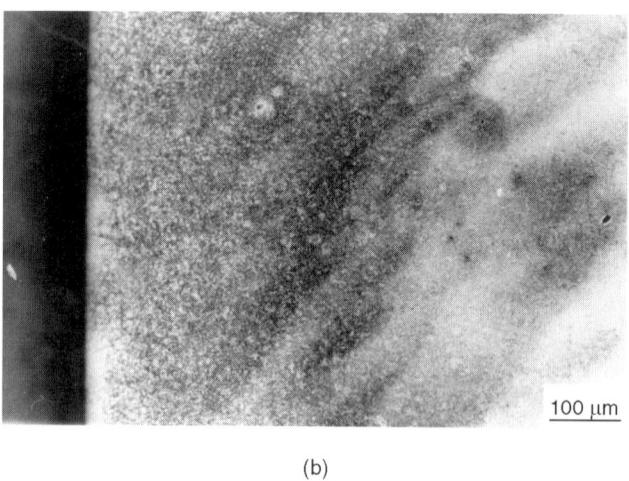

(b)

FIGURE C2.73
(a) Sulphide and oxide inclusion particles in the failed bevel gear. (b) Microstructural banding and acicular martensite in the case-hardened region.

TABLE C2.11

Inclusion Rating in the Failed Bevel Gear

Inclusion type	Rating
Sulphides	1.5 Thin
Alumina	1.0 Thin
Silicate	1.0 Thin
Globular oxides	2.5 Thin

Discussion

Fractographic examination revealed the presence of beach marks and striations suggestive of failure in fatigue mode. The fracture origin contained a stringer of nonmetallic particles that were found to be oxides of magnesium, aluminum and calcium, along with a few particles of MnS. The fatigue crack had initiated due to the stress concentration effect of the inclusion stringer located at the surface near a corner. The final failure in dimple rupture mode was due to the reduction of the cross-sectional area caused by the fatigue crack. The general inclusion content of the steel used for the manufacture of the gear was high. Even though the material standard BS S 157 does not specify the inclusion content, the inclusion rating indicated in the manufacturer's test certificate suggested the steel was much cleaner than it really was. The microstructure and hardness of the case and the core were in accordance with the specification requirements. Although microstructural banding was noticed, its role in initiating the fatigue crack was insignificant.

Conclusion

The failure of the gear was caused by fatigue, and the fatigue was initiated by a stringer of inclusions.

Part 3

Failures Due to Embrittlement

3.1 Delayed Cracking of Maraging Steel Billets

Key words: Maraging steel, Grinding cracks, Lath martensite, Hydrogen embrittlement

Introduction

Maraging steel (M-250) billets 590 mm in diameter and 400 mm long were forged in the temperature range of 1100 to 900°C from an ESR ingot 650 mm in diameter. Four billets were placed side by side and allowed to cool in air. The billet surface was ground to remove oxide scale. After about 2 months from the date of forging, cracks were noticed in two billets [22]. Factory records revealed that the cracks had originated at one end of the cylindrical surface and propagated over the entire length in the next 24 h.

Experimental Results

The failed billets were examined at the site. The surface of the billets contained a network of fine cracks that resembled grinding cracks. The main crack was straight with slight corrugations. It extended from one end to the other and to a depth of about half the radius of the billet. The fracture surface of a sample cut from the crack initiation region was found to be coarse and crystalline (Figure C3.1). SEM examination revealed that the fracture was transgranular with quasi-cleavage features and a few cavities (Figure C3.2). At some locations the fracture surface contained pockets of debris (Figure C3.3), which on analysis by EDS was found to contain Si, Al, Ca, Fe, Ni, and Ti, suggestive of slag inclusions.

A longitudinal section of the failed billet at the fracture origin was prepared for metallographic examination. A number of shallow intergranular cracks originating from the surface were noticed (Figure C3.4). The grain size was found to be about 3 to 4 mm. The microstructure consisted of lath martensite

FIGURE C3.1
Coarse and crystalline fracture features at the origin.

with fine precipitates (Figure C3.5). The chemical composition of the maraging steel billet is given in Table C3.1. The oxygen, hydrogen and nitrogen contents were 18, 10, and 2 ppm, respectively. The average hardness of the billet measured at a 10-kg load was 340 HV.

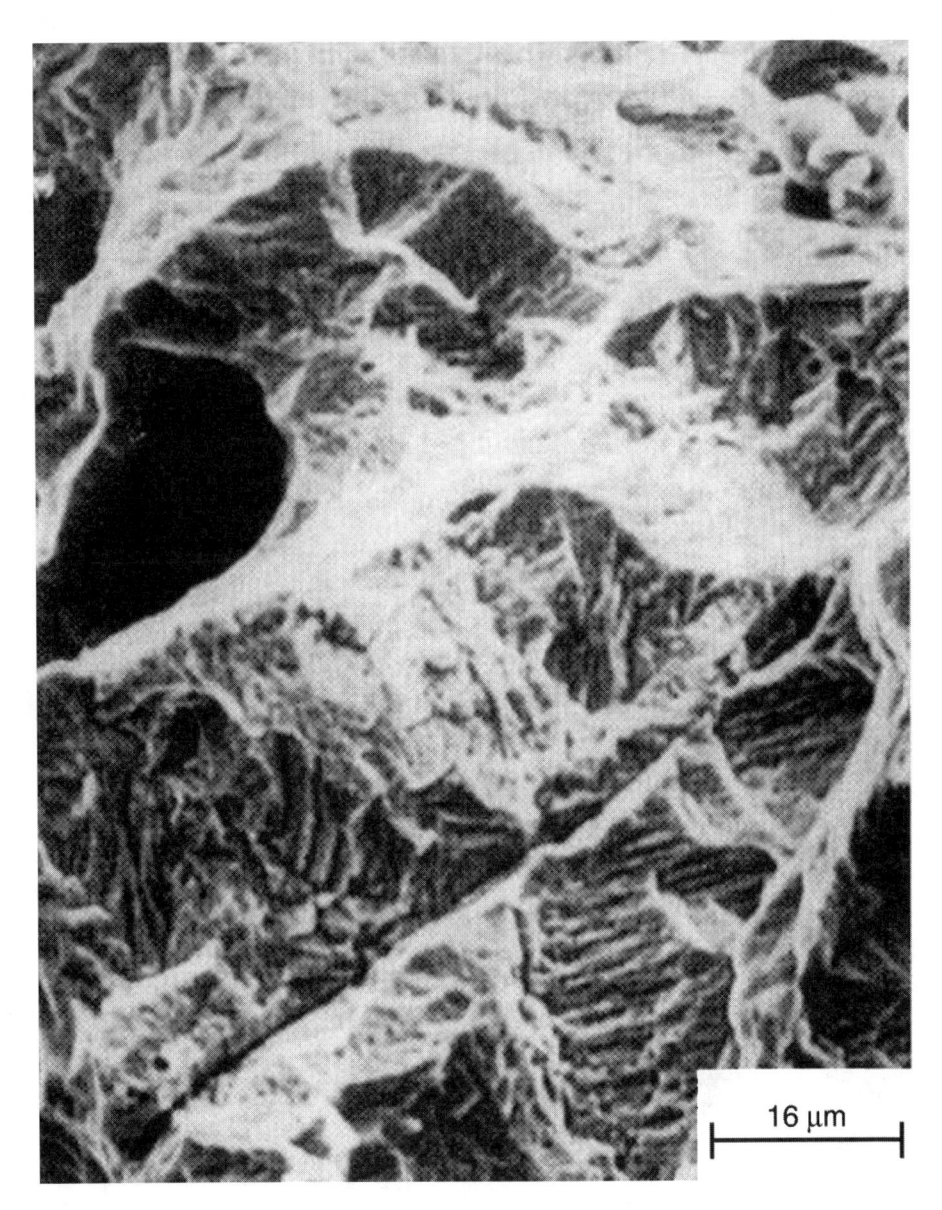

FIGURE C3.2
Fractograph showing transgranular quasi-cleavage fracture features at the origin.

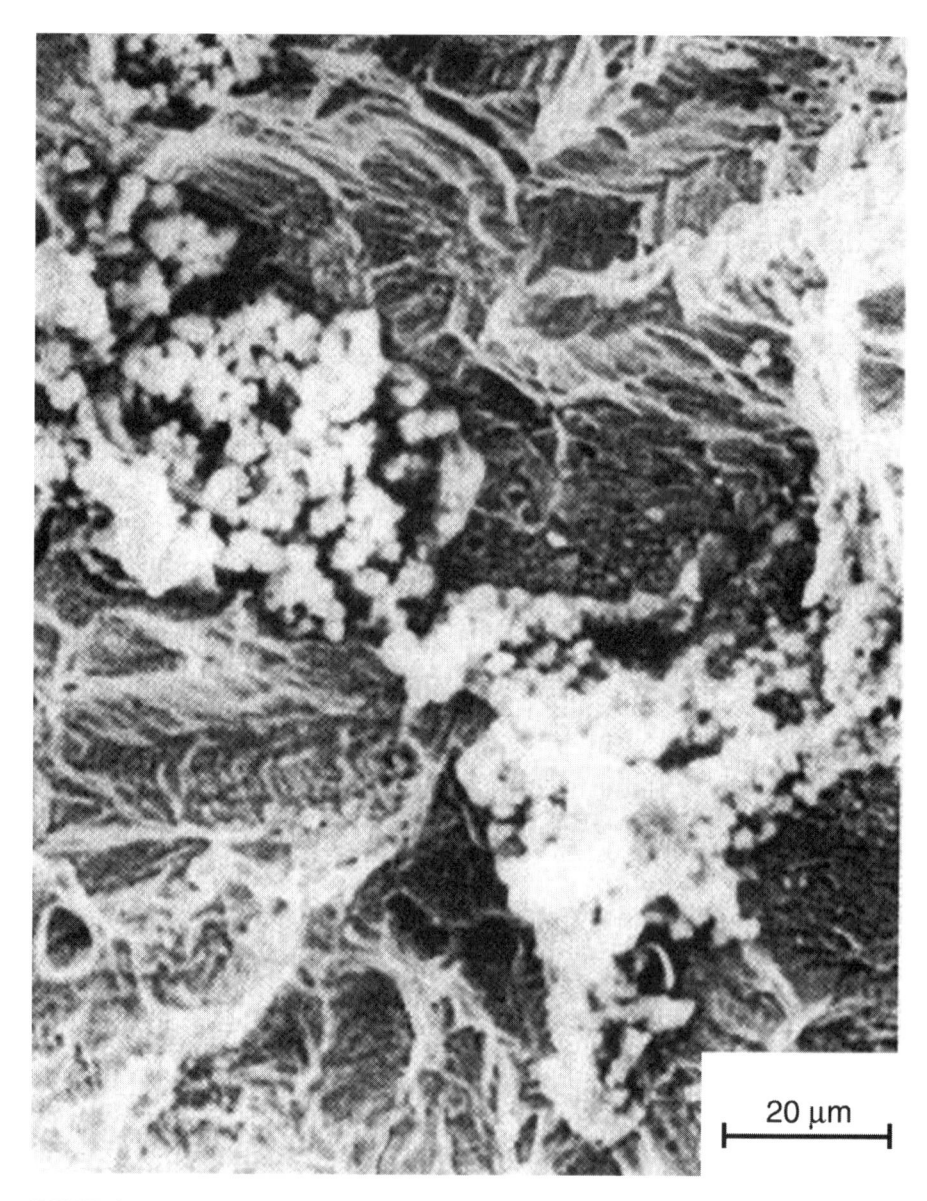

FIGURE C3.3
SEM fractograph showing quasi-cleavage fracture and chunks of slag.

Failures Due to Embrittlement

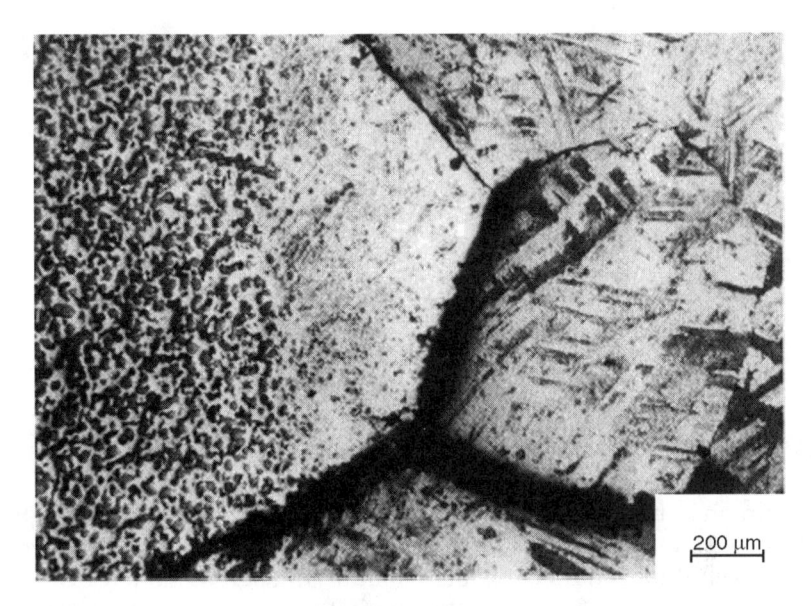

FIGURE C3.4
Intergranular grinding cracks originating from the billet surface.

FIGURE C3.5
Coarse lath martensite.

TABLE C3.1

Chemical Composition of the Maraging Steel Billet

Element	Wt%	Element	Wt%
C	0.005	Ti	0.46
Mn	0.02	Al	0.12
Ni	18.21	S	0.002
Mo	5.05	P	0.005
Co	7.98	Fe	Bal

Discussion

The chemical composition conforms to maraging steel grade M-250. The delay of about 2 months in failure of the billet after forging suggests that the failure may be due to hydrogen embrittlement or SCC. However, SCC can be ruled out as there is neither corrosive environment nor the tensile stress of magnitude required for such a failure. Transgranular quasi-cleavage fracture features, one of the characteristic features of hydrogen-assisted failure, were observed on the fracture surface of the billet. Quasi-cleavage fracture occurs due to fracture along the martensitic platelet boundaries to which hydrogen already present in the material diffuses and assists failure. The crack tips of numerous grinding cracks could have acted as ideal sites of high triaxial stress to which hydrogen diffused preferentially and initiated hydrogen-induced cracking. The long incubation period for hydrogen-induced failure appears to be reasonable in view of the low diffusivity of hydrogen in high-nickel maraging steels. The coarse martensitic platelet size might have aided the quasi-cleavage fracture due to hydrogen embrittlement.

Conclusion

The failure of the maraging steel billets in quasi-cleavage mode after a delay of 2 months after forging was due to hydrogen-assisted cracking.

3.2 Failure of Large-Caliber Gun Barrels

Key words: Ni-Cr-Mo steel, Intergranular fracture, Trace elements, Temper embrittlement

Introduction

Over a dozen large-caliber, smooth-bore gun barrels mounted on tanks burst into fragments during proof-firing [23]. A few barrels failed early in their

TABLE C3.2

The Specified and Revised Heat Treatments for Gun Barrels

Treatment	Revised	Specified
Hardening	860°C, 5 h, WQ for 5 min and OQ for 20 min	900°C, 5 h, WQ for 5 min and OQ for 20 min
Tempering I	420°C, 6 h, WQ	520°C, 6 h, WQ
Tempering II	430°C, 6 h, AC	530°C, 6 h, AC

first quarter of life in service. This led to quarantining of hundreds of barrels. The barrels were forged from ESR quality Ni-Cr-Mo-V steel conforming to GOST 5192-78 grade OXH3MQA. The barrels were quenched and double-tempered to achieve the specified mechanical properties. At the stage of heat treatment the length of the barrel was 6530 mm with outside diameters of 206 mm at the breech end and 215 mm at the muzzle end and a bore diameter of 125 mm. Scrutiny of process documents of the failed barrels indicated that the tempering temperatures were reduced in consultation with the collaborators to overcome the problem of lower strength when tempered as per the original schedule. The specified and revised heat treatment schedules are given in Table C3.2.

Experimental Results

Many of the failed barrels burst into fragments and scattered over a large area. The fragments were carefully examined to identify the fracture origin by tracing the chevron pattern evident on all the fragments. Though there were multiple crack origins in some cases, the fracture always originated at the breech end, where the pressure and temperature are the highest. A typical fragment of a failed barrel is shown in Figure C3.6. Over a length of about 180 mm, the fracture surface was flat, crystalline and free from shear lips. A chevron pattern originating from the boundaries of the flat fracture region was noticed over the rest of the fracture surface. The flat region of the fracture surface showed intergranular fracture with secondary cracks (Figure C3.7). The area showing a chevron pattern showed quasi-cleavage fracture features (Figure C3.8). Longitudinal sections slightly below the fracture surface showed a number of intergranular cracks (Figure C3.9). Extensive banding (Figure C3.10a) and tempered martensitic structure in both the light and dark etching bands were observed (Figure C3.10b).

The mechanical properties of the barrel along with the specified properties are given in Tables C3.3 through C3.5. The chemical composition of the failed barrel along with the specified limits is given in Table C3.6.

FIGURE C3.6
Photograph of one of the broken pieces of a failed gun barrel in as-received condition. The fracture region between arrow heads is flat, crystalline and free from shear lips.

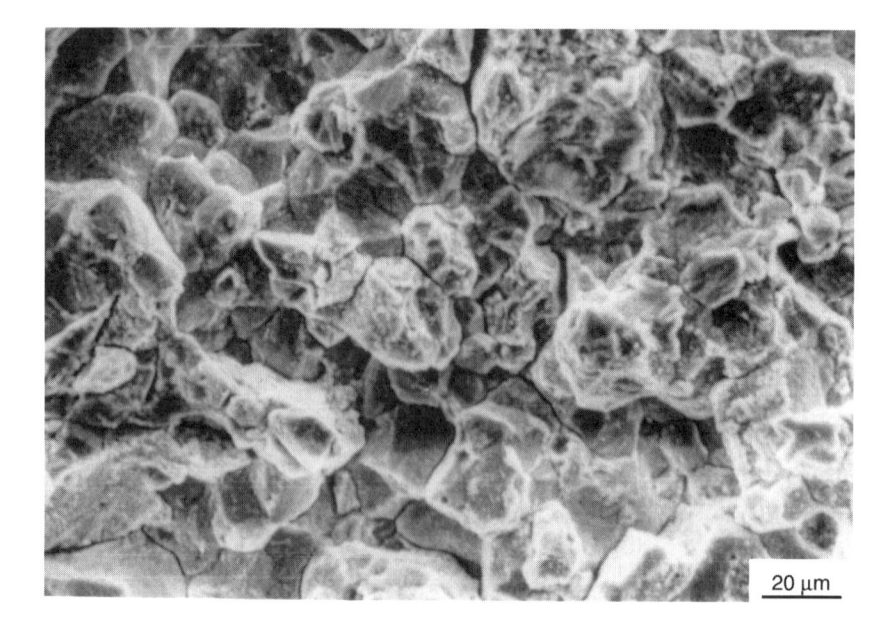

FIGURE C3.7
SEM fractograph showing intergranular fracture features with a few secondary cracks in the flat-faced region.

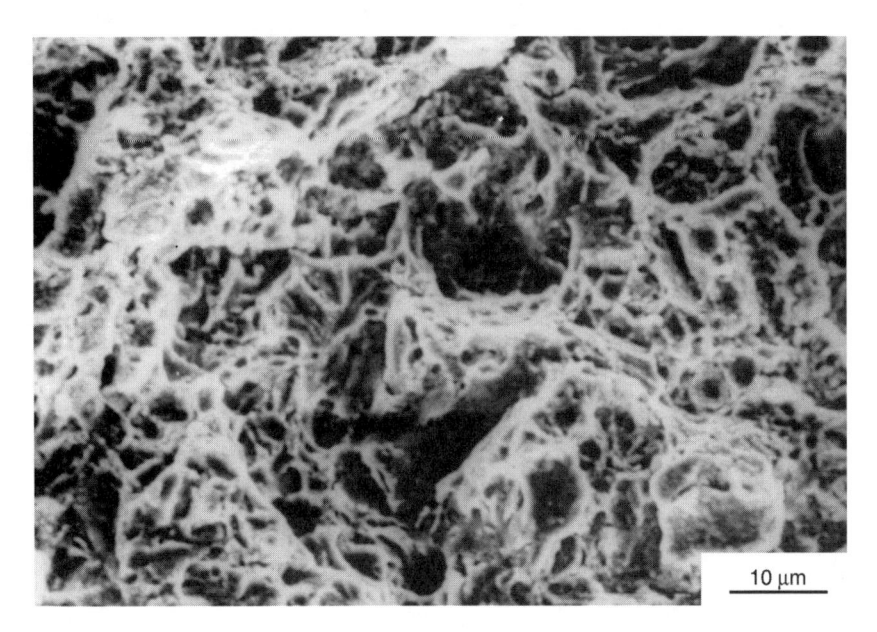

FIGURE C3.8
SEM fractograph in the region containing a chevron pattern showing quasi-cleavage fracture features.

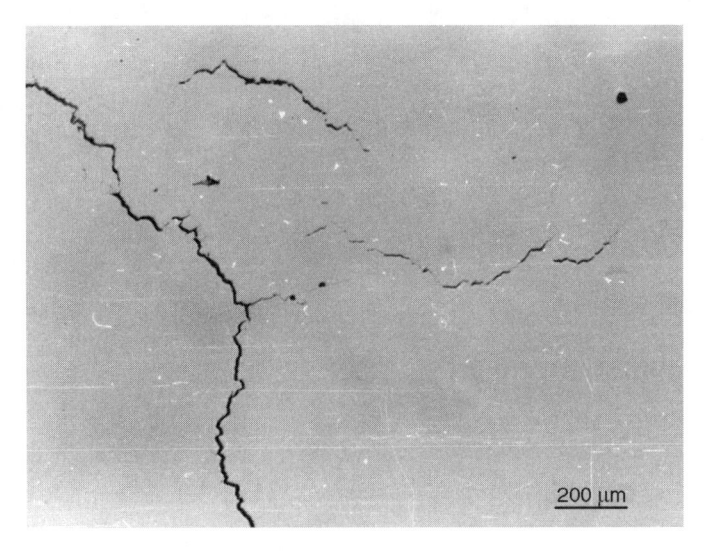

FIGURE C3.9
Photomicrograph showing intergranular cracks below the fracture surface.

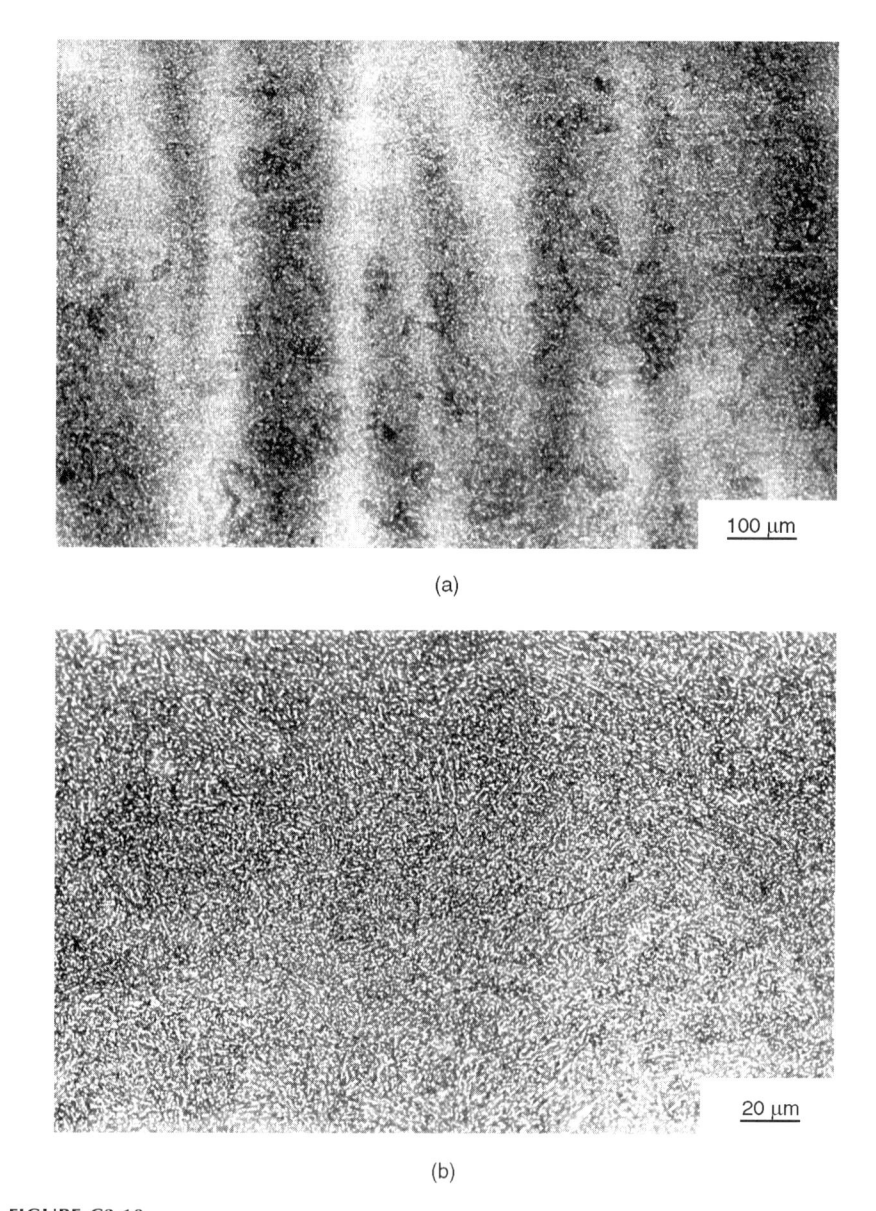

FIGURE C3.10
Photomicrographs of the longitudinal section showing (a) banded structure and (b) tempered martensite.

TABLE C3.3

Tensile Properties (Tangential)

Properties	Barrel	Specified (min)
Yield strength (MPa)	1240	1170
UTS (MPa)	1560	—
% El	10	—
% RA	35	20

TABLE C3.4

Impact Energy (Charpy U Notch) (Tangential)

| | Impact Energy (J) | |
Temperature	Barrel	Specified (min)
Room temperature	25	20
–50°C	18	16

TABLE C3.5

Hardness (HV)

Failed barrel	Specified
469	425–470

TABLE C3.6

Chemical Composition of the Failed Gun Barrels

Element	Specified	Barrel
C	0.33–0.40	0.39
Si	0.17–0.37	0.20
Mn	0.25–0.50	0.42
Ni	3.0–3.5	3.30
Cr	1.2–1.5	1.45
Mo	0.35–0.45	0.39
V	0.10–0.11	0.17
S	0.012	0.003
P	0.012	0.008
As	—	0.008
Sn	—	0.021
Sb	—	0.001
Fe	Bal	Bal

Discussion

The chemical composition, microstructure and mechanical properties are in conformance with the governing material standard. The barrel was tempered at 420°C and 430°C instead of 520°C and 550°C. The steel is susceptible to temper embrittlement (TE) when tempered between 350 and 500°C. The susceptibility to TE increases when the J factor ($J = (Si + Mn)(P + Sn) \times 10^4$ (amounts in wt%)) is more than 10. The J factor takes into account the synergistic effects of the trace elements and other alloying elements that promote segregation of trace elements to prior-austenite grain boundaries, causing embrittlement. In all the failed barrels examined, the J factor was in the range of 160 to 210, which indicates that the steel was prone to embrittlement when tempered in the susceptible temperature range. The observed intergranular fracture with secondary cracks and the presence of intergranular cracks at many locations are supportive evidence that the steel suffered temper embrittlement. The quasi-cleavage fracture at chevron pattern regions is indicative of crack propagation at high strain rates under the stress-raising effect of the intergranular cracks at the origin.

Conclusion

The gun barrels failed due to temper embrittlement.

Remedial Measures

The obvious remedy to temper embrittlement is to temper above the susceptible temperature range. By improving the quench efficiency by (a) agitating the quenching medium and (b) ensuring quenchant conditions such as temperature, viscosity, suspended particles, etc., the strength in as-quenched condition was increased for the same ruling section in order to ensure that the strength levels did not fall below the specified values, even after higher-temperature tempering. Tempering at 500°C (first stage) and 520°C (second stage) were implemented with success for all batches produced after the investigation.

Salvaging of hundreds of barrels already tempered in the TE range was an absolute necessity for economic reasons. The first choice was to heat-treat the barrels again, which was impractical for finished gun barrels. Thus, if the barrels were to be cleared for use, it was necessary to either ensure the absence of embrittlement by confirming that the trace element levels were too low to cause TE or to restore toughness by tempering above 520°C to diffuse out the trace elements from the prior-austenite grain boundaries. Accordingly, based on the trace element analysis, barrels were grouped into three categories as given in Table C3.7.

Based on multicentric analysis, barrels falling in Cat A were identified. The analysis showed that over 85% of the barrels quarantined belong to Cat A. Three barrels from Cat A were subjected to full life assessment trials. All the barrels survived the stipulated life without fracture. Considering the

TABLE C3.7
Grouping of Gun Barrels Based on Trace Element Analysis

Trace element concentration (wt%)	Cat A	Cat B	Cat C
P + Sn + As + Sb	<0.02	0.021–0.025	>0.026

small number of barrels in Cat B and the economics of retempering, no action was initiated to salvage Cat B, even though retempering at 500°C yielded a more than 30% enhancement in impact energy with less than a 5% drop in strength. Cat C barrels were rejected as unsatisfactory since it is not possible to reheat the finished barrel without warpage, and reheating is the only solution to eliminate TE that has already damaged the barrels. This case is an example of the failure analyst finding solutions and successfully implementing them to overcome the problem.

3.3 Failure of the Nose Fairing of an Aeroengine

Key words: Ti-Al-Mn alloy, Faceted fracture, Gaseous impurities, Welding, Embrittlement

Introduction

The nose fairing is located at the front of the engine to streamline the flow of air to the engine. The outer casing of the nose fairing was welded to the flange by gas tungsten arc welding. Cracks about 35 mm in length were observed along the weld joint of the casing and flange after completion of 97 h of life [24]. The nose fairing was made of 1-mm-thick titanium alloy sheet conforming to Russian grade OT4-1 (Ti-2.3Al-1.5Mn) and was welded to the flange made of BT 5-1 (Ti-5.5Al-2.5Sn) with filler wire conforming to BT-1 (CP titanium). The flange thickness at the weld joint was identical to that of the nose fairing.

Experimental Results

The failed component in as-received condition is shown in Figure C3.11. The component had two cracks perpendicular to each other, and both cracks were along the weld. On opening the cracks, it was found that fracture surfaces were discolored over a length of about 20 mm on all sides from the junction of the two cracks and the rest of the region was brighter. In general, the fracture surface was coarse and crystalline. The discolored region of the fracture surface showed faceted fracture (Figures C3.12a and b). The transition zone from the discolored region to the brighter region showed a change of fracture mode from faceted fracture to dimpled rupture (Figure C3.13).

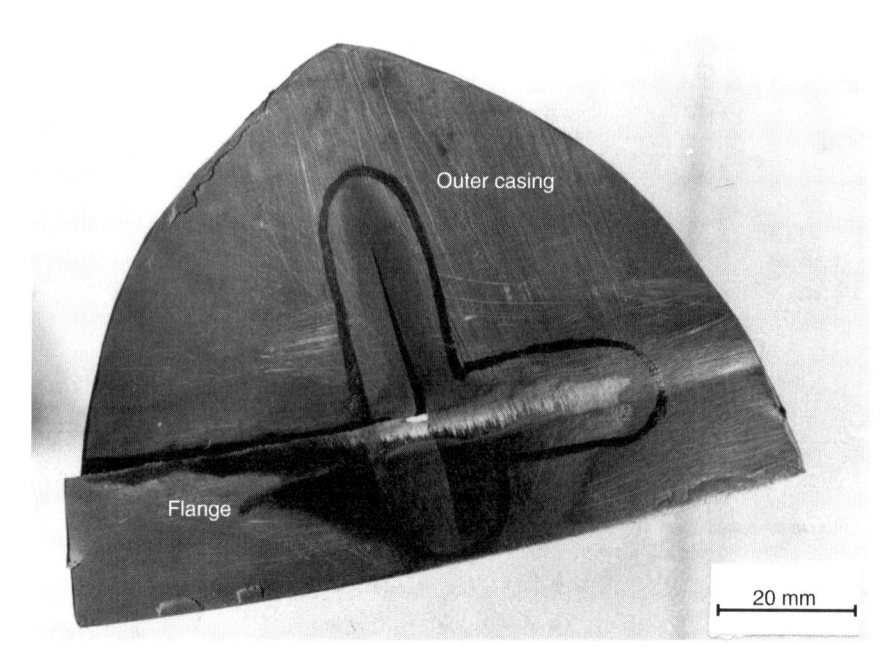

FIGURE C3.11
The failed component in as-received condition.

Sections containing the fracture surface across the weld were subjected to metallographic examination. The weld metal showed coarse acicular α and coarse prior-β grain size with a thin α casing at the surface and a few transgranular cracks (Figure C3.14a). The heat-affected zone showed relatively finer acicular α (Figure C3.14b). The parent metal (OT4-1) microstructure showed fine equiaxed α with a small amount of transformed β structure (Figure C3.14c). The details of chemical composition of the base metal (OT4-1) and the oxygen and nitrogen levels in the weld are given in Table C3.8. The hardness values of the base metal, HAZ and weld metal were 270 HV, 292 HV and 379 HV, respectively.

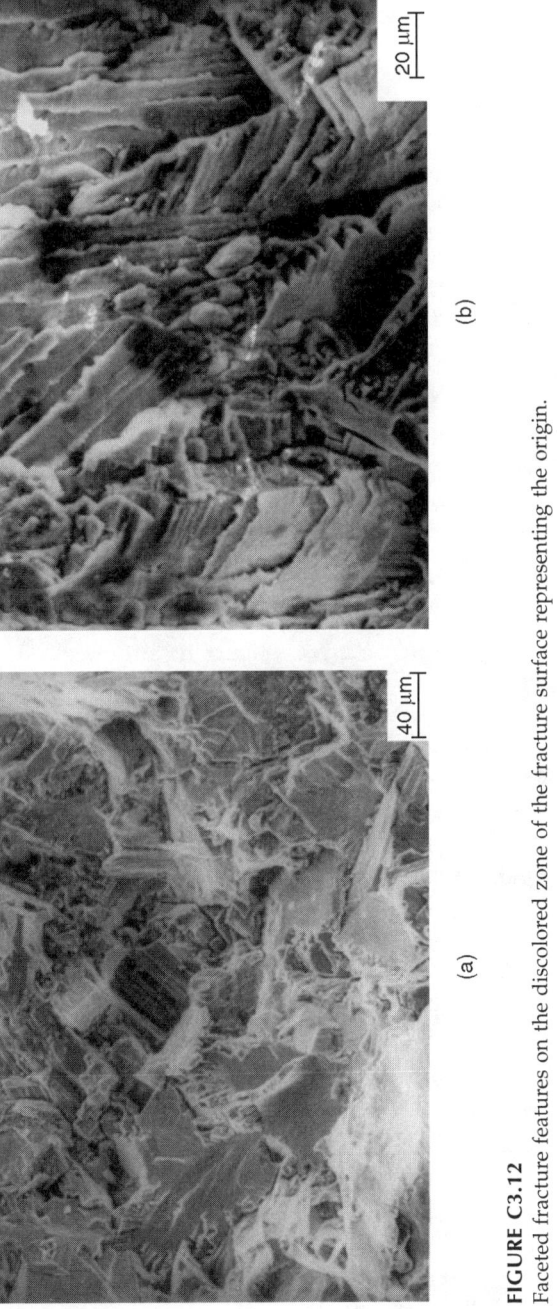

FIGURE C3.12
Faceted fracture features on the discolored zone of the fracture surface representing the origin.

FIGURE C3.13 Transition from faceted fracture to dimpled rupture.

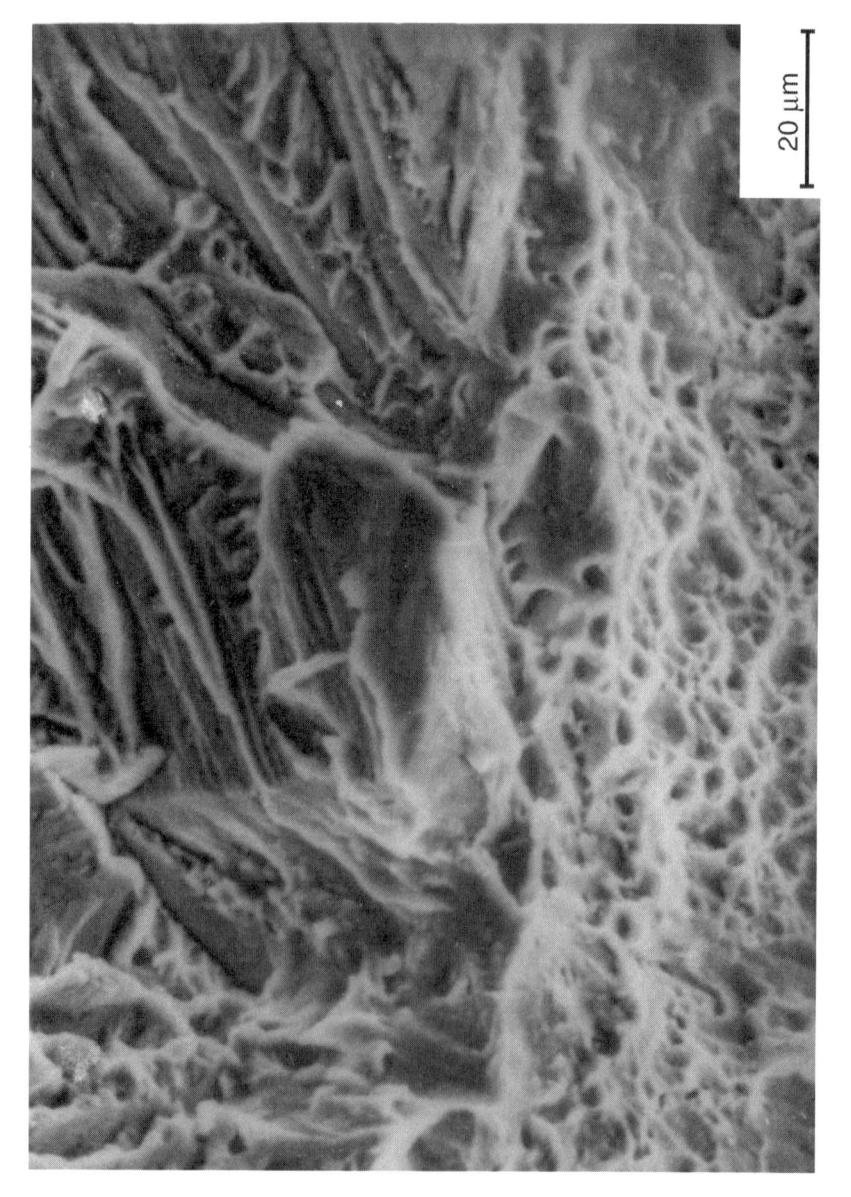

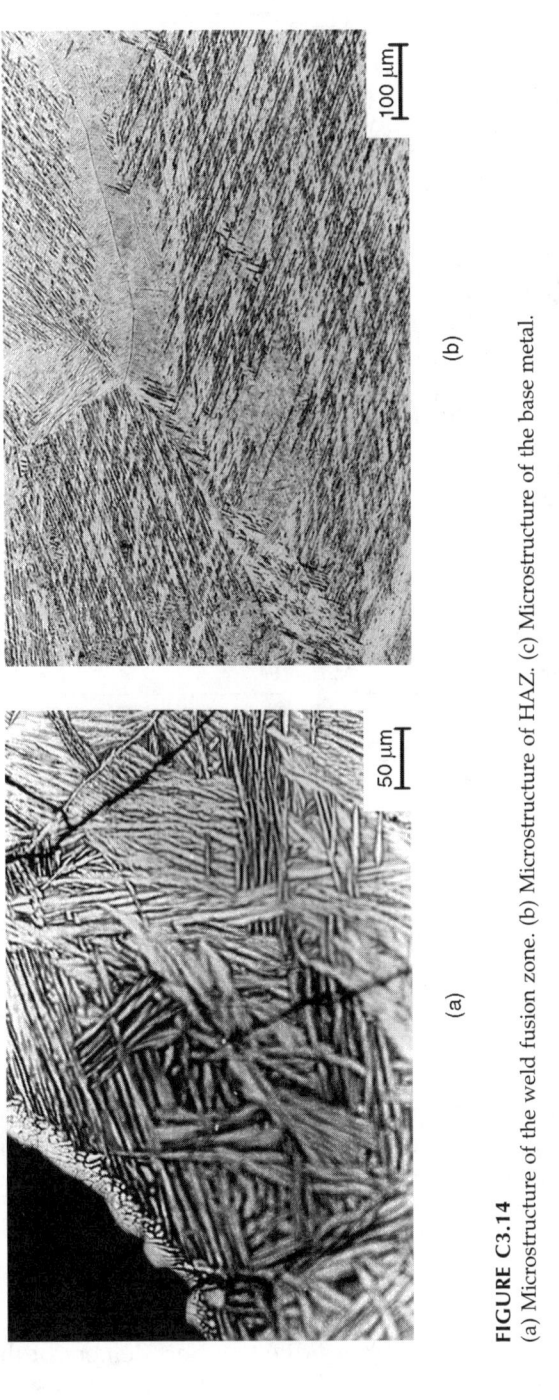

FIGURE C3.14
(a) Microstructure of the weld fusion zone. (b) Microstructure of HAZ. (c) Microstructure of the base metal.

(c)

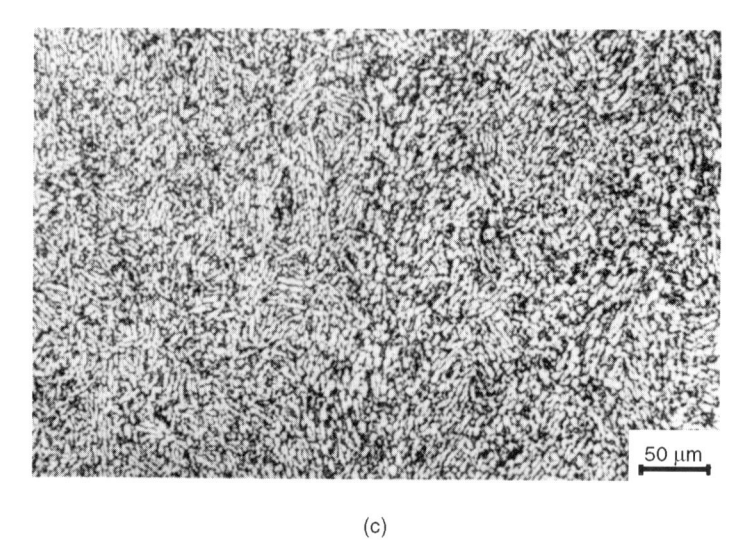

FIGURE C3.14 (continued)

TABLE C3.8

Chemical Composition of the Base and Weld Metals

Element (wt%)	Al	Mn	Fe	C	O	N	Ti
Base metal	1.4	1.2	0.05	0.02	0.09	0.02	Bal
Weld metal	—	—	—	—	0.25	0.17	—

Discussion

The chemical composition, microstructure and hardness of the material conformed to the specified grade OT4-1. Fractographic examination indicated that the failure occurred along the weld in a brittle mode. It is well known that the ductility of titanium alloys decreases with increasing concentration of oxygen and nitrogen. Further, the brittleness is enhanced by the observed coarse prior β grain size and acicular microstructure in the weld fusion zone. The abnormally high levels of oxygen and nitrogen, the presence of a large number of transgranular cracks and very high hardness in the weld metal pointed to a large ingress of gaseous elements into the weld pool, resulting in an increase in the strength and a decrease in ductility of the weld metal, leading to cracking of the weld in brittle mode.

Conclusion

The failure of the nose fairing was due to embrittlement of the weld metal by the increased levels of oxygen and nitrogen as a result of improper shielding during welding. Welding under the conditions of high external restraint and the stresses developed during weld metal solidification might have contributed to the initiation of the crack.

Part 4

Failures Due to Overheating

4.1 Failure of the HP Turbine Blade of an Aeroengine

Key words: *Nickel-based superalloy, Intergranular fracture, Gamma prime precipitates, Creep cavities, Overheating, Creep rupture*

Introduction

The high-pressure turbine blades of an aeroengine that was involved in an accident were found to have failed during strip examination [25]. The blades were forged from a Russian-origin Ni-based superalloy AE 109 B II. The blades completed a total service life of 659 h prior to failure. The operating temperature of the turbine was around 1000°C.

Experimental Results

A portion of one of the failed blades is shown in Figure C4.1. The blade failed at about two thirds of the blade length from the root. The failure appeared to have originated at the leading edge and propagated across the aerofoil towards the trailing edge and also toward the root. The fracture surface and the aerofoil region of the blade were heavily oxidized. The entire fracture surface showed intergranular fracture with cavities on a few grain facets and secondary cracks (Figure C4.2).

Samples from just below the fracture surface and from the root portion were metallographically examined. A few round cavities at the grain boundaries and intergranular cracks perpendicular to the blade axis were observed below the fracture surface (Figure C4.3). While coarse grains devoid of matrix precipitates with clusters of carbides were observed at the failed region (Figure C4.4a), the root area showed equiaxed grains with precipitates (Figure C4.4b). Transmission electron microscopic examination showed fine spheriodal γ' precipitates at the failed zone (Figure C4.5a) and bimodal γ' precipitate distribution at the root (Figure C4.5b). The chemical composition of the failed turbine blade is given in Table C4.1. The hardness values at the failed area and root area were 336 HV and 392 HV, respectively.

FIGURE C4.1
Photograph of the failed blade showing surface oxidation and the fracture path.

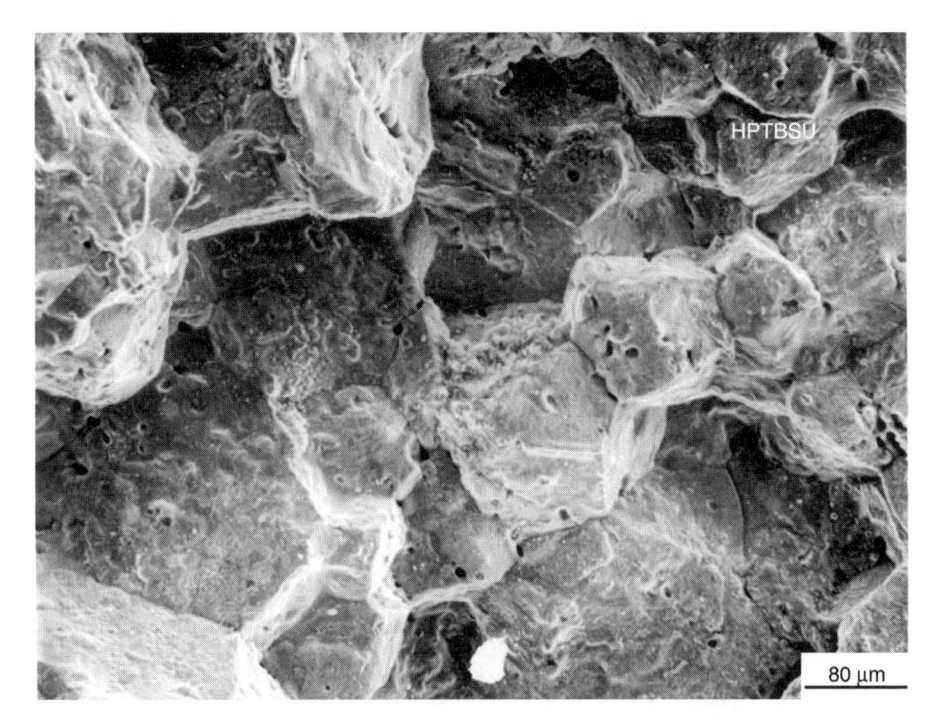

FIGURE C4.2
Fractograph showing intergranular fracture and cavities on the grain facets.

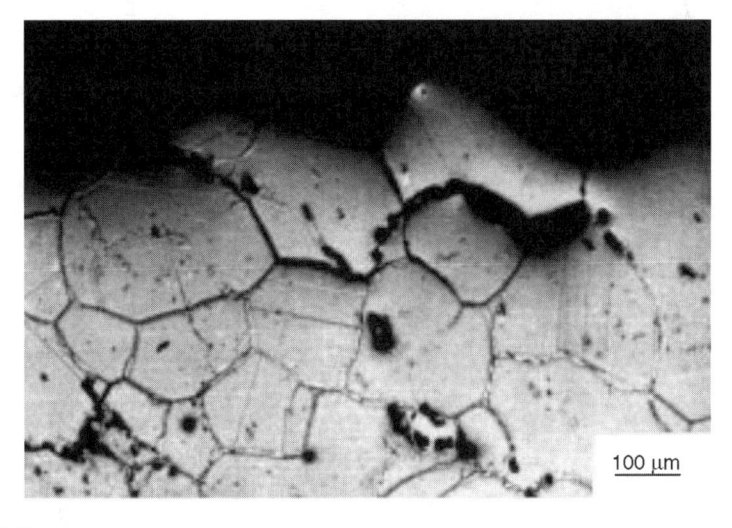

FIGURE C4.3
Photomicrograph showing intergranular cracks and cavities close to the fracture surface.

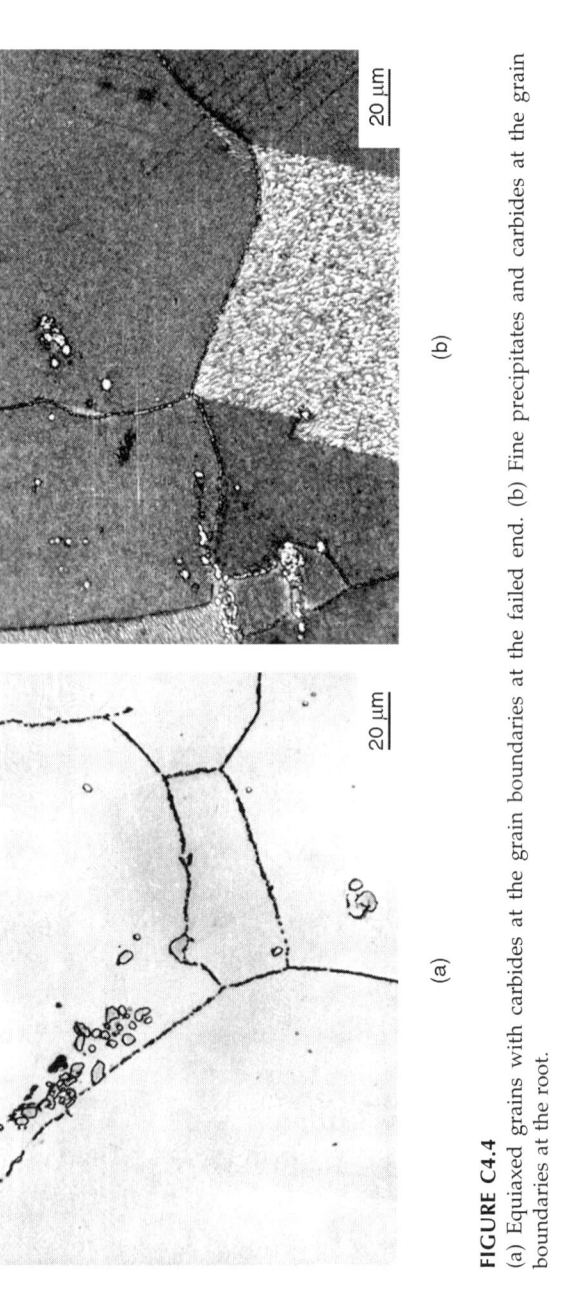

FIGURE C4.4
(a) Equiaxed grains with carbides at the grain boundaries at the failed end. (b) Fine precipitates and carbides at the grain boundaries at the root.

Failures Due to Overheating

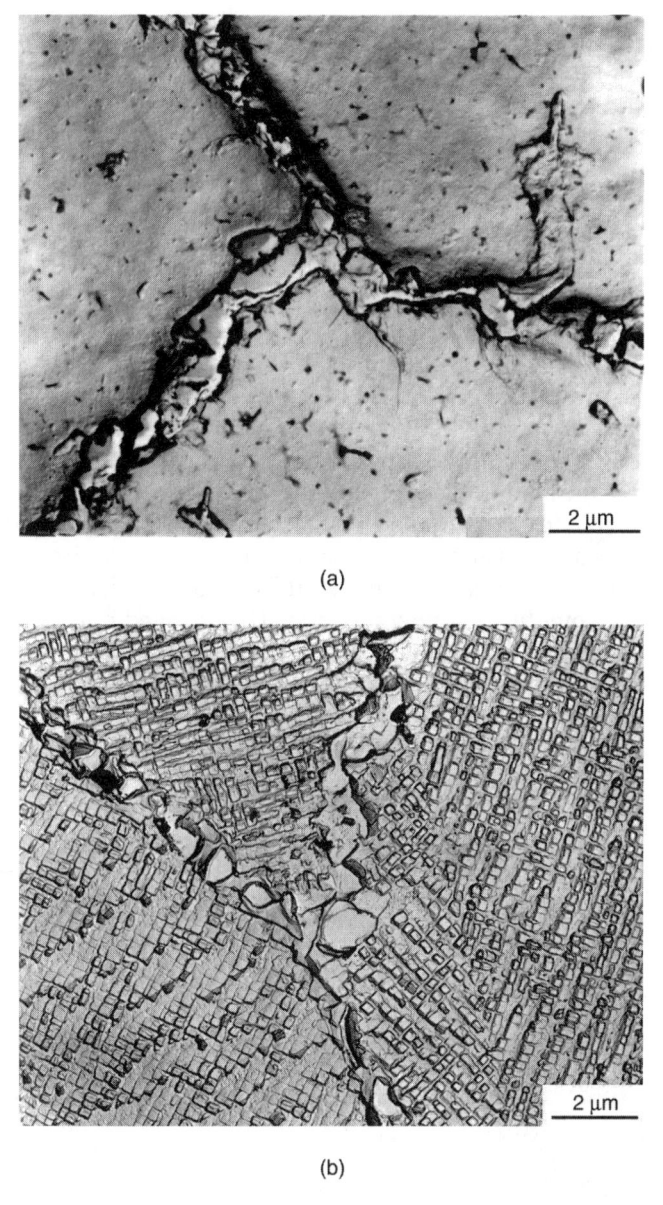

FIGURE C4.5
Electron micrograph showing (a) very fine precipitates and carbides at the grain boundaries at the failed regions. (b) γ′ precipitates and carbides at grain boundaries at the root.

TABLE C4.1

Chemical Composition of the Failed Turbine Blade

Element	Wt%	Element	Wt%
C	0.12	W	6.8
Cr	9.5	Al	4.35
Mo	6.0	Ni	Bal

Discussion

The composition conforms to the specified grade. The presence of fine spheroidal γ' precipitates at the failed zone, in contrast to coarse cuboidal and fine spheroidal γ' precipitates at the root, was considered to be the effect of overheating of the blade at the failed region above the γ' solvus temperature. The fine precipitates at the failed end might have formed during cooling after the overheating incident. The changes in hardness and microstructure at the failed region compared to the unaffected root region are confirmative evidence of the overheating phenomenon. The changed microstructure due to overheating had poor creep and stress rupture properties. The intergranular fracture and presence of round cavities and cracks at the grain boundaries perpendicular to the stress axis near the fracture surface suggested creep failure.

Conclusion

The failure of the turbine blades was caused by overheating and consequent creep damage.

4.2 Failure of a Mounting Bolt of the Second-Stage NGV of an Aeroengine

Key Words: Stainless steel, Intergranular fracture, Overheating, Stress rupture

Introduction

Mounting bolts used for fixing the nozzle guide vanes to the casing were found broken during scheduled servicing of an aeroengine [26]. The failed bolt was manufactured from a heat-resistant, precipitation-hardened austenitic stainless steel conforming to Russian grade 10Kh11N23T3. The component had completed a service life of 349 h.

Experimental Results

A portion of the failed bolt and a new bolt are shown in Figure C4.6. The failure took place at the first thread from the bolt shank. The fracture surface appeared coarse and crystalline. The fracture surface of the failed bolt showed intergranular fracture over the entire surface, and the cracks had originated from the thread roots (Figure C4.7).

Longitudinal sections of the samples from both the failed and a new bolt were prepared for metallographic examination. On etching with aquaregia, the failed bolt sample showed coarse equiaxed grains of austenite and intergranular secondary cracks adjacent to the fracture surface (Figure C4.8). Significant grain coarsening was observed in the used bolt compared to the unused bolt. The grain size of the failed bolt was 110 μm and that of the new bolt was 50 μm (Figure C4.9). The hardness of the failed bolt was 383 HV and that of the new bolt was 406 HV. The chemical composition of the failed bolt is given in Table C4.2.

(a)

(b)

FIGURE C4.6
(a) New mounting bolt. (b) A portion of the failed bolt. Arrow indicates fracture surface.

Failures Due to Overheating

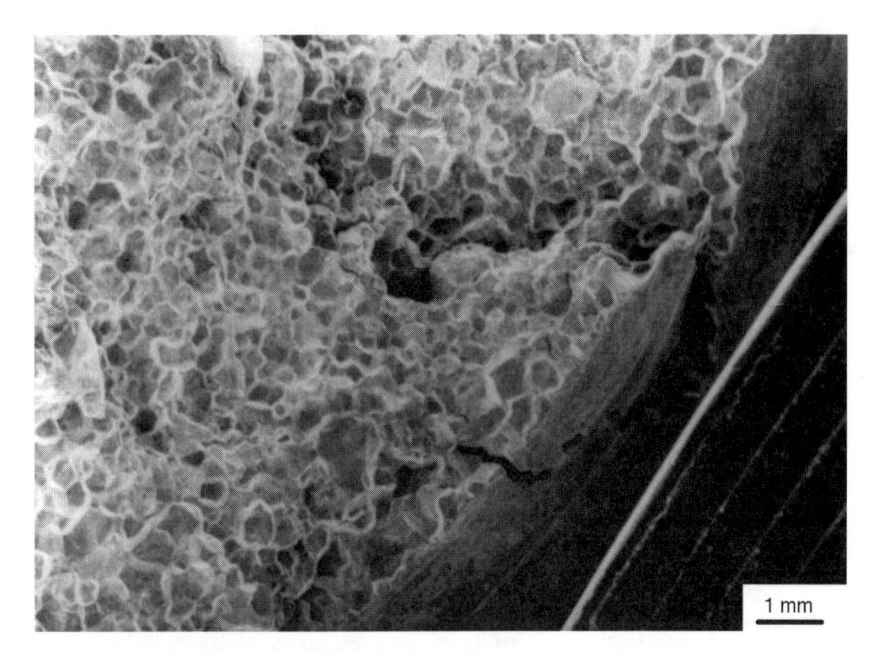

FIGURE C4.7
SEM fractograph showing intergranular fracture over the entire fracture surface of the bolt.

FIGURE C4.8
Photomicrograph showing intergranular secondary cracks close to the fracture surface.

(a)

(b)

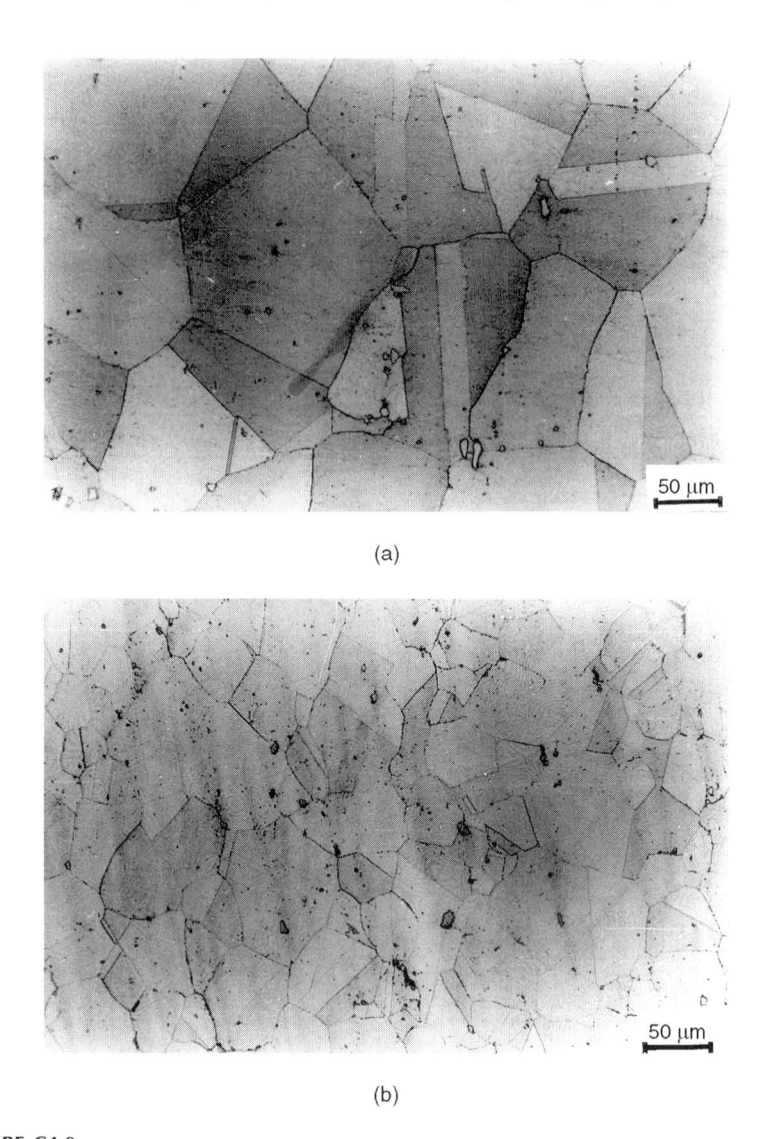

FIGURE C4.9
(a) Coarse grains and annealing twins in the used bolt. (b) Fine grain structure in the new bolt.

TABLE C4.2

Chemical Composition of the Failed Bolt

Element	Wt%	Element	Wt%
C	0.05	Cr	12.2
Si	0.001	Mo	1.6
Mn	0.21	Ti	2.0
Ni	19.0	Fe	Bal

Discussion

The chemical composition of the failed bolt conformed to the specified grade. Hardness and microstructure of the new blade sample indicated that it was in solution-annealed and precipitation-hardened condition. The coarsening of the austenite grains with a drop in hardness in the failed bolt suggested that it had experienced a temperature above that of the aging temperature during its service. The reduction in strength of the bolt due to precipitate dissolution and grain coarsening due to overheating led to intergranular fracture of the bolt by stress rupture. The thread root could have acted as a stress raiser.

Conclusion

The mounting bolt failed due to overheating in stress rupture mode.

4.3 Failure of the Nozzle Guide Vane of an Aeroengine

Key words: Ni-based superalloy, γ' precipitates, Overheating, Incipient melting

Introduction

An aircraft was involved in an accident while performing a rolling exercise. Preliminary investigations suggested that malfunctioning of the aeroengine fitted to the aircraft could have caused the accident. Detailed examination of the engine wreckage was conducted and damaged nozzle guide vanes (NGV) were taken for study to find out whether the damage was caused in flight or was the result of postaccident events [27]. The NGV was cast with a Ni-based superalloy conforming to Russian grade ZC-6YBE, heat-treated and aluminide-coated. The normal operating temperature of the guide vane is 1050°C.

Experimental Results

The trailing edge of the NGV was found badly damaged and a large portion was observed to be missing near the midlength of the aerofoil (Figure C4.10). Deep cracks were seen initiating at the damaged region and propagating inward. The damaged surface of the NGV showed a thick adherent oxide layer masking the features even after cleaning with oxide remover. Sections of the NGV both from the damaged region and the root were prepared for metallographic examination. The damaged region showed a number of deep craters filled with aluminide (Figure C4.11a) and dendritic structure (Figure

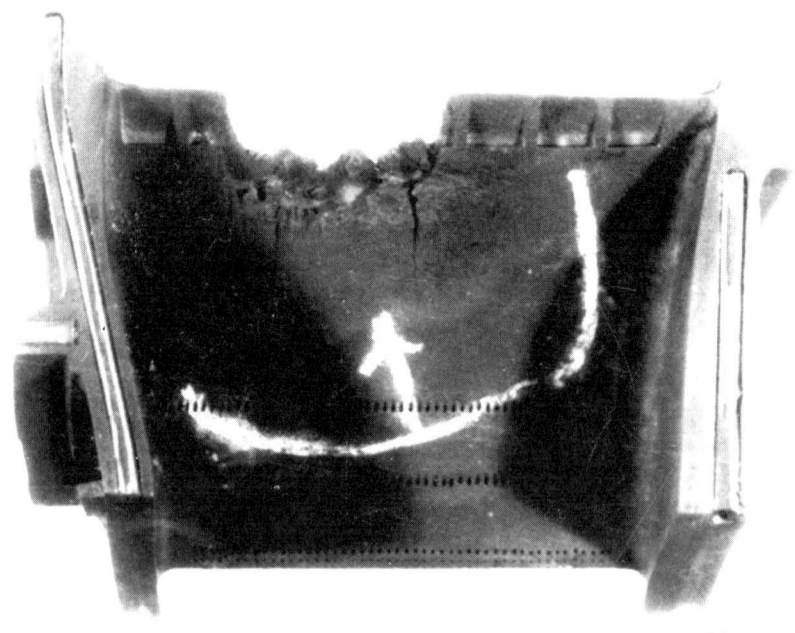

FIGURE C4.10
Damaged NGV.

C4.11b). The root region also showed cast microstructure with slightly coarser dendrites (Figure C4.12).

Sections near and away from the damaged region of the NGV aerofoil and the root region were examined for precipitate morphology. The sample near the damaged region showed cavities at grain boundaries, a large number of carbides both along the grain boundaries and within grains and no γ' precipitates (Figures C4.13a and 4.13b). The area away from the damaged region showed cuboidal γ' precipitates in the γ matrix and carbides along the grain boundaries (Figure C4.14). The root region showed a slightly coarser γ' in γ matrix (Figure C4.15). The chemical composition and hardness of the NGV at different locations are given in Tables C4.3 and C4.4, respectively.

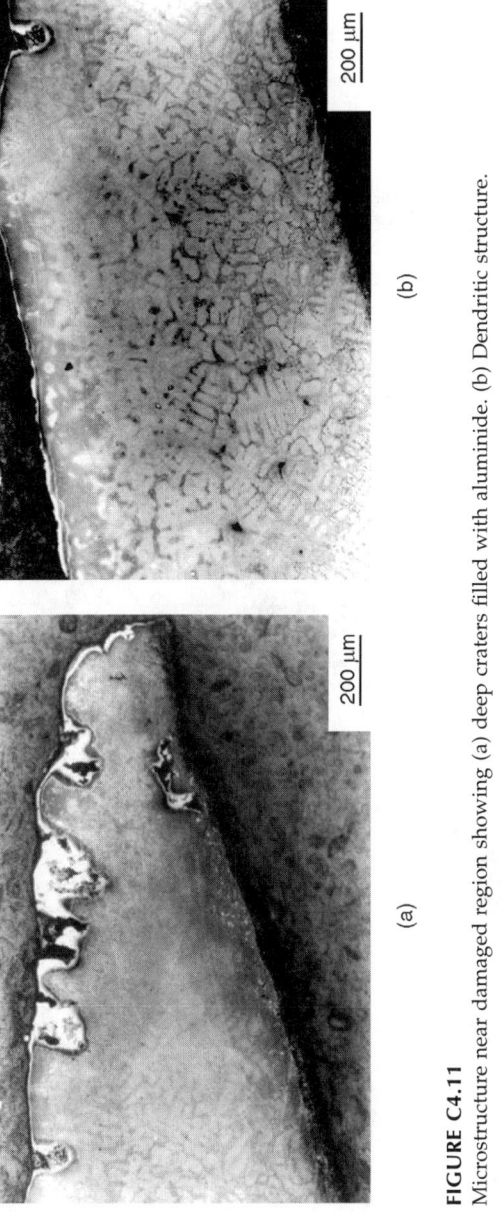

FIGURE C4.11
Microstructure near damaged region showing (a) deep craters filled with aluminide. (b) Dendritic structure.

FIGURE C4.12
Photomicrograph of the root region showing cast microstructure with slightly coarser dendrites.

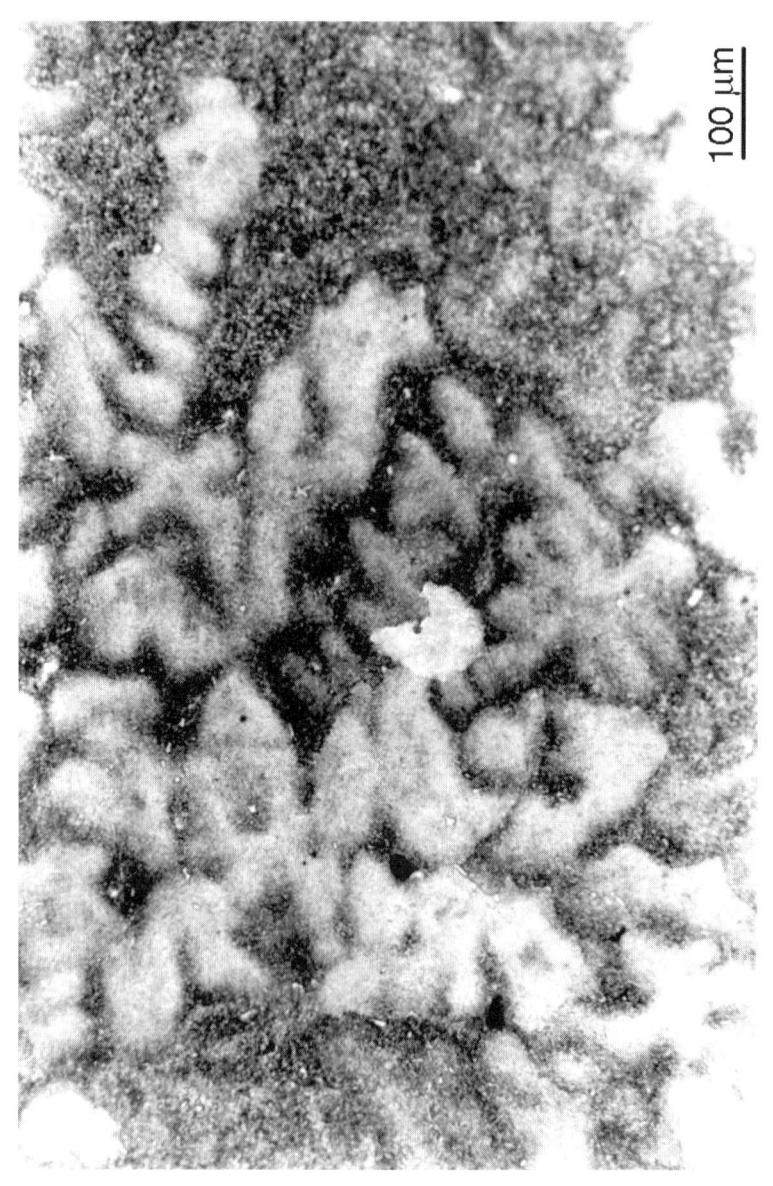

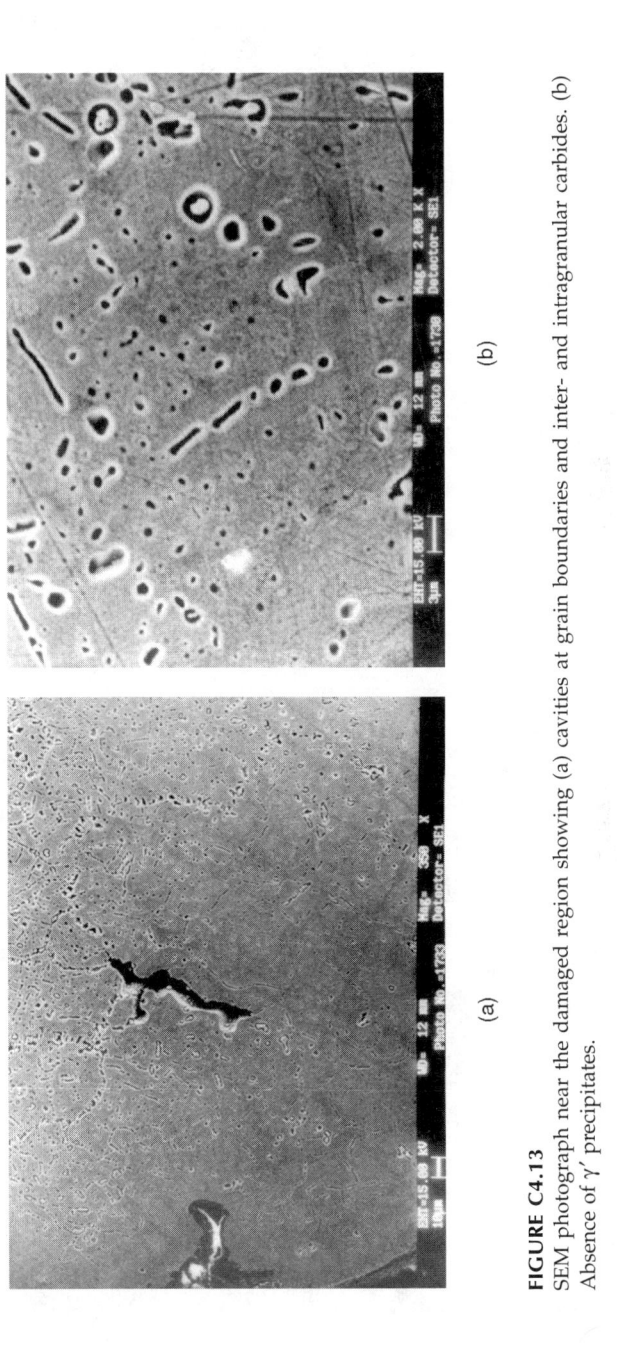

FIGURE C4.13
SEM photograph near the damaged region showing (a) cavities at grain boundaries and inter- and intragranular carbides. (b) Absence of γ' precipitates.

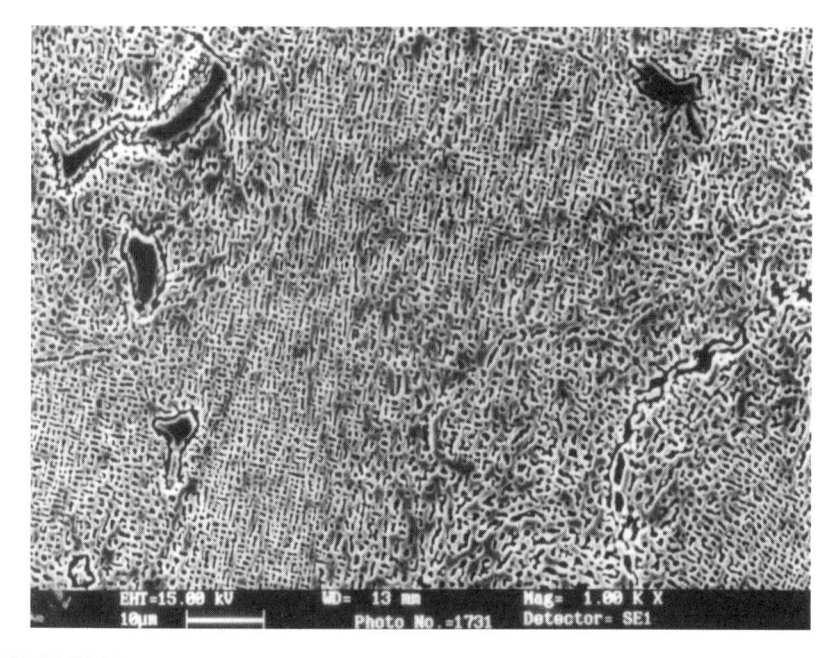

FIGURE C4.14
Microstructure of the aerofoil region away from the damaged portion showing cuboidal γ' precipitates and carbides along the grain boundaries in austenite matrix.

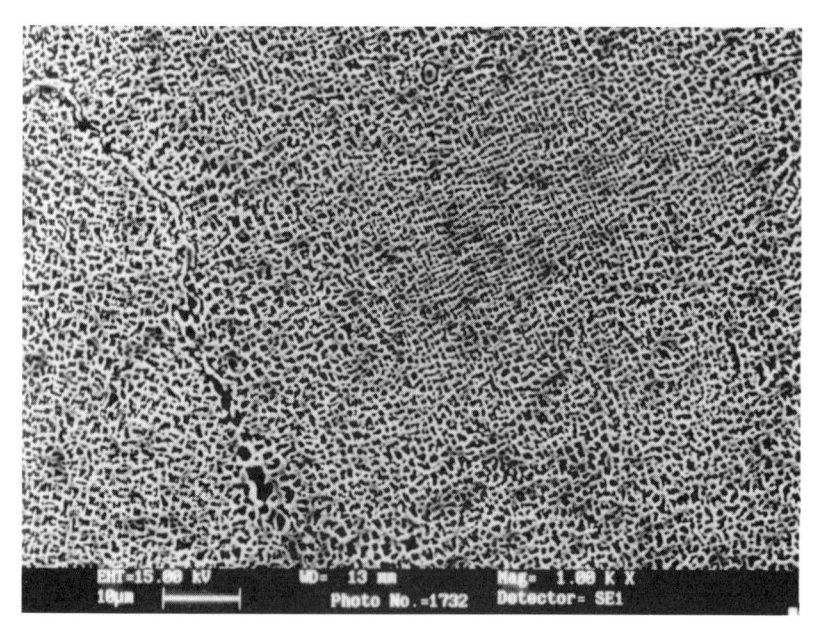

FIGURE C4.15
Microstructure in the root region showing slightly coarser γ' in austenite matrix.

TABLE C4.3

Chemical Composition of the Failed Component

Element	Wt%	Element	Wt%
C	0.15	W	10.24
Cr	9.01	Co	9.75
Ti	2.46	Nb	1.10
Al	5.52	Mo	2.17
Ni	Bal		

TABLE C4.4

Hardness of the Failed NGV at Aerofoil and Root Regions

Location	Hardness (HV)
Aerofoil near damaged region	430
Aerofoil away from damaged region	383
Root	357

Discussion

The chemical composition, microstructure and hardness at the root region were in conformity with the material standard. Microstructural examination of the damaged region and root region of the failed NGV indicated that (a) the surface of the vane became molten and resolidified, (b) the microstructure at the damaged region was significantly different from that of the root and (c) the γ' precipitates at the damaged region were not resolvable even at 2000×, while cuboidal γ' precipitates were observed at the root region. The hardness at the damaged region was higher by 70 HV compared to the root region. The higher hardness at the damaged region is due to (a) dissolution of original precipitates caused by overheating and (b) reprecipitation of extremely fine γ' precipitates during the cooling stage after the accident. These results indicated that the trailing edge of the NGV experienced very high temperature dissolution of γ', incipient melting and material loss. The presence of aluminide in the craters is probably due to flowing of semimolten aluminide into the craters created by substrate melting and resolidification. As the root region showed normal microstructure, it is evident that the damage and microstructural changes at the trailing edge of the NGV were caused in flight and were not due to postcrash events.

Conclusion

The damage to the NGV was due to overheating caused by the operation of the engine at temperatures far in excess of the defined operating temperature.

4.4 Failure of an Aeroengine Center Support Bearing

Key Words: *Bearing Steel, Aluminum bronze, Overheating, Lubrication failure*

Introduction

The center support bearing of an aeroengine failed after about 300 h of service [28]. At the commencement of take-off, the oil pressure and metal chip detector systems gave warning signals. On strip examination, metal chips were detected in the oil filter in large quantities and the bearing was found to be severely damaged. The outer and inner races and the balls were made from steel AISI 52100 and the cage was made from an aluminum-bronze conforming to DTD 197A. The initial hardness of the ball and the races were in the range of R_C 62 to 65 and R_C 59 to 62, respectively. The operating temperature of the bearing is maintained around 150°C by forced circulation of lubricating oil.

Experimental Results

The ribs of the cage of the bearing were found to be broken and bent. The balls were also found to be scored and deformed. The inner surface of the outer race contained a deposit of the cage material. A part of the damaged bearing is shown in Figure C4.16. One of the severely damaged balls and a section of the outer race were prepared for metallographic study. Fine, undissolved carbide particles uniformly dispersed in a matrix of fine pearlitic matrix and ferrite were noticed in the steel ball (Figure C4.17).

A two-stage plastic-carbon replica of the 2% Nital-etched ball was examined in TEM. The lamellar pearlitic matrix as observed in TEM is shown in Figure C4.18. Metallographic examination of the outer race sample showed that the aluminum-bronze cage material was bonded to the inner surface of the outer race (Figure C4.19). The microstructure of the outer race consisted of fine, undissolved carbide particles dispersed in a matrix of tempered martensite. The average hardness of the steel ball was R_C 30. The outer race had a hardness of R_C 49. The chemical composition of the ball and outer race are given in Table C4.5.

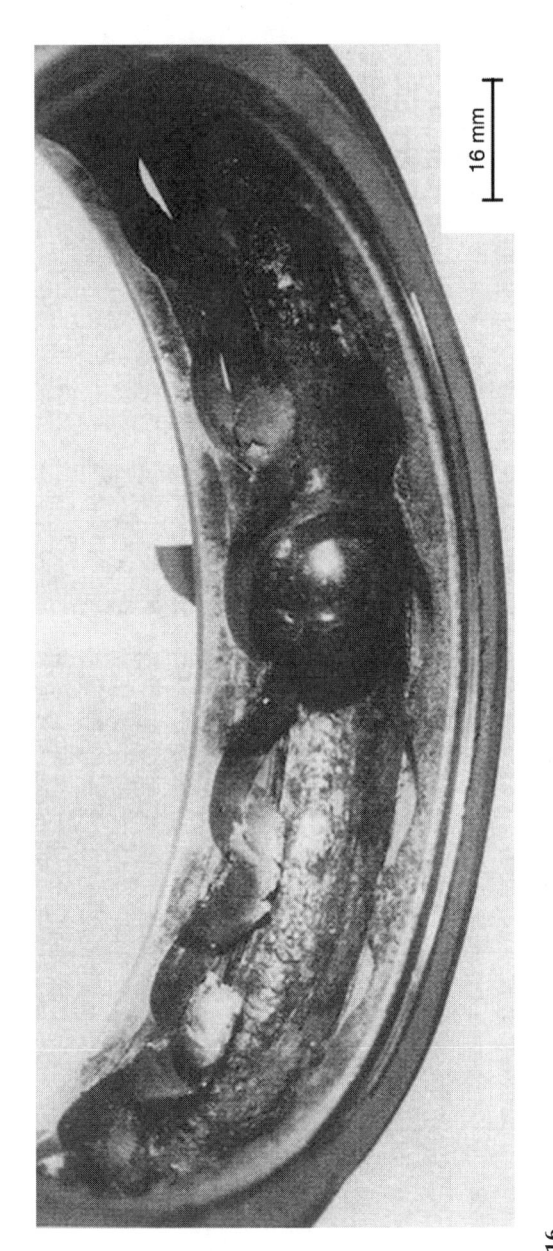

FIGURE C4.16
Failed bearing with broken cage ribs, deposit on the inner surface and deformed ball.

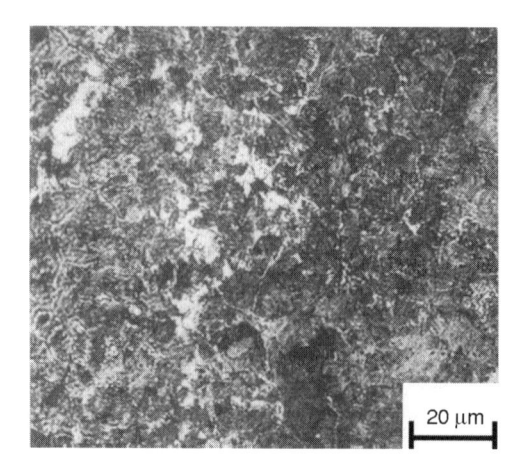

FIGURE C4.17
Lamellar pearlite and ferrite in the damaged steel ball.

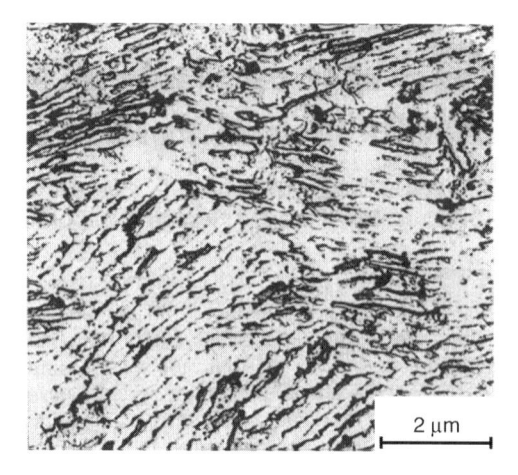

FIGURE C4.18
TEM micrograph showing lamellar pearlite in the ball.

Failures Due to Overheating

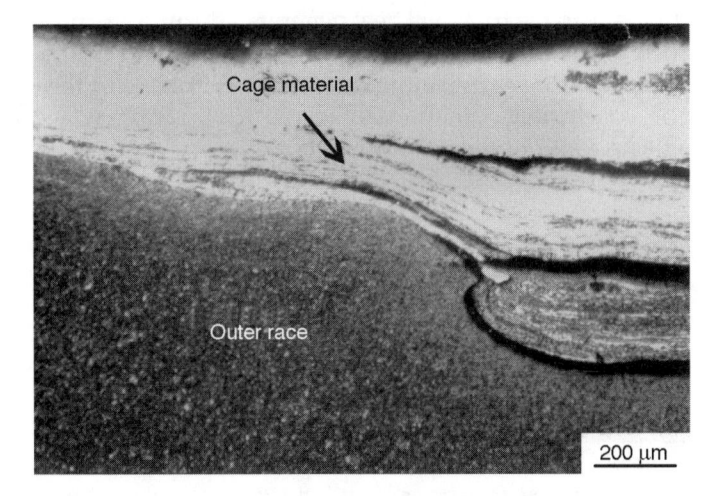

FIGURE C4.19
Cage material (light) welded to the inner surface of the outer race.

TABLE C4.5

Chemical Composition of the Ball and Outer Race Materials

Elements	Component	
	Ball	Race
C	1.0	1.0
Si	0.27	0.36
Mn	0.32	0.29
Ni	0.22	0.07
Cr	1.40	1.38
S	0.003	0.002
P	0.006	0.008
Fe	Bal	Bal

Discussion

The chemical composition of the ball and outer race conformed to AISI 52100. Metallographic examination of the deformed ball showed a ferrite network and fine pearlitic structure containing very fine carbide particles. The steel when hardened and tempered to the specified hardness would show a lightly tempered martensitic structure containing undissolved carbides. Such a structure can transform to the observed structure only if it is heated to a temperature in the two-phase austenite + carbide field and cooled relatively slowly. The minimum temperature for the formation of austenite in the steel is about 730°C. Thus, the deformed ball must have been heated at least to this temperature to develop the observed pearlitic structure, which is possible only if there is complete stoppage of the flow of lubricating oil, resulting in overheating due to the frictional heat generated during dry running. Other evidence such as the reduction in hardness of the ball and the race and the pressure bonding of the cage material on the inner surface of the outer race supports the conclusion that the damage to the bearing was caused by overheating as a result of failure of lubrication.

Conclusion

The deformation of the balls and consequent failure of the bearing was due to overheating caused by failure of lubrication.

4.5   Failure of a Drive Shaft

Key words: Ni-Cr-Mo steel, Intergranular cracks, Frictional overheating, Bearing seizure

Introduction

A drive shaft of the cooling air pack of an aircraft failed after a service life of 288 h [29]. The shaft was designed for unlimited life subject to inspection during overhaul at the end of 800 h of service. The maximum operating temperature was stated to be 45°C. The shaft was machined from Ni-Cr-Mo steel and heat-treated to a hardness level of 360 HV. Strip examination conducted by the user revealed that the bearing fitted on one of the sides of the drive shaft could be removed, while the inner race of the bearing fitted to the broken end could not be taken out from the shaft.

Experimental Results

Visual examination of the broken drive shaft indicated that it was of stepped construction with a threaded portion at both ends (Figure C4.20). The inner race of the bearing fitted on the right side of the shaft was found to be damaged and jammed. The stepped shaft was found to have failed at the

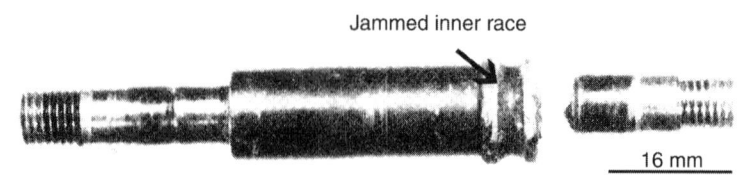

FIGURE C4.20
Failed drive shaft with jammed inner race.

FIGURE C4.21
Mating fracture surface of failed shaft.

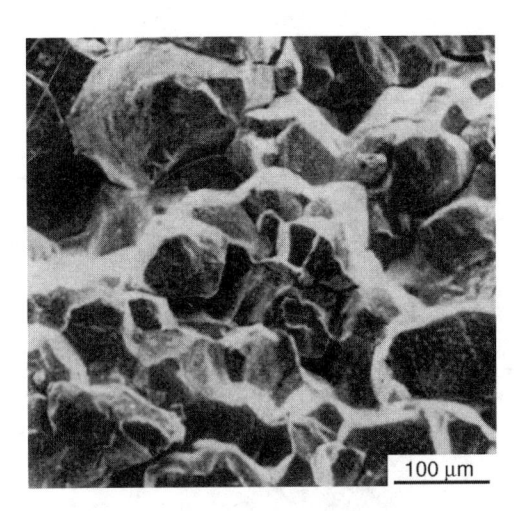

FIGURE C4.22
SEM fractograph showing intergranular fracture and secondary cracks.

circumferential groove, located immediately after the bearing inner race. The fracture surfaces were coarse and crystalline and contained severe rub marks in some places (Figure C4.21). The fracture surface of the shaft was examined in SEM. Wherever the fracture surface was free of rub marks, it showed intergranular fracture features with secondary cracks (Figure C4.22).

A longitudinal section of the shaft along with the jammed inner race of the bearing was prepared for metallographic examination. On etching with 2% Nital, the shaft showed a banded structure at low magnification and tempered martensitic structure with prominent acicularity at higher magnification (Figure C4.23). The inner race was unaffected by 2% Nital. Cavities of different sizes were found in the inner race close to the inner race/shaft interface. On the shaft side of the interface, unetched material containing a

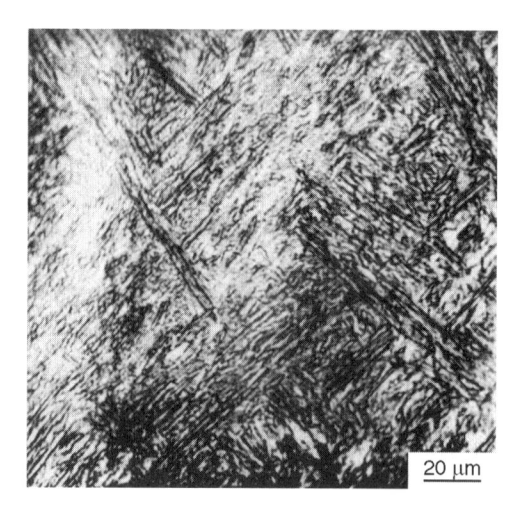

FIGURE C4.23
Acicular martensite near failed region in the shaft.

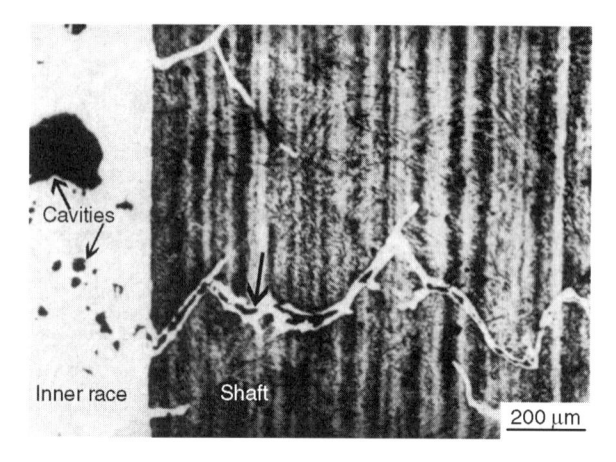

FIGURE C4.24
Microstructure at the inner race/shaft interface. Note the unetched material penetrated into the shaft.

few cavities was found in zig-zag pathways inside the shaft (Figure C4.24). The extraneous material found inside the drive shaft near its interface with the inner race of the bearing was analyzed qualitatively in EPMA. The electron and X-ray images for Cr and Ni indicated that the extraneous material in the shaft was of similar composition to the inner race material.

The inner race/shaft interface was found to be free of any discontinuity. On etching the inner race with aquaregia, nearly equiaxed austenite grains and

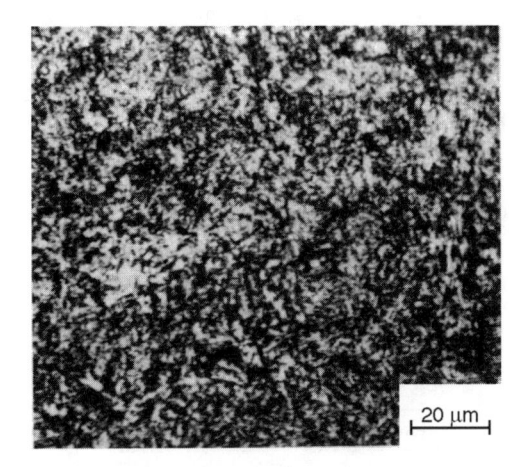

FIGURE C4.25
Lamellar structure and austenite grains in the inner race.

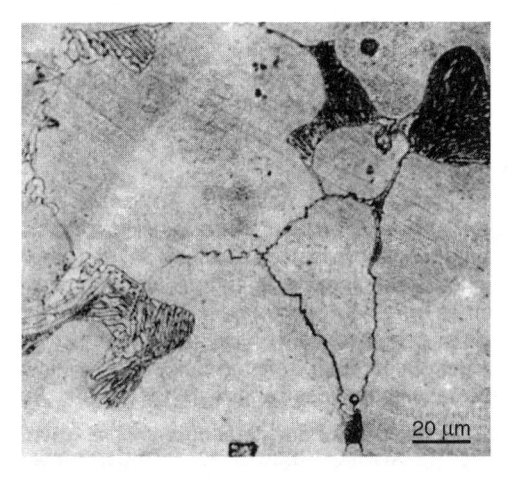

FIGURE C4.26
Tempered martensite in shaft (away from failed end).

colonies of lamellar structure were seen (Figure C4.25). Metallographic examination of a longitudinal section of the shaft taken far away from the failed region revealed a tempered martensitic structure without any acicularity (Figure C4.26). Vickers hardness was measured at a 30-kg load on samples used for metallographic examination. The results of hardness measurements are given in Table C4.6. X-ray diffraction analysis carried out on the inner race of the bearing showed that it contained about 85% retained austenite. The chemical composition of the drive shaft and the inner race of the bearing are given in Table C4.7.

TABLE C4.6

Results of Hardness Measurements

Location	Hardness (HV)
Shaft near the broken end	550
Shaft away from the broken end	360
On inner race of bearing	320

TABLE C4.7

Chemical Composition (wt%) of the Shaft and Inner Race Materials

Elements	Drive shaft	Inner race
C	0.35	1.17
Si	0.33	0.95
Mn	0.48	0.41
Ni	3.9	Trace
Cr	1.24	17.5
Mo	0.49	0.46
Fe	Bal	Bal

Discussion

The drive shaft was made from a Ni-Cr-Mo steel that was hardened and tempered to 360 HV, whereas its hardness was 550 HV at the broken end. The hardness of the inner race was expected to be 700 HV, whereas the hardness of the failed race was only 320 HV. The hardness of both inner race and shaft were changed considerably from their original values. The increase in hardness of the shaft was associated with a change in microstructure, which could be produced only by re-austenitization of the steel and its subsequent fast cooling. The altered microstructure of the shaft near the broken end was thus suggestive of its reheating to at least 800°C.

Metallographic examination and EPMA studies confirmed that the material penetrated into the prior-austenite grain boundaries of the shaft was from the inner race. The presence of a large number of shrinkage cavities in the inner race near its interface with the shaft suggested that melting of the inner race took place near this region. The molten material diffused along the austenite grain boundaries of the shaft.

The experimental evidence, therefore, indicates that uncontrolled frictional heat was generated at the inner race/shaft interface, resulting in localized melting of the inner race. The molten material penetrated into the shaft along prior-austenite grain boundaries. The shaft material was heated to its austenitization temperature, resulting in formation of martensite during cooling, thus justifying the higher hardness. Dissolution of the chromium carbide at high temperature in the race resulted in formation of austenite enriched with carbon, which remained untransformed on cooling due to substantial lowering of M_s temperature. The drastic reduction of the hardness on the inner race from about 700 HV to 320 HV supports this conclusion. The final intergranular failure of the drive shaft was due to high temperature overload conditions caused by the seizure of the bearing at the failed end.

Conclusion

The failure of the drive shaft was associated with overheating caused by frictional heat generated due to relative motion of the inner race over the shaft, leading to bearing seizure.

4.6 Failure of the Center Main Bearing of an Aeroengine

Key words: *High speed steel, Intergranular cracks, Untempered martensite, Overheating*

Introduction

The center main bearing fitted to an aeroengine failed during the ground run [30]. The bearing was manufactured by a reputed aeroengine manufacturer. The maximum working temperature of the component was stated to be 120°C. The component had done a service of 1508 h prior to failure.

Experimental Details

The failed component in as-received condition contained fragments of cage, cracked inner race, outer race and rollers, as shown in Figure C4.27. Careful examination of all the failed parts indicated that failure of the inner race was the primary failure. The inner raceway of the bearing containing a wide crack is shown in Figure C4.28. On opening the crack, the fracture surface showed coarse crystalline fracture with radial marks pointing to the origin located at the outer surface (Figure C4.29a). At the origin, intergranular fracture features were observed (Figure C4.29b) and the rest of the fracture surface showed quasi-cleavage fracture features (Figure C4.29c).

Sections from the inner (fractured) and outer raceways were prepared for metallographic examination. The inner raceway showed a white etching layer and intergranular cracks on the outer surface (Figure C4.30a). The inner surface of the outer raceway also showed intergranular cracks originating from the white etching layer. Away from the damaged region both the samples showed tempered martensite with carbide banding (Figure C4.30b).

FIGURE C4.27
Broken pieces of the bearing.

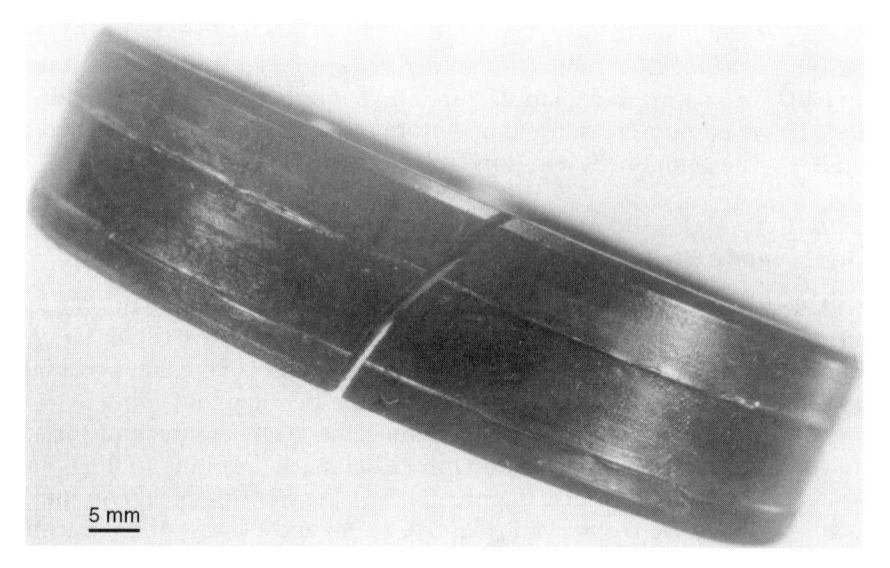

FIGURE C4.28
Inner raceway of the bearing showing a crack.

Hardness measurement carried out at a load of 200 g indicated that the hardness of the white etching layer was 902 HV and that of the undamaged region was 781 HV. The chemical composition of the steel used for the center main bearing is given in Table C4.8.

Failures Due to Overheating

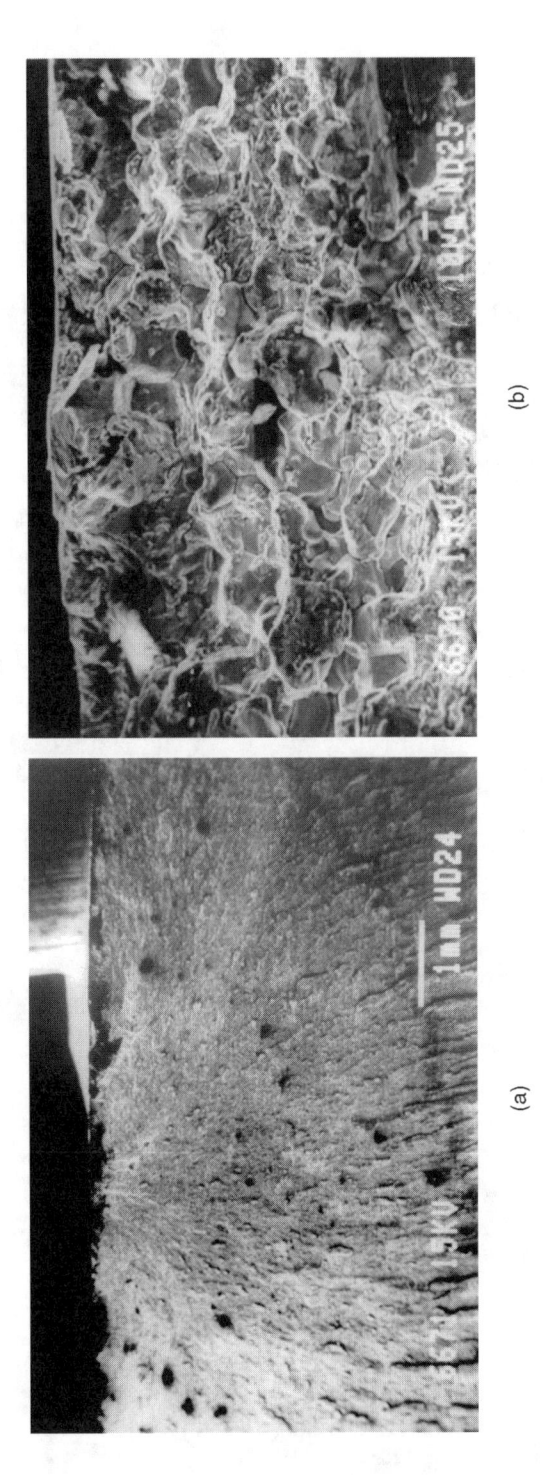

FIGURE C4.29
(a) Fracture surface showing the radial marks pointing to the fracture origin at the outer surface. (b) Intergranular fracture features at the origin. (c) Quasi-cleavage fracture features away from the fracture origin.

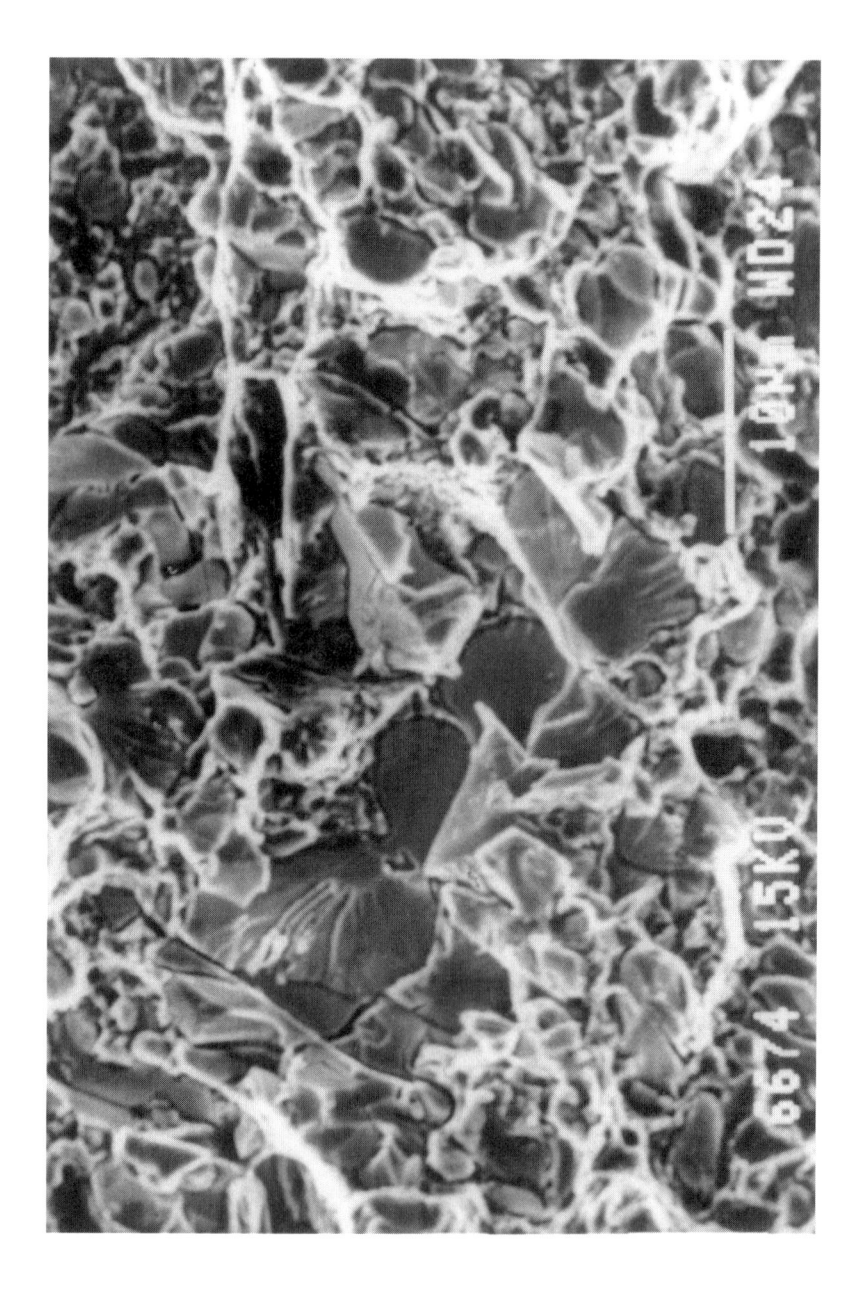

FIGURE C4.29 (continued)

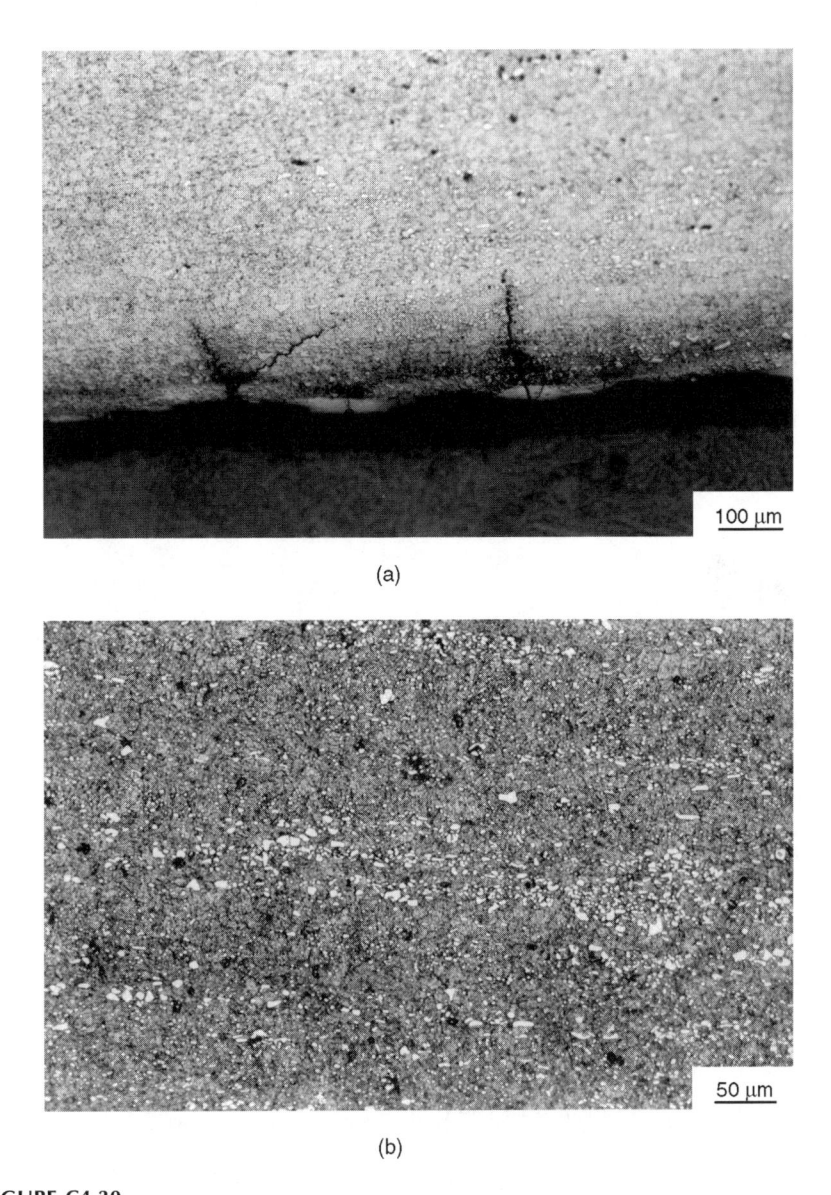

FIGURE C4.30
(a) Microstructure at the outer surface of the inner raceway showing a white etching layer and intergranular cracks. (b) Tempered martensitic structure with carbide banding at regions away from the damaged region.

TABLE C4.8

Chemical Composition of the Steel Used for the Center Main Bearing

Element	Wt%
C	0.82
Si	0.48
Mn	0.28
Cr	4.0
Mo	0.62
V	1.30
W	18.30
S	0.008
P	0.012
Fe	Bal

Discussion

Chemical composition indicated that the inner raceway was made from 18W-4Cr-1V type high-speed steel. The microstructure and hardness indicated that the component was used in quenched and tempered condition. Banding of undissolved carbides, though not responsible for failure, was observed. The outer surface of the inner race and inner surface of the outer race showed a white etching layer with a large number of intergranular cracks. The intergranular fracture at the origin represented the intergranular cracks in the white etching layer that were directly responsible for the failure. The final failure in quasi-cleavage mode is the result of the intergranular crack acting as a stress raiser.

The featureless white etching, untempered martensite layer, with very high hardness, on the contact surfaces can only form as a result of friction heating up the contact surfaces to temperatures above AC_3 of the steel and subsequent quenching action of the colder substrate. Such high frictional heat in a bearing could have been generated only under conditions of failure of lubrication and dry running of the bearing.

Conclusion

The center main bearing failed due to overheating caused by ineffective lubrication.

Part 5

Failures Induced by Corrosion

5.1 Failure of the Blow-off Vanes of an Aeroengine

Key words: Ti-Al-Sn alloy, Quasi-cleavage fracture, SCC, Residual stresses

Introduction

Thirty-three of the 120 blow-off vanes in the bypass casing of an aeroengine failed within 40 h of operation [31]. The maximum operating temperature was 60°C. The blow-off vanes were forged from a Ti-Al-Sn alloy conforming to Russian grade BT-5-1 and heat-treated.

Experimental Results

The blow-off vane with a crack at the U-section is shown in Figure C5.1. Deep milling marks were observed within the U grooves and on the surfaces. The surface was bluish in color. The crack was opened and the fracture surface was examined. The crack had originated at the bottom surface of the U section and propagated inward through the thickness. The fracture surface of the failed samples showed quasi-cleavagelike features (Figure C5.2). Corrosion debris was observed at the origin, which on qualitative analysis using an electron probe microanalyzer was found to contain Ca, Mg, Cl, and O (Figure C5.3).

Longitudinal sections at and away from the fracture region were examined. A number of secondary transgranular branching cracks were observed near the fracture (Figure C5.4). The microstructure consisted of nonuniform grains with acicular α and transformed β (Figure C5.5). The chemical composition of the failed component is given in Table C5.1. The hardness of the failed samples was 312 HV. Residual stresses were estimated at critical locations using an X-ray stress analyzer. On the face perpendicular to the crack, a tensile stress of 310 MPa and on the flat face at midlength a tensile residual stress of 111 MPa were measured.

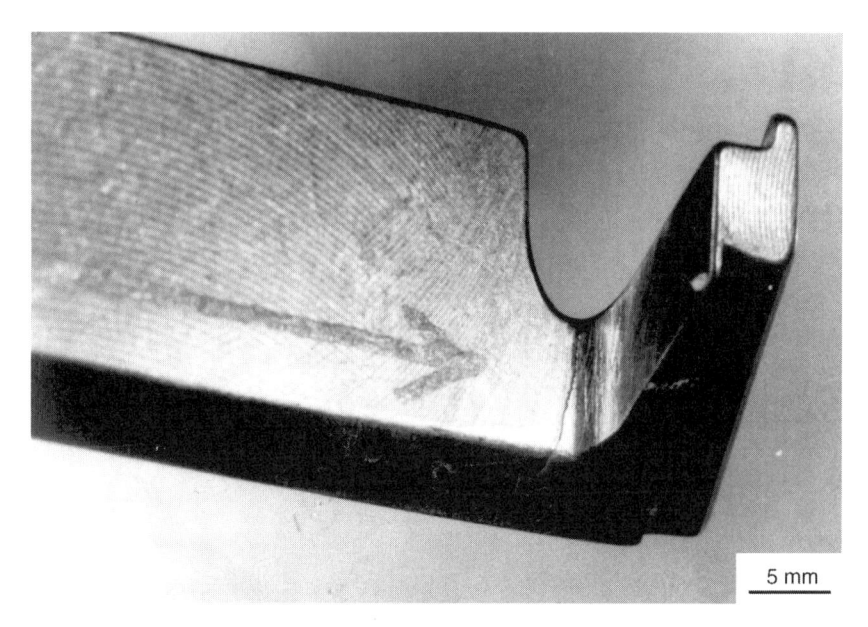

FIGURE C5.1
Blow-off vane with a crack at the U section.

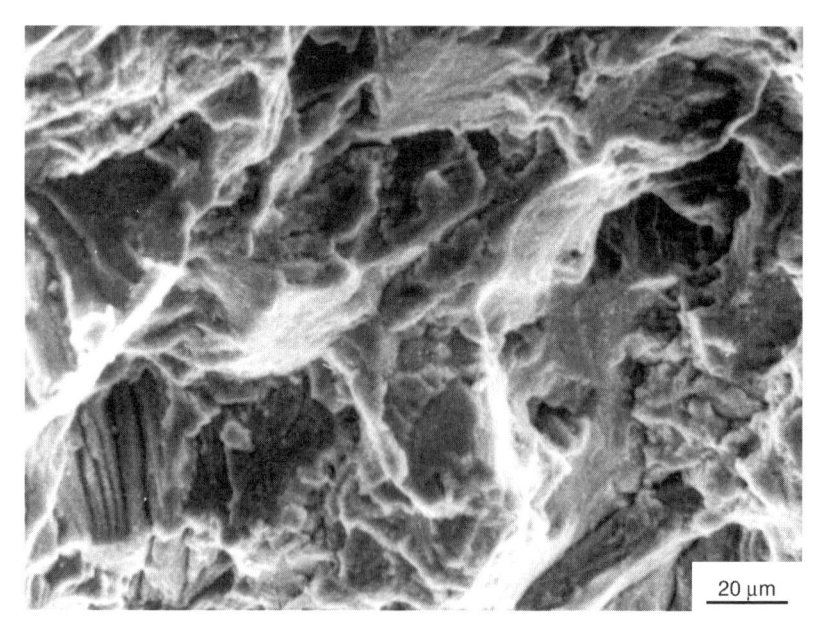

FIGURE C5.2
SEM fractograph showing quasi-cleavage features with secondary cracks.

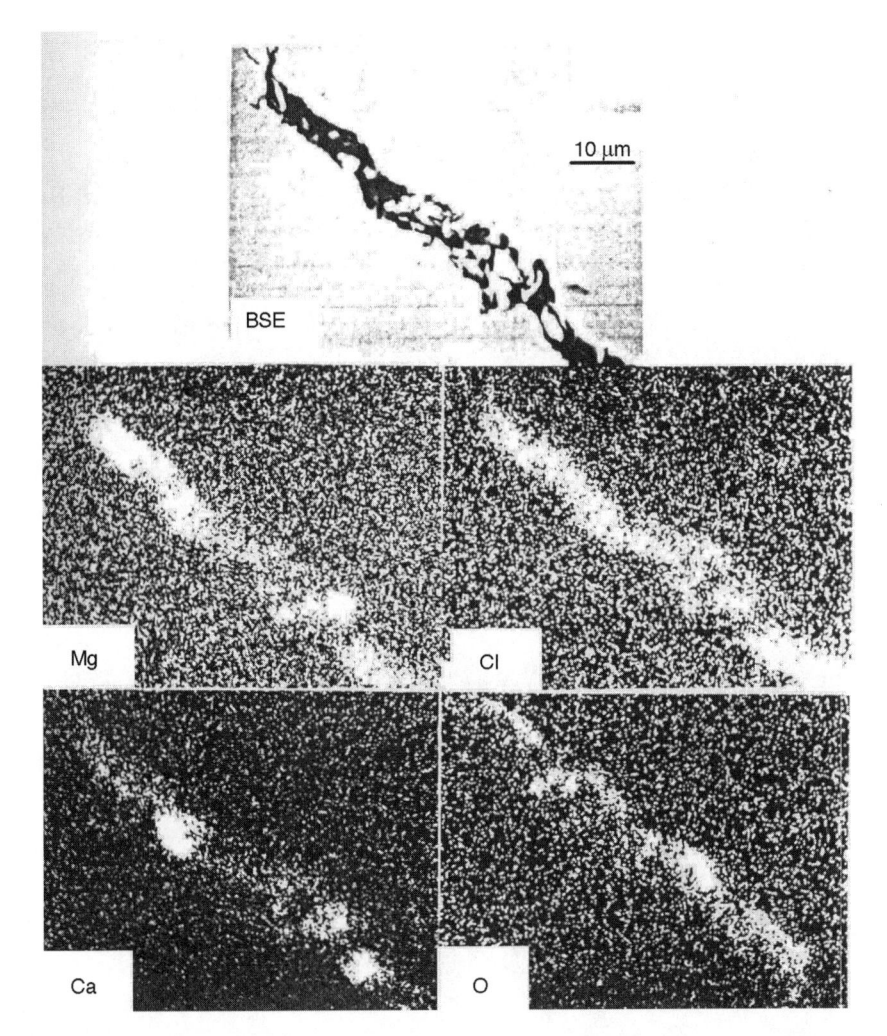

FIGURE C5.3
BSE and X-ray images at a crack in the failed vane.

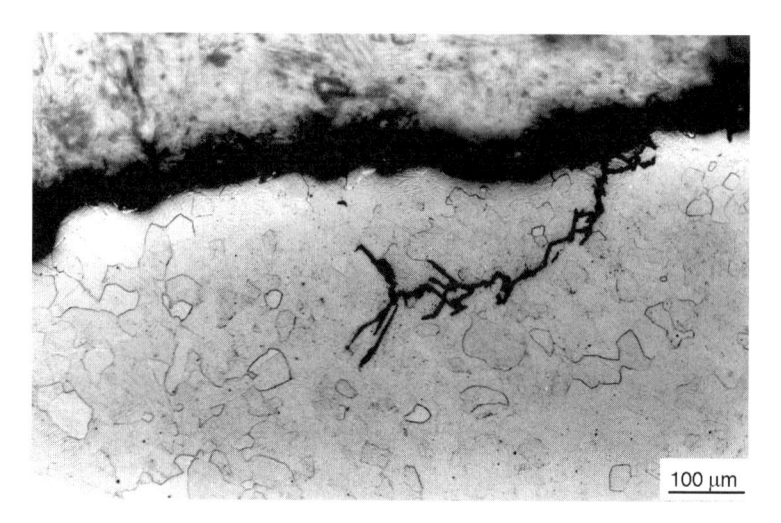

FIGURE C5.4
Photomicrograph showing transgranular branching cracks.

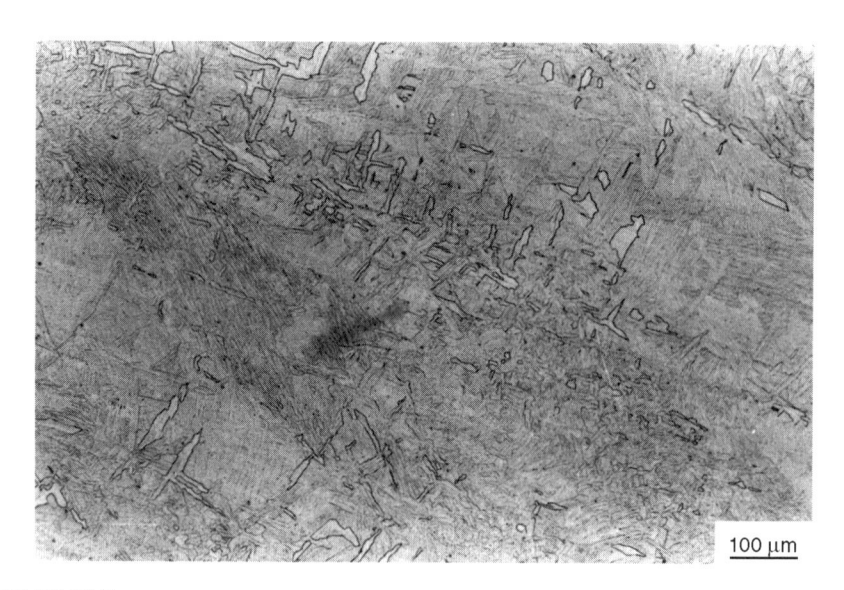

FIGURE C5.5
Photomicrograph showing nonuniform grain structure with acicular α and transformed β.

TABLE C5.1
Chemical Composition of the Failed Blow-off Vane

	Al	Sn	Fe	O	H	N	Ti
Wt%	5.7	3.4	0.05	0.15	0.015	0.05	Bal

Discussion

The chemical composition and hardness conformed to the specified grade. The bluish color on all the used samples was confirmed to be due to a "stabilization" treatment imparted after heat treatment. The microstructure was found to vary from batch to batch. The presence of transformed β and acicular α makes it prone to easy crack initiation and hence is not normally preferred. Ti-Al-Sn alloys are known to be susceptible to SCC in a chloride-rich environment. The presence of transgranular secondary cracks filled with corrosion debris consisting of Cl, Ca, Mg, and O and a significant amount of residual stress indicated that the crack initiation was due to stress corrosion. The stress concentration effect of the machined grooves might have aided SCC.

Conclusion

Failure of the blow-off vanes occurred due to stress corrosion cracking.

5.2 Failure of the Undercarriage Cylinder of an Aircraft

Key words: Al-Zn-Mg alloy, Intergranular cracking, SCC

Introduction

A long crack was noticed in the inner cylinder of the nose wheel undercarriage of an aircraft during overhaul [32]. The service life of the undercarriage at the time of the failure was 1000 h. The cylinder was forged from Al-Zn-Mg alloy and precipitation-hardened and anodized.

Experimental Results

The forged cylinder (Figure C5.6) was found to have a 170-mm-long crack along the parting line. One end of the crack was about 20 mm away from the flat region at the fork end and the other end was found to have extended up to the cylindrical region. The crack was opened to examine the fracture surface. The fracture pattern indicated that the fracture had originated at the edge of a curved step on the inner surface (Figure C5.7). The anodized layer was found to be locally damaged at the fracture origin. The fracture surface contained a bright region R_1 and a dark region R_2 (Figure C5.8).

The fracture surface was examined in SEM. In the region R_1, intergranular fracture was observed along with a number of secondary intergranular cracks (Figure C5.9). Corrosion debris was found on the fracture surface in this region. In region R_2, dimpled rupture was observed on the fracture

FIGURE C5.6
Photograph of the cracked inner cylinder.

surface without any corrosion product (Figure C5.10). The region R_3 representing shear lips showed elongated dimples.

The corrosion product found in region R_1 was qualitatively analyzed by energy-dispersive spectroscopy. The corrosion product was found to contain Ca, S, Cl, and K in addition to the elements in the alloy (Figure C5.11). The chemical composition of the failed component is given in Table C5.2.

The edge from which the crack had originated was prepared for metallographic examination after a very light grinding. The sample showed nearly equiaxed grains with some undissolved particles and a few intergranular cracks (Figure C5.12). The average hardness of the component at 2.5 kg load was 186 HV.

Failures Induced by Corrosion

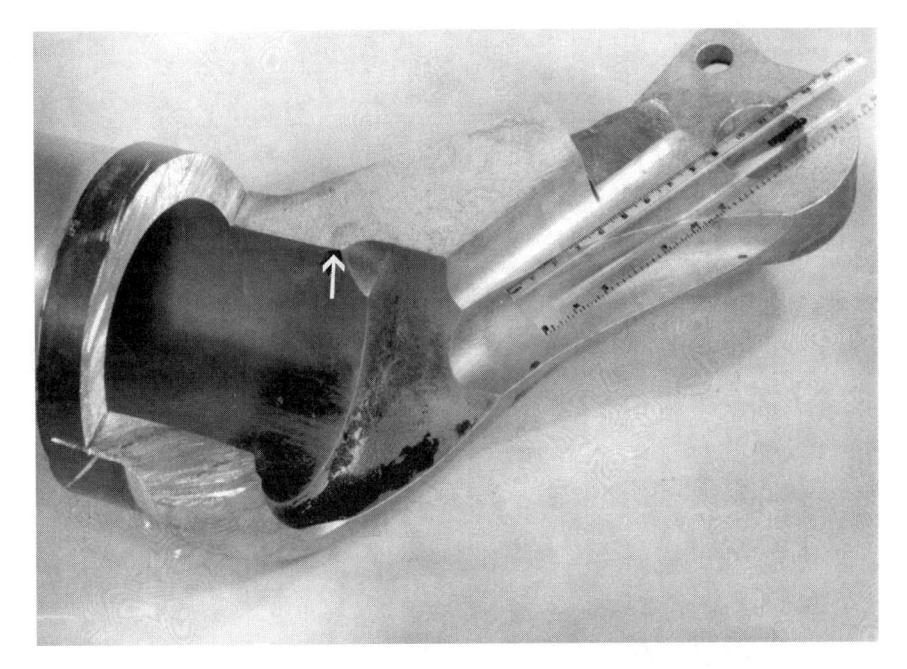

FIGURE C5.7
Photograph showing the fracture origin (arrow).

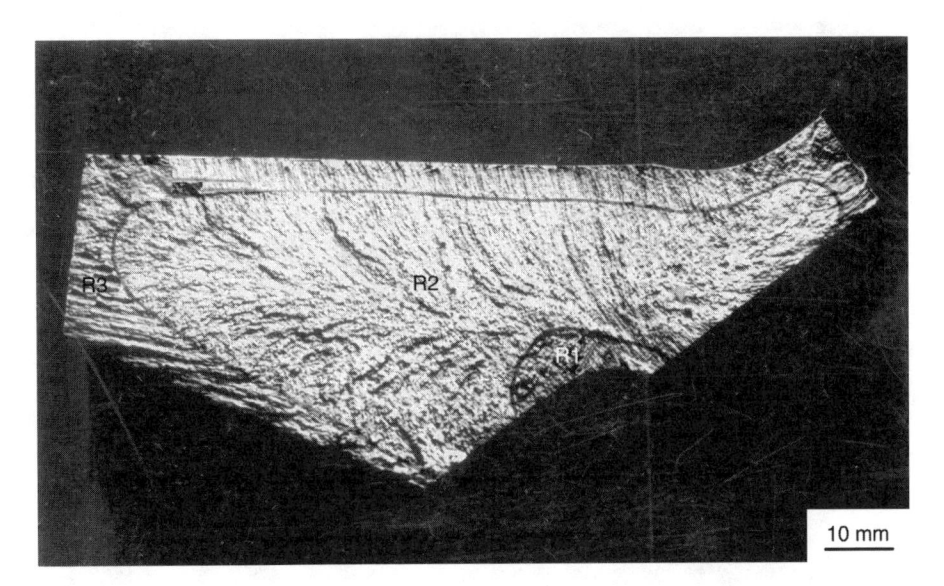

FIGURE C5.8
Photograph showing the fracture surface with a dark region (R_1) and bright region (R_2). Region R_3 represents shear lips.

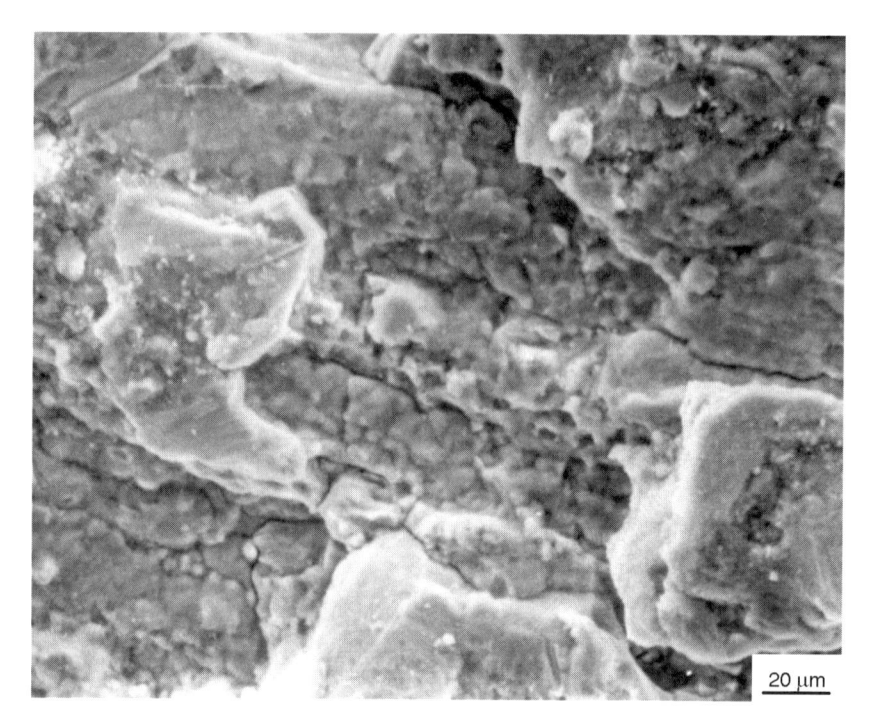

FIGURE C5.9
Fractograph showing intergranular fracture with secondary cracks in region R_1.

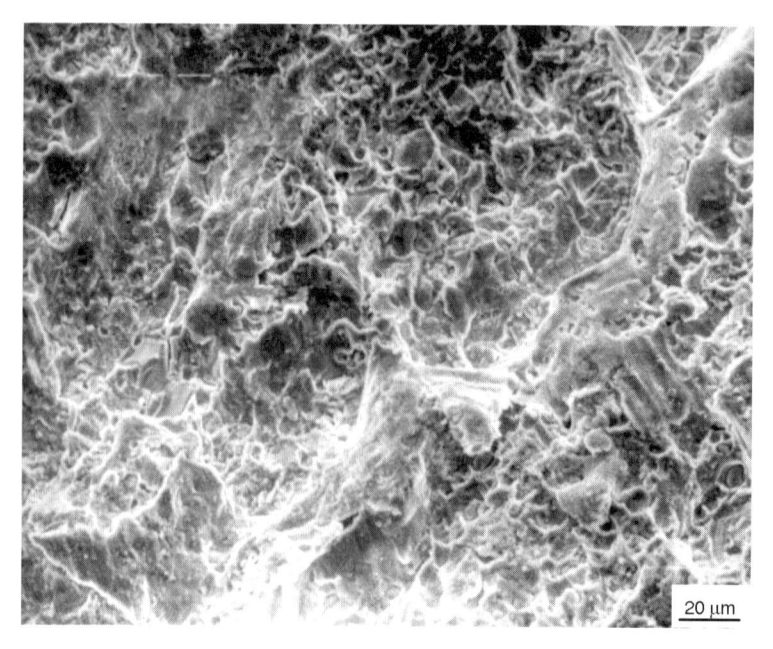

FIGURE C5.10
Fractograph showing equiaxed dimples in the region R_2.

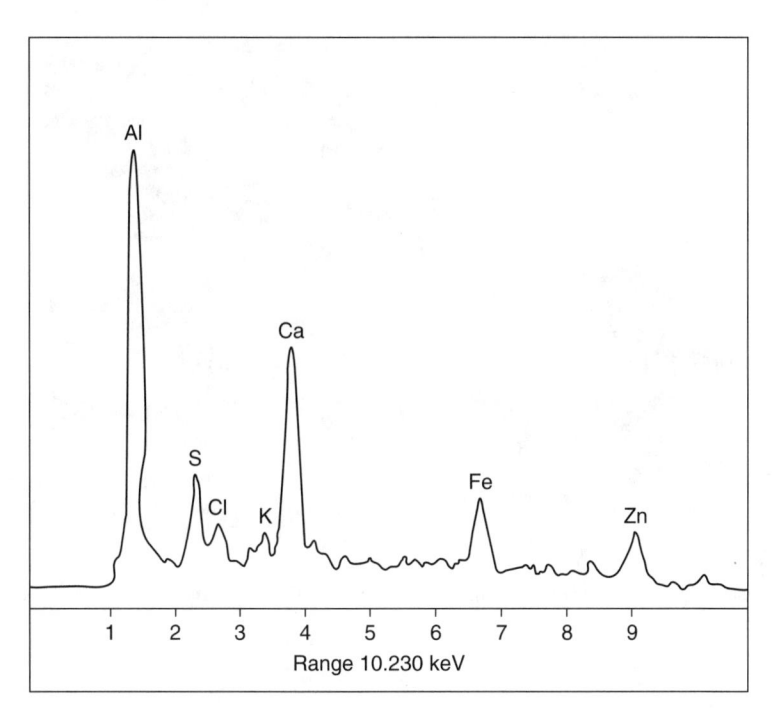

FIGURE C5.11
EDAX spectrum of the corrosion debris.

TABLE C5.2
Chemical Composition of the Failed Undercarriage Cylinder

	Zn	Mg	Cu	Mn	Si	Fe	Ti	Al
Wt%	6.0	3.0	0.32	0.51	0.12	0.31	0.07	Bal

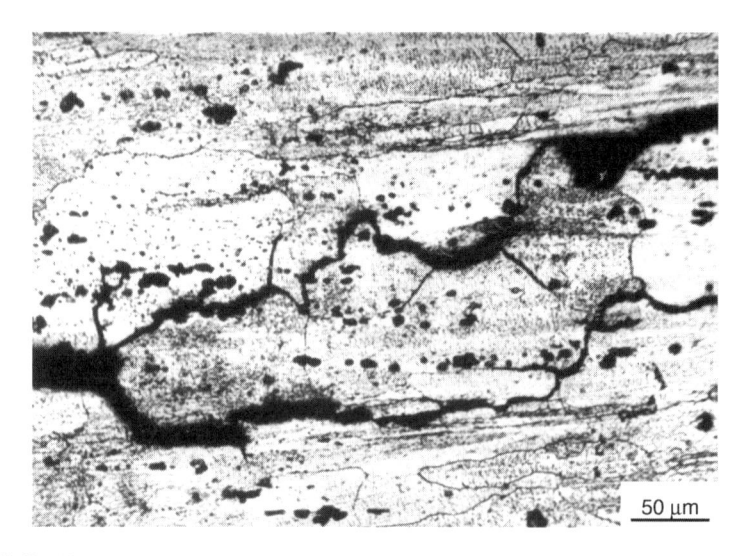

FIGURE C5.12
Photomicrograph at the fracture origin showing intergranular cracks and nearly equiaxed grains.

Discussion

Chemical analysis confirmed that the material of the forging was an Al-Zn-Mg-Cu-Mn alloy. The hardness value indicated that the component was solution-treated and aged to peak hardness. These alloys in the peak-hardened condition are susceptible to intergranular and stress corrosion. The parting plane of a forged component is more susceptible to failure by SCC because of the presence of residual tensile stresses. SCC is further enhanced by SO_2 present in air as a pollutant.

The presence of intergranular cracks near the fracture origin indicated that the protective anodized layer was damaged in that region. Chlorine and SO_2 present in the air could thus react with susceptible material to cause SCC, as evidenced by intergranular fracture at the origin. The crack propagation in ductile mode took place when the remaining wall thickness was unable to withstand the stress imposed on the component.

Conclusion

The failure of the undercarriage cylinder was due to SCC. The high-strength aluminum alloy component was hardened to peak hardness, making it susceptible to SCC in an environment containing chlorine and SO_2.

5.3 Failure of the Flame Tube Retainer Bolts of an Aeroengine

Key Words: Stainless steel, Transgranular fracture, Intergranular fracture, SCC

Introduction

The flame tube retainer bolts of an aeroengine operated in a coastal area failed after 367 h of service [33]. The normal working temperature of the bolts was around 600°C. The bolts were manufactured from precipitation-hardening, austenitic stainless steel conforming to Russian alloy AE 481 and heat-treated to a hardness level of 241 to 302 HB.

Experimental Results

Visual examination showed that the bolts failed at the thread region. The failure originated at a thread immediately next to the nut, as shown in Figure C5.13. The fracture surfaces were smooth and covered with corrosion and oxidation products. There was no evidence of general corrosion of the bolts.

The fracture surface was examined in a scanning electron microscope. The fracture surface was found to be masked by corrosion/oxidation products (Figure C5.14). EDAX spectra were taken at the peripheral and central regions of the fracture surface. The peripheral region indicated the presence of chlorine, sulfur and calcium along with iron, chromium and nickel, whereas in the central region iron, chromium and nickel were found to be present (Figure C5.15).

The cleaned fracture surface revealed transgranular fracture features at the periphery (Figure C5.16) and intergranular fracture and secondary cracks at the central region (Figure C5.17). A longitudinal section of the failed bolt showed transgranular cracks originating at the thread roots followed by intergranular cracks (Figure C5.18). The intergranular cracks contained a few large grain boundary cavities (Figure C5.19). Away from the fracture surface, a number of branching transgranular cracks were seen at the thread roots up to a depth of 0.35 mm (Figure C5.20). The sample showed equiaxed grains of austenite with precipitates. At the thread roots slightly deformed grains were noticed. The average hardness was 298 HB. The chemical composition of the failed bolt is given in Table C5.3.

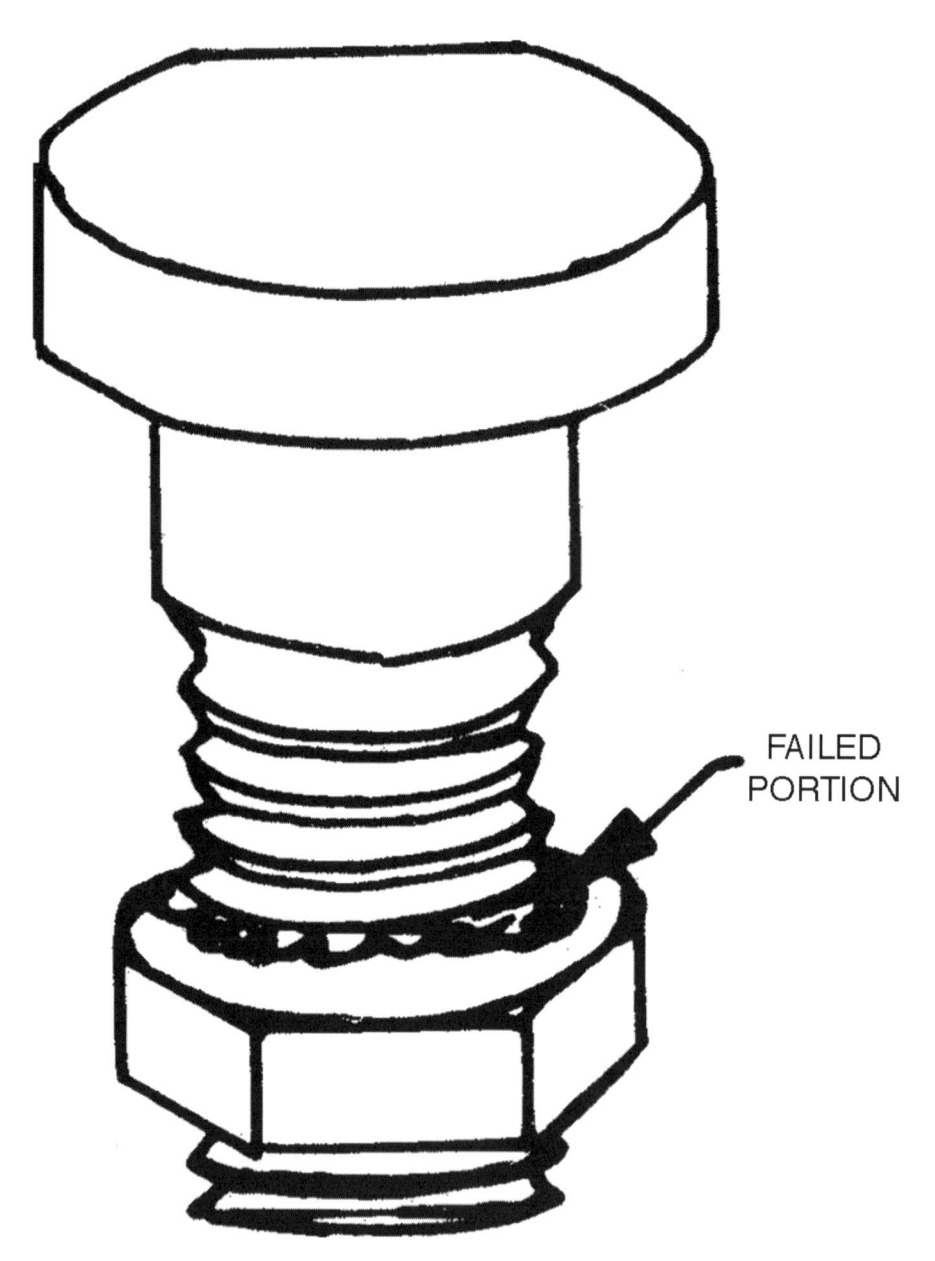

FIGURE C5.13
Sketch showing the location of failure of the bolt.

FIGURE C5.14
SEM picture of the fracture surface.

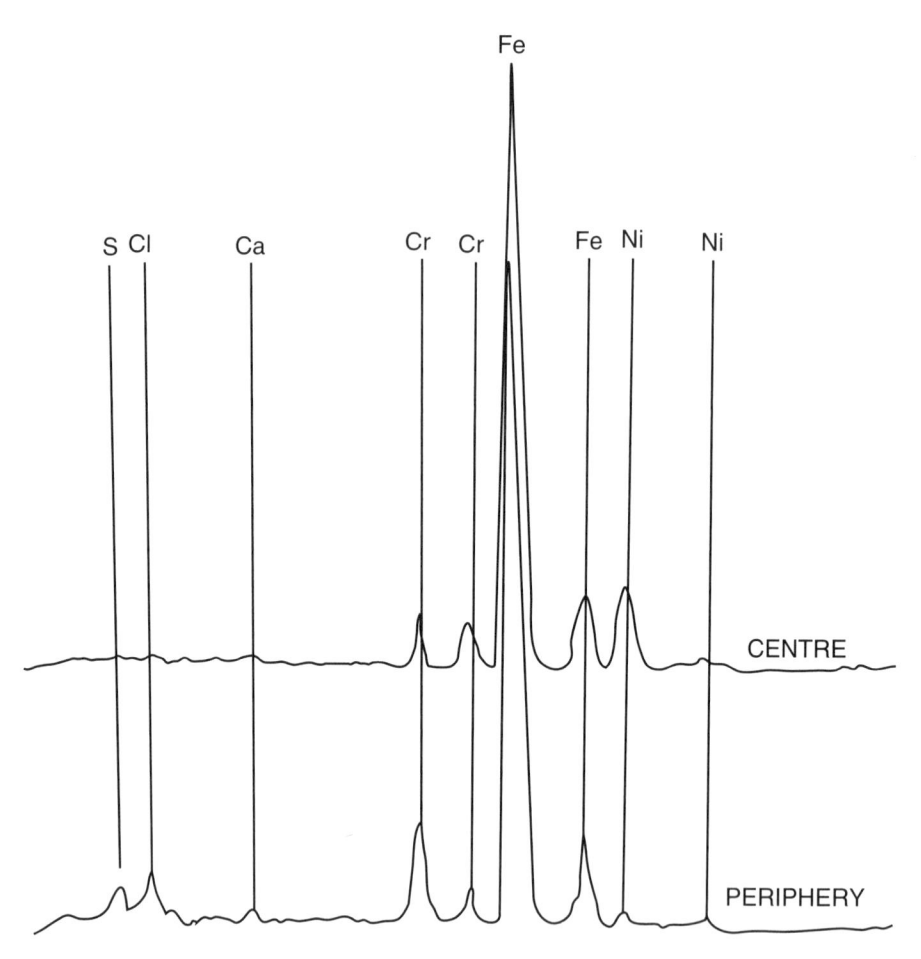

FIGURE C5.15
EDAX spectra at the periphery and central region of the bolt fracture surface.

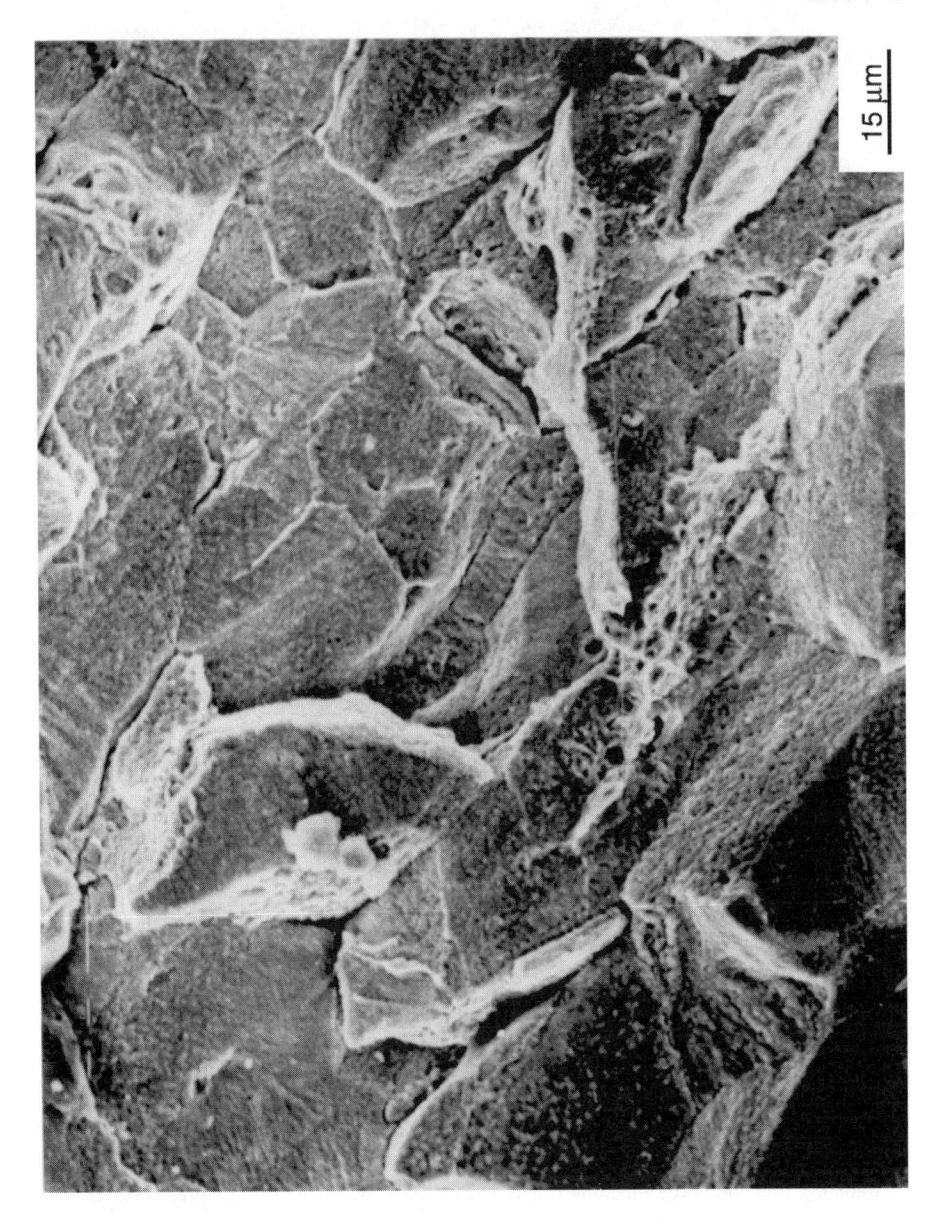

FIGURE C5.16 Fractograph showing transgranular fracture at the periphery.

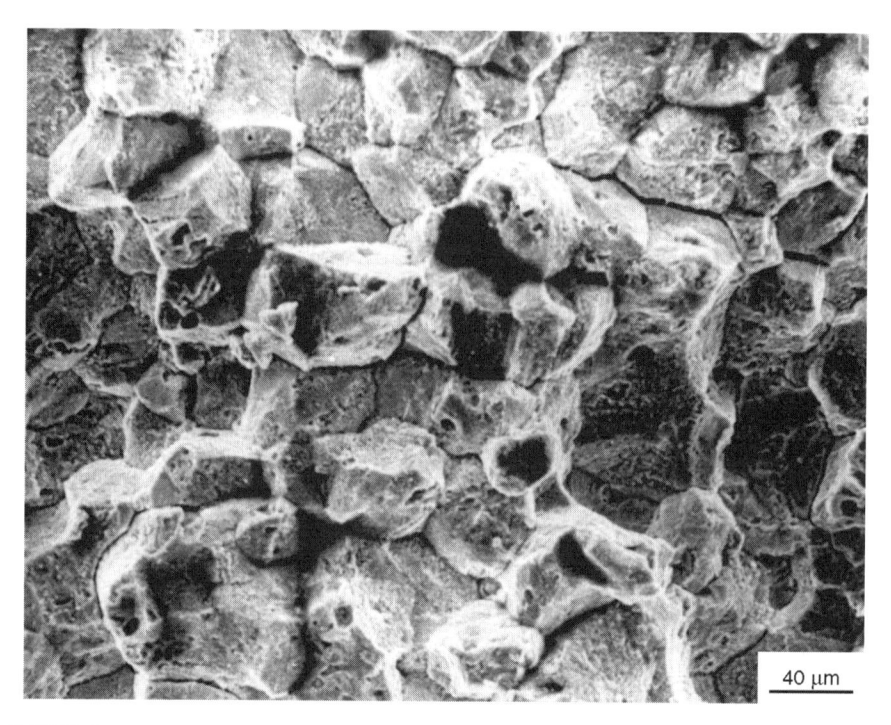

FIGURE C5.17
Fractograph showing intergranular fracture with secondary cracks at the central portion.

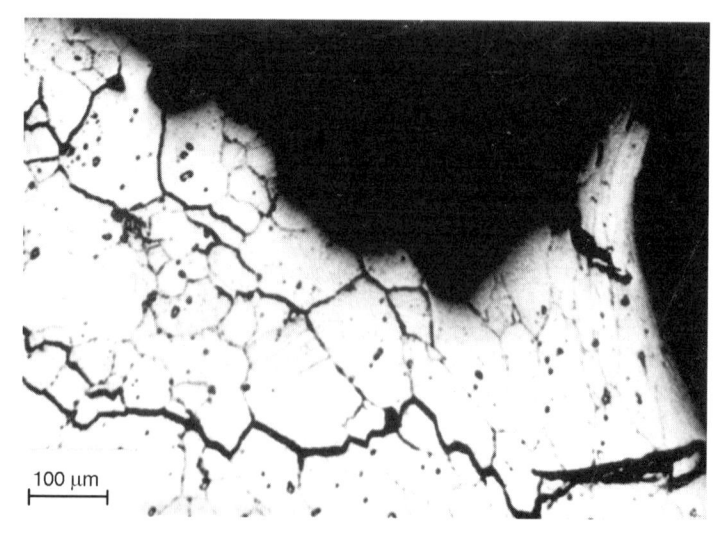

FIGURE C5.18
Photomicrograph showing branching transgranular cracks followed by intergranular cracks.

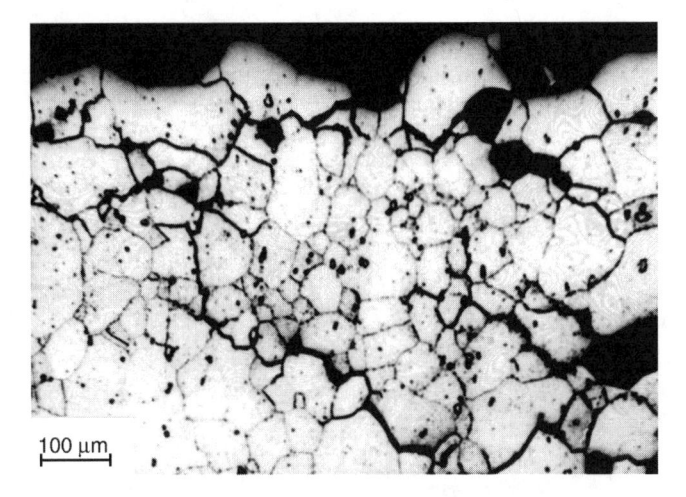

FIGURE C5.19
Photomicrograph showing large cavities on the intergranular crack paths.

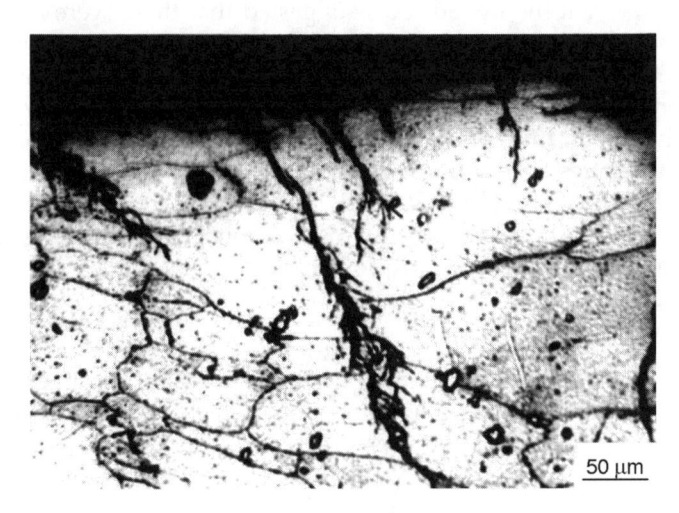

FIGURE C5.20
Photomicrograph showing transgranular branching cracks originating from thread roots.

TABLE C5.3

Chemical Composition of the Failed Flame Tube Retainer Bolt

Elements	Wt (%)
C	0.38
Mn	7.96
Ni	7.86
Cr	11.73
Mo	0.95
V	1.50
Nb	0.32
Si	0.5
S	0.02
P	0.03
Fe	Bal

Discussion

The flame tube retainer bolts were made of precipitation-hardening, austenitic stainless steel conforming to grade AE 481. Transgranular fracture at the periphery with a chloride-bearing corrosion product and transgranular branching cracks at the thread roots suggested that these were formed due to SCC. The SCC took place at ambient temperature when the engine was stationed at a coastal base. The nut-component crevice could have helped in retention of the corrodant and promotion of SCC at the thread root above the nut. The intergranular fracture covered with oxides and the presence of cracks and cavities at the grain boundaries close to the fracture surface indicated that the central region failed at high temperature under overload conditions due to the reduction in cross-sectional area resulting from the presence of peripheral stress corrosion cracks.

Conclusion

The failure of the flame tube retainer bolts was initiated by stress corrosion cracking in a chloride-containing atmosphere. The final failure was due to high temperature overload rupture.

5.4 Failure of the Side Strut of a Helicopter

Key words: Ni-Cr-Mn-Si Steel, Corrosion pits, Intergranular cracks, SCC

Introduction

A helicopter fell onto its starboard side while landing during its first flight, before which it was kept in a hangar for over 1 year [34]. The helicopter's

Failures Induced by Corrosion

life was extended beyond its service life, considering the fewer number of landings performed during its specified service. Telltale evidence indicated that the side strut had failed, causing the accident. The strut was manufactured with Ni-Cr-Si-Mn steel tubes with a paint finish on the outer surface.

Experimental Results

The failed side strut, along with other parts of the assembly, is shown in Figure C5.21. The fracture pattern suggested that the crack origin was at the inner surface of the tube. The fracture surface of the side strut at the origin showed crystalline fracture features over 30 mm of the circumferential fracture, while the remaining portion showed fibrous fractures with shear lips.

The fracture surface containing the origin showed corrosion debris near the inner surface and predominantly intergranular fracture with secondary cracks (Figure C5.22). The remaining portion of the fracture surface showed dimpled rupture.

Longitudinal sections of the side strut samples both near and away from the fracture surface were metallographically examined. The inner surface of the sample showed corrosion pits and intergranular cracks. On etching with 2% Nital solution, the sample revealed a decarburized layer to a depth of about 15 µm, corrosion pits, and tempered martensitic structure (Figure C5.23). The microstructure of the strut away from the fracture region also contained tempered martensitic structure. EPMA analysis of the debris was carried out and the electron and X-ray images for Cl, Mg, Ca, O, and Fe are shown in Figure C5.24. The chemical composition of the failed component is given in Table C5.4. The hardness of the strut was 530 HV.

FIGURE C5.21
Side strut assembly. Arrow head shows the failed side strut.

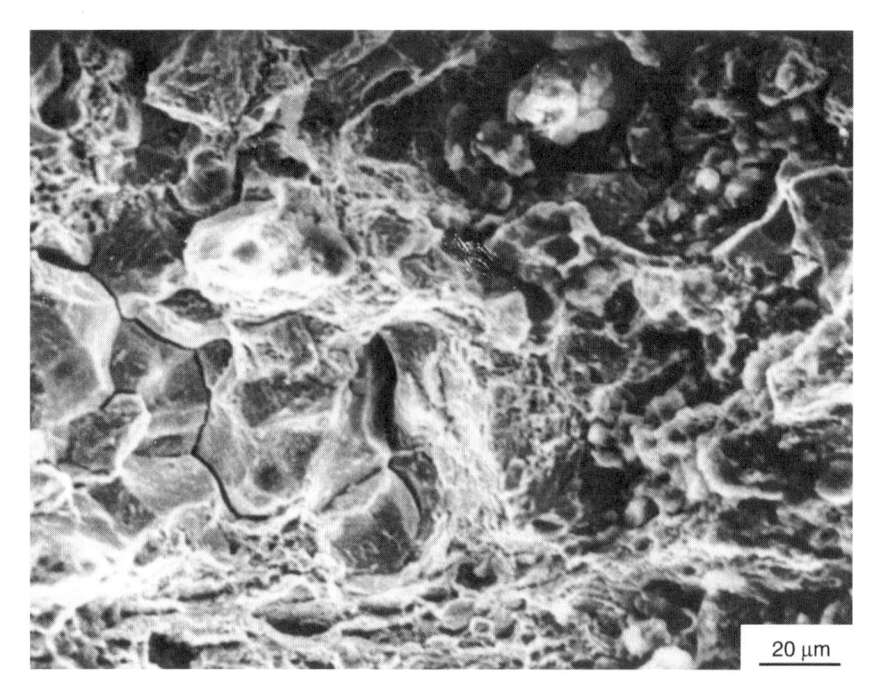

FIGURE C5.22
SEM fractograph showing corrosion debris near the inner surface and intergranular fracture with secondary cracks.

FIGURE C5.23
Photomicrograph showing the decarburized layer and pits.

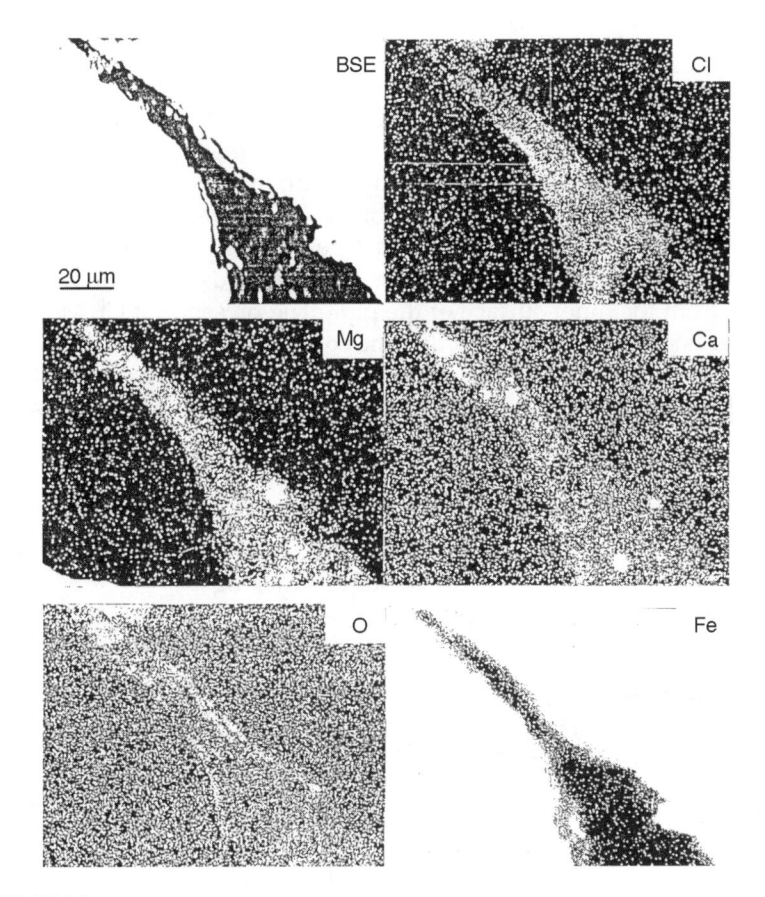

FIGURE C5.24
BSE and elemental X-ray images of the corrosion debris.

TABLE C5.4

Chemical Composition of the Failed Side Strut

Elements	Wt%
C	0.30
Si	0.80
Mn	1.23
Ni	1.86
Cr	1.0
S	0.006
P	0.016
Fe	Bal

Discussion

Chemical analysis, microstructure and hardness results suggested that the side strut sample was made from Ni-Cr-Si-Mn steel and was used in hardened and tempered condition. Microstructure of the strut sample revealed that the inner surface was severely decarburized and corroded. Pits and intergranular cracks filled with corrosion debris were also observed all along the inner surface. The intergranular fracture, corrosion debris-filled pits and cracks suggested that the failure of the strut was caused by stress corrosion cracking from the inner surface of the pipe. The presence of Cl, Mg, and Ca in the corrosion debris indicated that the corrosion had occurred in a marine environment probably during prolonged storage in the hanger. The final failure during landing was due to overload consequent to a reduction in the load-bearing cross-sectional area caused by stress corrosion cracks.

Conclusion

The side strut failed due to stress corrosion cracking.

5.5 Failure of Fourth-Stage Stator Casing Bolts

Key words: Martensitic stainless steel, Intergranular fracture, Corrosion debris, SCC

Introduction

The stator blades fitted to an outer stator casing shroud ring are tightened to the inner shroud ring with the help of bolts [35]. Two of the bolts in one segment of a stator casing failed within 16 h since the last visual examination (Figure C5.25), and the failure was noticed during scheduled servicing. The operating temperature was below 200°C. The bolts completed a total life of 556 h. The bolts were made from Russian-grade martensitic stainless steel conforming to AP 866W and heat-treated to a strength level of 1100 MPa.

Experimental Results

The failure took place at the step near the first thread root. The failed samples showed a dark-brownish deposit at the fracture origin. The entire fracture surface appeared coarse and crystalline. The fracture features suggested that the crack initiation was at the root of the first thread, corresponding to the step on the fracture surface (Figure C5.26). At higher magnification, the fracture surface showed predominantly intergranular fracture with a number

Failures Induced by Corrosion

FIGURE C5.25
A segment of stator casing showing the two failed bolts.

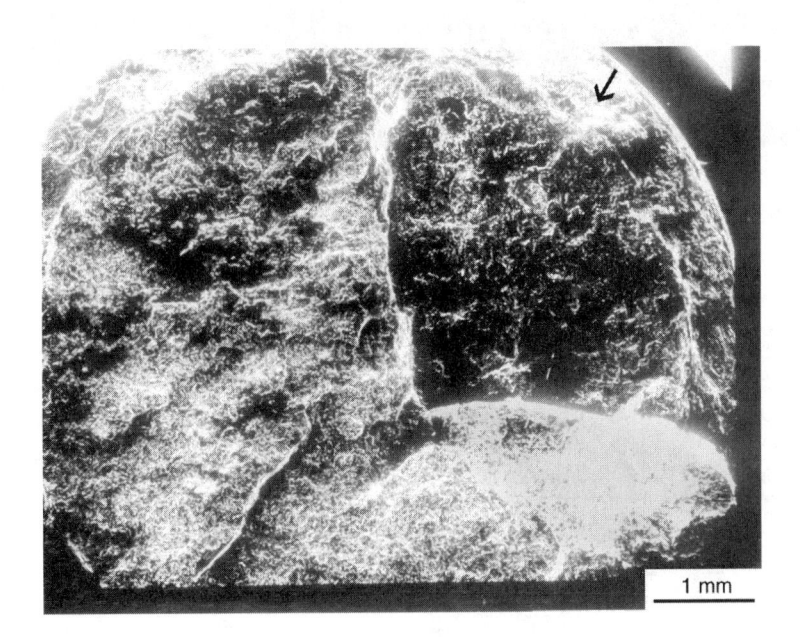

FIGURE C5.26
SEM fractograph showing the fracture origin from the thread root (arrow).

of secondary cracks (Figure C5.27). The fracture features were masked with corrosion debris at the step where the fracture had originated. The corrosion product observed in pits and inside the cracks was analyzed by EPMA. The BSE and X-ray images of Fe, O, Na, Cl, Cr, S, Ca, and Mg present in the corrosion product are presented in Figure C5.28.

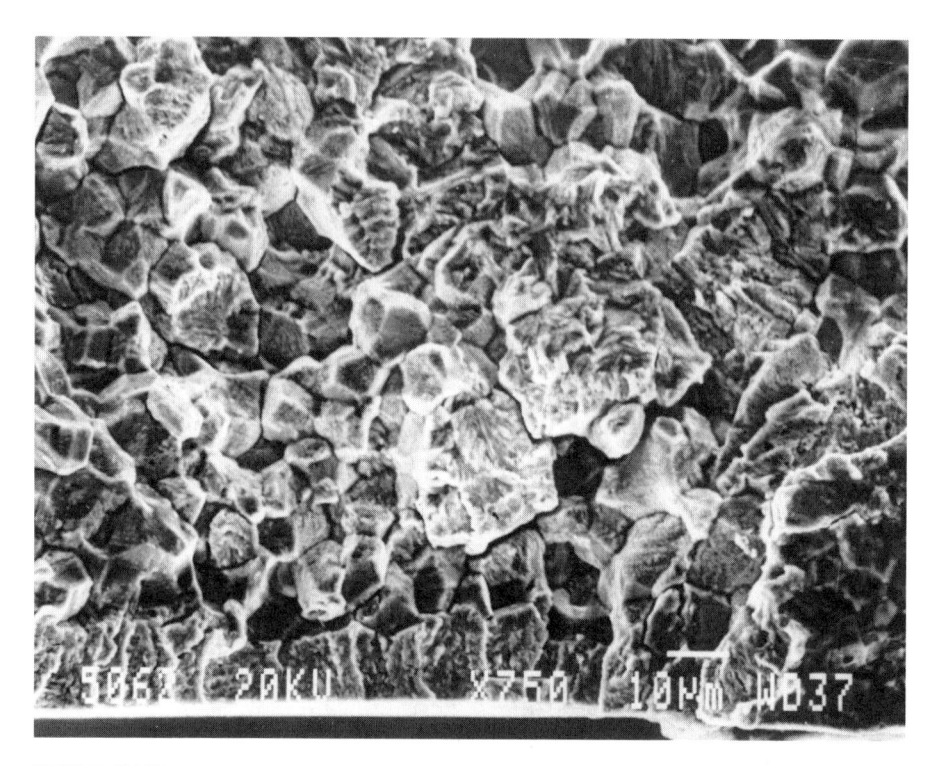

FIGURE C5.27
SEM fractograph showing intergranular fracture with secondary cracks.

The longitudinal section of the failed bolt was examined. In the unetched condition, the sample showed deep pits filled with debris and cracks originating from the root (Figure C5.29). On etching with aqua regia, the failed and new samples showed tempered martensite (Figure C5.30). The chemical composition of the bolt is given in Table C5.5. The average hardness was 386 HV.

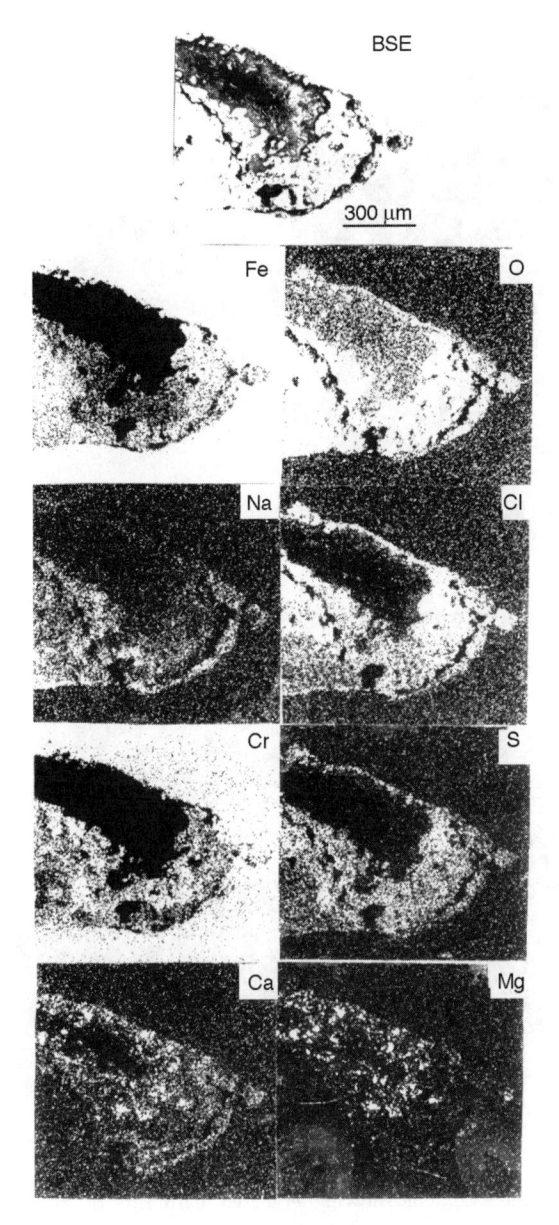

FIGURE C5.28
BSE and X-ray images of the elements present in the corrosion product of the failed bolt.

FIGURE C5.29
Photomicrograph showing corrosion in the groove. Note the crack originated from the root (arrow).

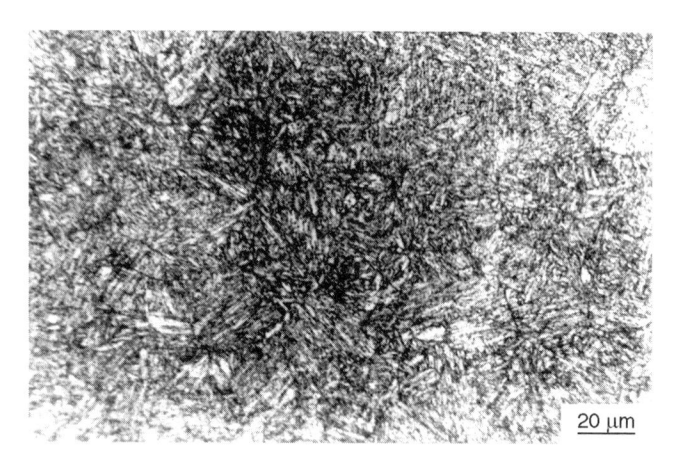

FIGURE C5.30
Photomicrograph of the failed bolt showing tempered martensite.

TABLE C5.5

Chemical Composition of the Failed Bolt

Elements	Wt%
C	0.14
Mn	0.3
Si	0.5
Ni	1.82
Cr	17.0
Mo	1.92
Co	4.85
V	0.20
Nb	0.25
S	0.01
P	0.015
Fe	Bal

Discussion

The chemical composition, microstructure and hardness conformed to the requirements of the relevant material standards. The predominantly intergranular fracture with secondary cracks and corrosion debris rich in Na, Cl, S, and O indicated that corrosion played a dominant role in the failure. The presence of deep pits from which intergranular cracks were found to originate indicates that corrosion damaged the bolt. The fracture mode, the composition of the corrosion debris and the operating conditions suggest that the failure occurred due to stress corrosion cracking initiated near the first thread root under a chloride- and sulphide-containing atmosphere and service stresses.

Conclusion

The failure of the bolts occurred due to stress corrosion in a marine environment.

5.6 Failure of the NGV Bolts of an Aircraft

Key words: PH stainless steel, Corrosion pits, Intergranular cracks, Hot corrosion

Introduction

During strip examination of an aeroengine, 1 of the 12 bolts used to attach the first-stage nozzle guide vane locating flange was found to be missing

FIGURE C5.31
Bolt in as-received condition.

[36]. The missing bolt caused secondary damage to the hot end parts and the turbine rotor assembly. As the failed bolt could not be located, the remaining 11 bolts were examined to assess the nature of the damage that led to the bolt failure. The bolts were manufactured using Russian precipitation-hardened stainless steel AE 481 and coated with copper. The bolts were used with a static torque of 1.5–1.7 kg at temperatures varying between ambient and 1000°C in an environment of burning gases and hot air.

Experimental Results

One of the bolts in as-received condition is shown in Figure C5.31. The bolt surfaces were rough and oxidized. Longitudinal sections of the bolt samples showed pits with corrosion debris and intergranular cracks with extensive branching initiated near the outer surface of the bolt (Figures C5.32a and C5.32b). On etching with aquaregia, the sample showed equiaxed grains of austenite and carbide precipitates both along the grain boundaries as well as within the grains.

The debris in the pits on analysis using EPMA showed the presence of chlorine. A back-scattered electron image of a pitted area along with X-ray mapping for chlorine is shown in Figure C5.33. The chemical composition of the failed bolt is given in Table C5.6. The hardness of the bolts was 348 HV.

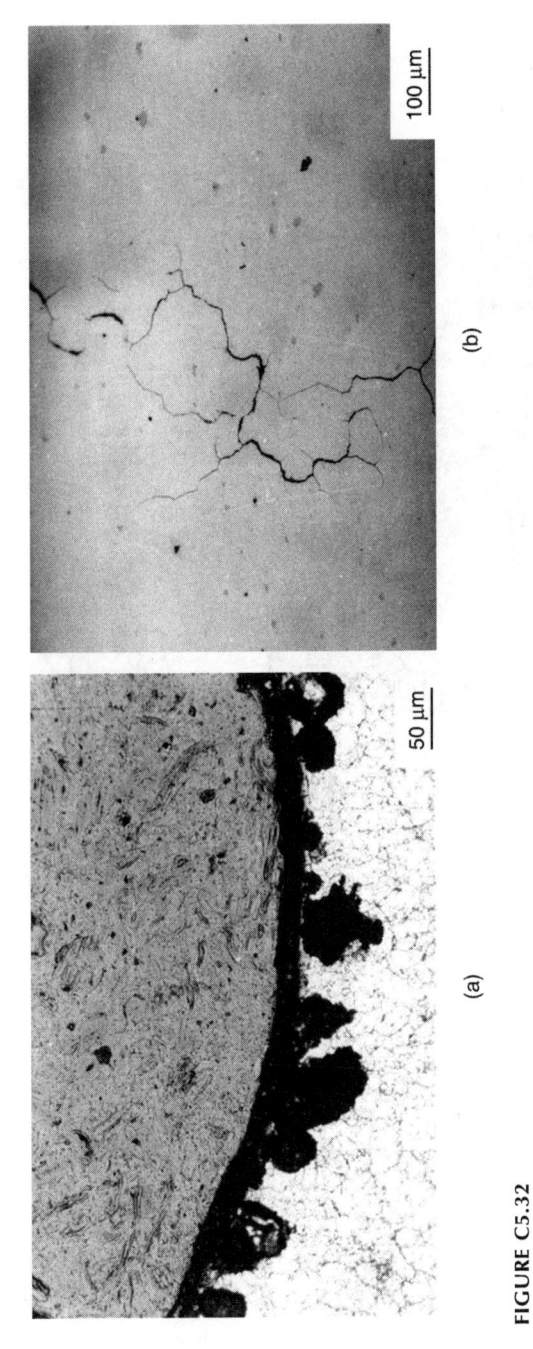

FIGURE C5.32
Photomicrographs showing (a) severe pitting and corrosion at the surface of the bolt sample and (b) intergranular cracks with extensive branching that originated at the surface of the bolt.

BSE

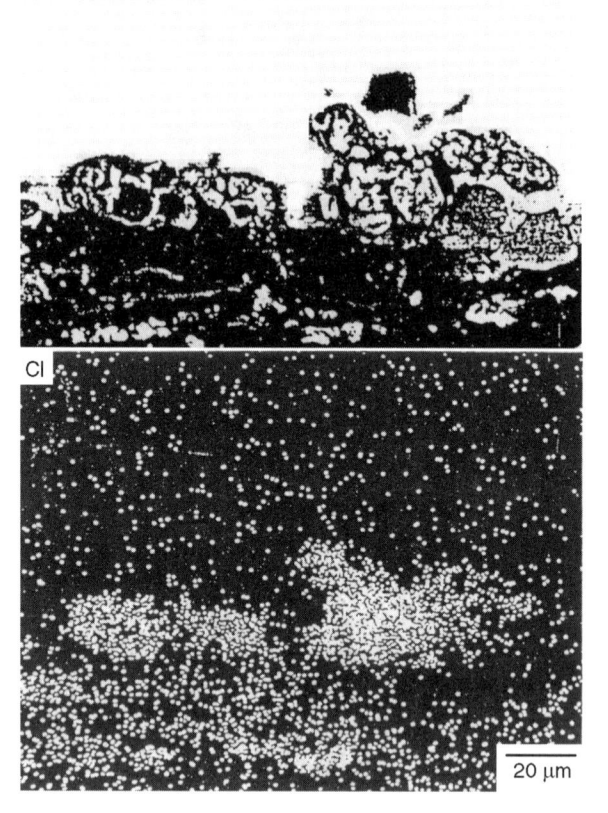

FIGURE C5.33
BSE and chlorine X-ray images of a corrosion pit.

TABLE C5.6

Chemical Composition of the Failed NGV Bolt

Elements	Wt%
C	0.41
Si	0.50
Mn	8.9
Ni	9.1
Cr	13.2
V	1.8
S	0.007
P	0.011
Fe	Bal

Discussion

Chemical analysis showed that the bolts conformed to the specified grade. The hardness indicated that the bolts were used in solution-treated and aged condition. Metallographic examination revealed extensive pitting and corrosion at all locations. Cracks were observed to have initiated from the surface pits at a few locations and propagated into the material along the grain boundaries. The evidence suggested that the failed bolt also must have been damaged due to hot corrosion and cracking in a chloride-bearing atmosphere at high operating temperatures and service loads.

Conclusion

The failure of the missing NGV bolt was probably due to damage induced by hot corrosion.

5.7 Exfoliation Corrosion in an Aircraft Structural Member

Key Words: *Al-Cu-Mg alloy, Delamination, Fibrous grain structure, Exfoliation corrosion*

Introduction

The rear spar of the stabilizer of a transport aircraft with an expired life of 12 years was found to have been corroded over an area of 150 × 60 mm [37]. The rear spar was an extruded T-section made of aluminum alloy grade D16 of GOST-4784 and was used in solution-treated and aged condition. A skin made of the same alloy was fixed to the T-section with rivets as shown schematically in the Figure C5.34.

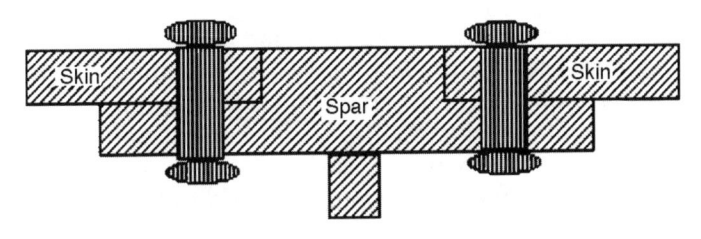

FIGURE C5.34
Schematic of the rear spar.

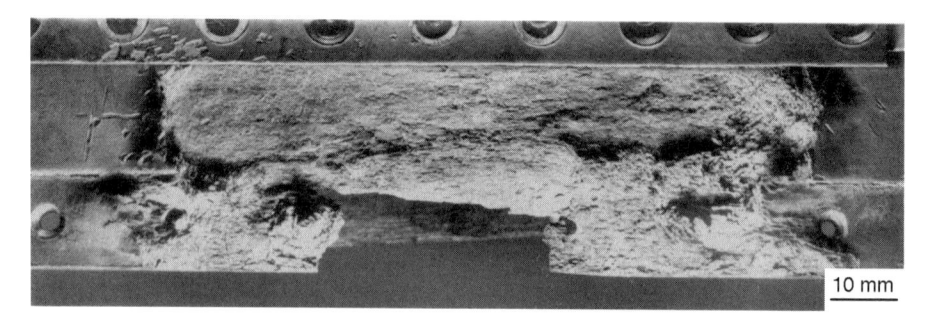

FIGURE C5.35
Photograph of the damaged portion of the T-section.

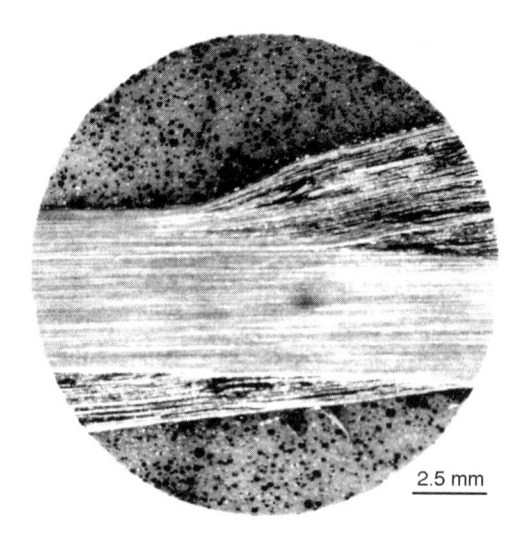

FIGURE C5.36
Macrograph of the T-section showing the fibrous structure, bulging and delamination.

Experimental Results

Visual and stereomicroscopic examination indicated that one arm of the T-section was completely corroded; six rivets were found to be broken and the skin was lifted up. The other arm also showed evidence of corrosion, but to a lesser extent. The material in the corroded region appeared flaky and was separated into layers (Figure C5.35). Extensive bulging and delamination were observed at the corroded region. Undamaged regions were well protected by corrosion-resistant paint.

Longitudinal sections from the T-section close to the damaged region and the skin were prepared for microstructural studies. Macroexamination of the extruded section revealed a fibrous structure, bulging and delamination (Figure C5.36). Microstructural examination showed banded grain structure

FIGURE C5.37
Photomicrograph of the T-section at the uncorroded region showing the fibrous structure.

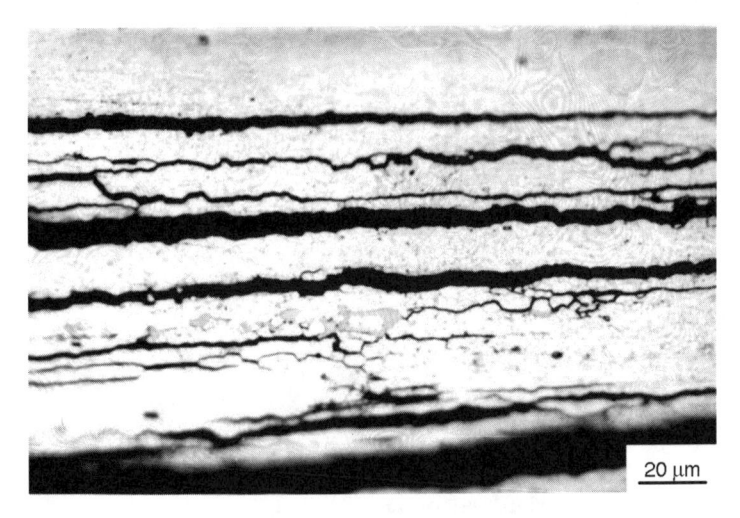

FIGURE C5.38
Photomicrograph of the longitudinal section showing cracks along the grain boundaries of the fibrous structure.

and cracks along the grain boundaries (Figures C5.37 and C5.38). The skin sample showed equiaxed grain structure with aluminum cladding (Figure C5.39). The chemical compositions of the extruded section and skin are given in Table C5.7. As the cladding material was very thin, its composition could not be evaluated. The hardness of the extruded section and the skin are given in Table C5.8.

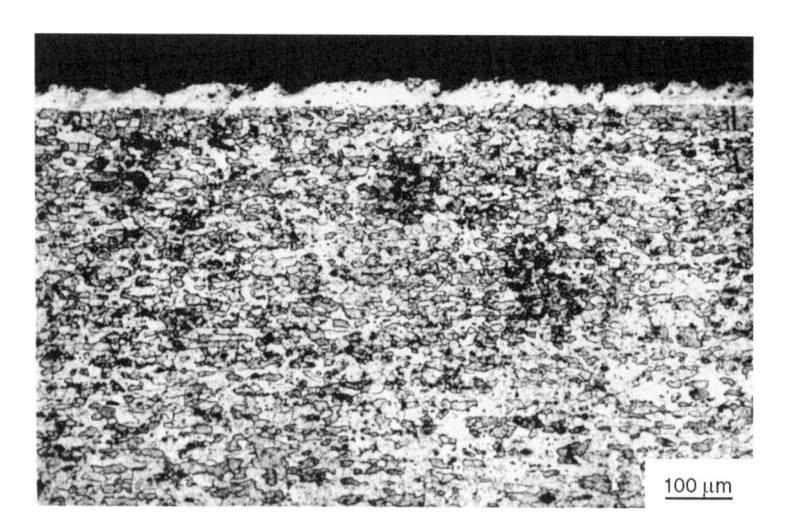

FIGURE C5.39
Photomicrograph of the longitudinal section of the clad sheet showing equiaxed grain structure. The clad layer is on the top.

TABLE C5.7

Chemical Composition of the Extruded Section and Skin

Elements	Extruded section	Skin
Cu	3.82	4.80
Mg	1.23	1.37
Mn	0.54	0.56
Fe	0.35	0.2
Si	0.28	0.30
Al	Bal	

TABLE C5.8

Hardness of the Extruded Section and Skin

Part	Hardness (HV)
Extruded section	153
Skin	143

Discussion

The chemical composition of the extruded section and sheet and strip samples conformed to grade D16 of GOST 4784-49. The hardness of the extruded section was found to be higher than that of the sheet and strip, even though it contained lower amounts of Cu and Mg. Being a leaner alloy, the extruded section would have a slower response to natural aging. The hardness of the extruded section was near the peak value obtainable in such alloys. On the other hand, in the sheet with a higher amount of Cu and Mg, lower hardness indicated that it was in overaged condition.

Visual and metallographic examination confirmed that the corrosion took place along the elongated grain boundaries, resulting in splitting of the extruded section into layers. However, the clad sheet with an equiaxed grain structure was not corroded. The type of damage observed in the T-section is classified as layer, laminar or exfoliation corrosion. In exfoliation-susceptible alloys, the corrosion takes place along the grain boundaries of the matrix phase, as these are more anodic to the grain interior when the alloy is in either underaged or peak aged condition. In a riveted structure, the corrosive electrolyte can have easy access to the directional grain structure through the rivet holes, leading to exfoliation corrosion in susceptible alloys. As corrosion proceeds along multiple narrow paths parallel to the surface, the insoluble corrosion products occupy a larger volume than the metal consumed in producing them. These voluminous corrosion products exert a wedging action, leading to splitting, flaking or delamination.

The sheet was found to be immune to exfoliation corrosion even though the sheets were cut transversely at the rivet holes. This immunity is thought to be due to cathodic protection offered by the cladding, the presence of equiaxed grain structure and the overaged condition of the alloy.

Conclusion

It was concluded that the extruded T-section had undergone exfoliation corrosion due to underaged elongated grain structure cut transversely at the rivet holes. Failure by this mode can be minimized or prevented by using artificially overaged material and sealing the rivet holes with suitable sealants.

Remedial Measures

Overaging can prevent exfoliation corrosion in Al-Cu-Mg alloys. The effect of the overaging treatment is to lessen the tendency for intergranular corrosion to occur by providing a more equipotential situation.

5.8 Failure of a High-Pressure Oxygen Cylinder

Key Words: *Al-Mg alloy, Pitting corrosion, Intergranular cracks, Stress concentration*

Introduction

High-pressure oxygen cylinders, parts of divers' harnesses, were made of an Al-Mg alloy conforming to grade AG-5 of a French standard. The cylinders consisted of 1.5 L of oxygen at a pressure of 200 kg/cm^2. All the cylinders were anodized and painted. During periodic servicing and proof testing, it was observed that some of the cylinders were leaking at a pressure of 150 kg/cm^2, even though the specified proof pressure was 300 kg/cm^2 [38].

Experimental Results

A large number of pits were observed on the outer surface of the cylinders. Some of the pits were as large as 1 cm^2, extending to a depth of over 1 mm. The cylinders were sectioned longitudinally to examine the inner surfaces. There was no evidence of any damage on the inner surface.

The longitudinal and transverse sections of the cylinders were prepared for metallographic examination. The sample revealed a heavily corroded outer surface with a large number of corrosion pits and a few intergranular

FIGURE C5.40
Photomicrograph showing corroded outer surface with a large number of corrosion pits and a few intergranular cracks that originated from the pits.

FIGURE C5.41
Photomicrograph of the longitudinal section showing fine, elongated grains.

cracks originating from the pits (Figure C5.40). Fine, elongated grains were observed on the longitudinal section (Figure C5.41).

The pits had the characteristic appearance of corrosion damage and were filled with corrosion debris. EDAX analysis of the corrosion debris showed that it contained Ca, Cl, S, Ti, and Si in addition to Al and Mg contained in the base material (Figure C5.42). The chemical composition of the failed cylinder is given in Table C5.9. The hardness of the cylinder was 98 HV. The tensile properties of the cylinder are given in Table C5.10.

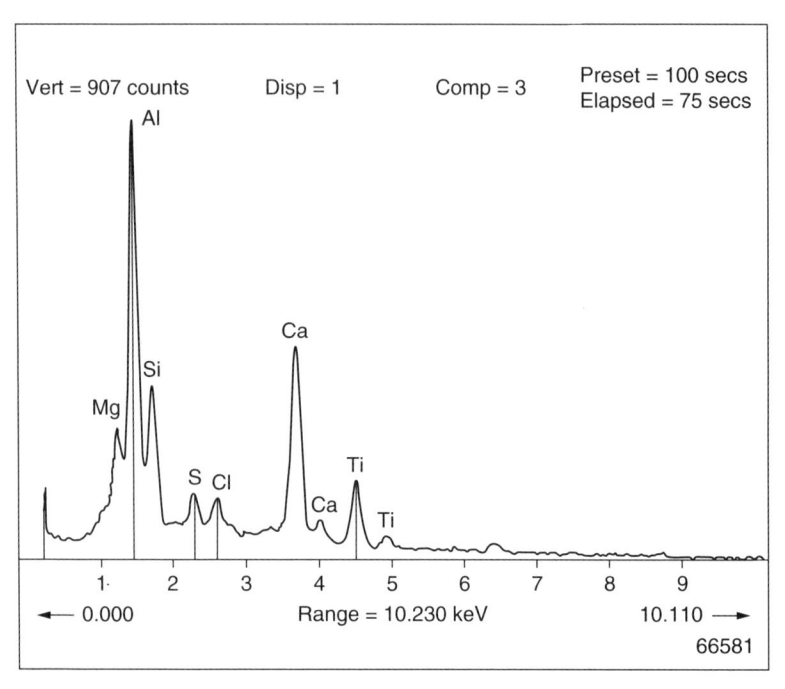

FIGURE C5.42
EDAX spectrum obtained on corrosion debris.

TABLE C5.9

Chemical Composition of the Failed High-Pressure Cylinder

	Mg	Mn	Si	Fe	Cu	Al
Wt%	4.88	0.88	Trace	Trace	Trace	Bal

TABLE C5.10

Tensile Properties of the Failed Cylinder

YS (MPa)	UTS (MPa)	Elongation (%)
238	344	19

Discussion

The chemical composition conformed to the specified grade. The hardness and tensile properties were in agreement with those expected from the material. The experimental results clearly suggested that the outer surface of the cylinder suffered extensive pitting corrosion. Localized damage to the protective anodized layer had exposed the susceptible base material to corrosion in a marine environment. The absence of the anodized layer due to local damage coupled with the presence of Cl and S in the corrosion debris in the pit confirmed corrosion damage. If one assumes a minimum stress concentration factor of 2.2 for a pit (a reasonable assumption for a pit ignoring the intergranular cracks, whose stress concentration factor is much higher), the calculated stress (based on the Von-Mises yield criterion) imposed at a pressure of 150 kg/cm^2 (the pressure at which leakage was observed during proof testing) exceeds the yield stress of the material, thus confirming that the pits were responsible for the failure of the cylinders at half the proof pressure.

Conclusion

The failure of the oxygen cylinders was due to the stress-raising effect of corrosion pits and associated intergranular cracks.

5.9 Failure of the Boiler Tube of a Thermal Power Station

Key words: Plain carbon steel, Galvanic corrosion, Perforation

Introduction

A boiler tube connecting the bottom drum to the bottom header in the rear water wall of a thermal power station failed due to perforation at the bend near the nose baffle after 11 years of service [39]. The boiler tube was made of low-carbon steel. The tube operated at a temperature of 515°C and at a pressure of 97 kg/cm^2.

Experimental Results

A perforation about 10 mm in diameter was observed at the bent portion of the tube (Figure C5.43). The longitudinal section of the tube showed the presence of a large chunk of deposit on the inner surface only near the perforation. Part of the deposit was found dislodged and removed from the inner surface around the perforation (Figure C5.44). The tube was found to be locally thinned near the perforation. Such a deposit was not observed at any other location either on the inner or the outer surface of the tube.

A transverse section of the tube containing a part of the perforation and the deposit was prepared for metallographic examination. The sample showed that the deposit contained reddish areas with metallic luster (Figure

FIGURE C5.43
Photograph showing a perforation in the failed tube.

FIGURE C5.44
Photograph showing a deposit on the inner surface near the perforation.

C5.45). The bright reddish, metallic areas in the deposit remained unattacked by 2% Nital, the etchant used for etching steels. The microstructure of the tube consisted of pearlite and ferrite in widmanstatten distribution (Figure C5.46). There were no microstructural changes between the perforated zone and the rest of the tube. The deposit on the inner surface of the tube was analyzed by EPMA. The electron and elemental X-ray images of the deposit are given in Figure C5.47. The deposit was found to contain mostly Cu and Zn.

The chemical composition of the failed tube is given in Table C5.11. The average hardness of the tube at 5 kg load was 158 HV.

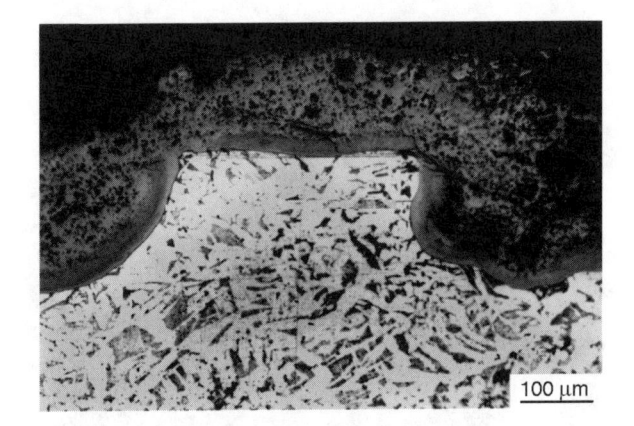

FIGURE C5.45
Photomicrograph showing the deposit on the inner surface of the tube. The deposit is on the top.

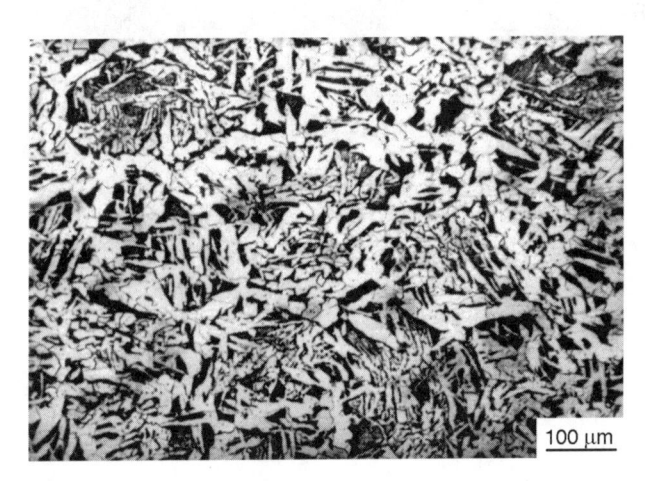

FIGURE C5.46
Photomicrograph showing ferrite and pearlite.

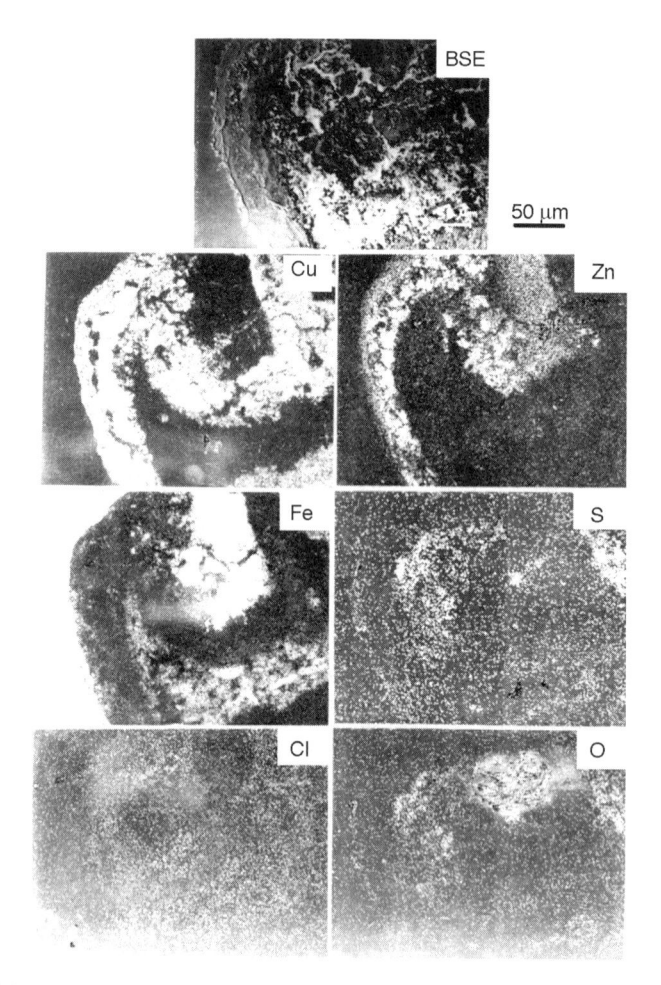

FIGURE C5.47
Electron and elemental X-ray images of the deposit.

TABLE C5.11

Chemical Composition of the Failed Boiler Tube

	C	Si	Mn	Cr	S	P
Wt%	0.21	0.16	0.81	Trace	0.018	0.015

Discussion

Visual and metallographic examinations indicated the presence of a deposit on the inner surface of the tube near the perforation. The presence of copper and zinc at identical locations in the deposit indicated that the debris originated from brass. Depletion of zinc at a few locations in the deposit suggested that brass had suffered dezincification. Since brass deposition on a small area of the inner surface of the tube is a most unlikely event, the probable source could be failure of some component of the control valves made of brass in the water circuit and the entanglement of the failed piece at the bent portion of the boiler tube. Since steel is anodic to brass, entanglement of a brass piece appeared to have resulted in the formation of a galvanic cell leading to anodic dissolution of the steel tube with water as the electrolyte. Since no structural change of the tube material could be detected near the perforation, the failure of the tube was not associated with any overheating effects. Galvanic corrosion between the brass piece and steel tube resulted in reduction in the wall thickness of the tube, leading finally to perforation under the operating pressure.

Conclusion

The failure of the tube was due to galvanic corrosion caused by entanglement of a brass piece in the steel tube.

Part 6

Failures Initiated by Wear

6.1 Failure of a High-Pressure Turbine Rotor Blade

Key Words: *Ni-base superalloy, Beach marks, Fatigue striations, Fretting*

Introduction

Two high-pressure turbine rotor (HPTR) hollow blades, made of Russian nickel-based superalloy ZC6K-BD by investment casting, were found to have failed at the root region [40]. The blade aerofoils were given aluminide coating after heat treatment. The fir-tree root regions were stipulated to be shot-peened. The failure was observed after the engine had completed 796 h since it was new and 98 h since its last overhaul.

Experimental Results

The fracture surface of the failed HPTR blade on stereomicroscopic examination revealed that (a) the fracture origin was at the first root groove region below the trailing edge on the concave surface and (b) the crack had propagated into the aerofoil region toward the leading edge, resulting in final failure. Suitable sections of the blade were examined in SEM. The fracture pattern clearly suggested that the crack origin was at the first root below the trailing edge. A few cavities and dark patches, aligned along the contour, were observed a few micrometers below the fracture origin (Figure C6.1). The fracture surface near the trailing edge showed distinct beach marks (Figure C6.2a) and fatigue striations (Figure C6.2b). The fracture surface at and near the leading edge revealed fracture along interdendritic boundaries (Figure C6.3).

The longitudinal section at the root region had a typical cast dendritic structure consisting of γ' precipitated in γ matrix and carbides (Figure C6.4). The average hardness of the blade at the root region was 385 HV. The composition of the blade is given in Table C6.1.

FIGURE C6.1
SEM fractograph showing a few cavities and dark patches below the fracture origin (arrow).

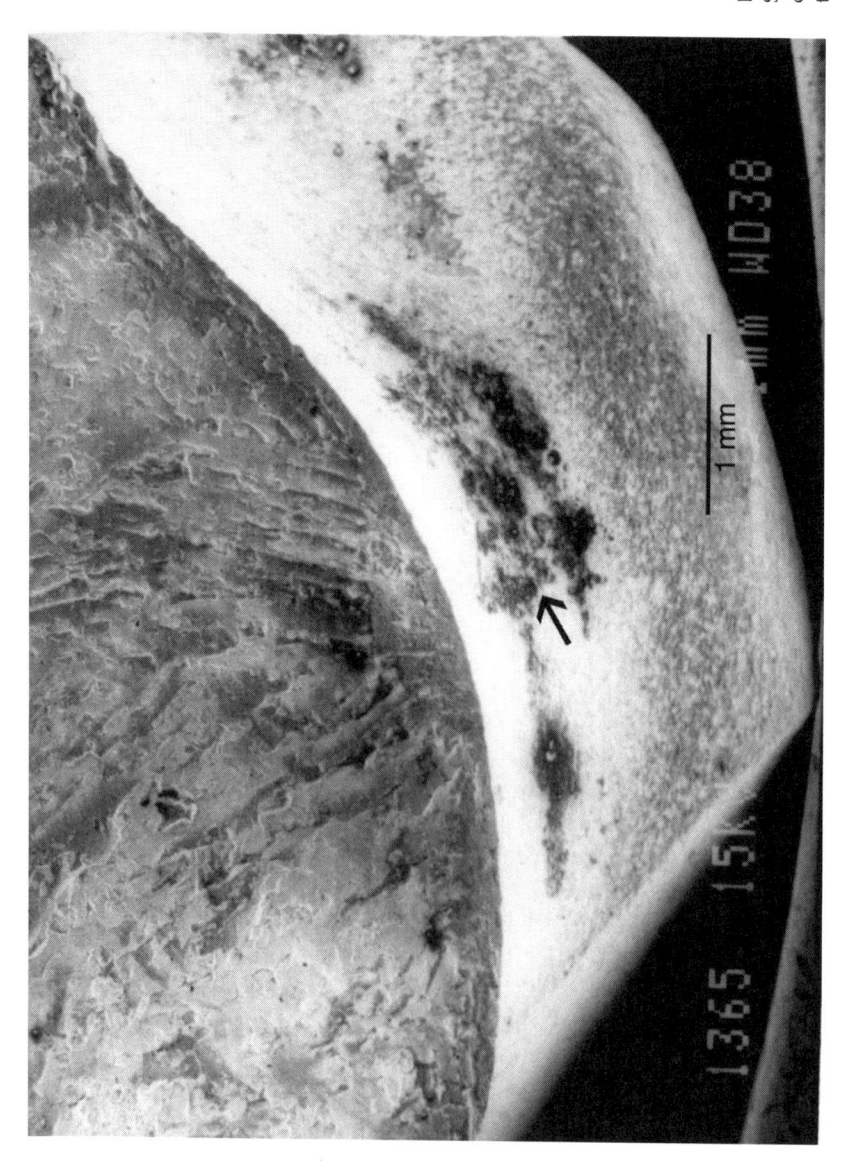

Failures Initiated by Wear

FIGURE C6.2
SEM fractographs showing (a) distinct beach marks and (b) fatigue striations.

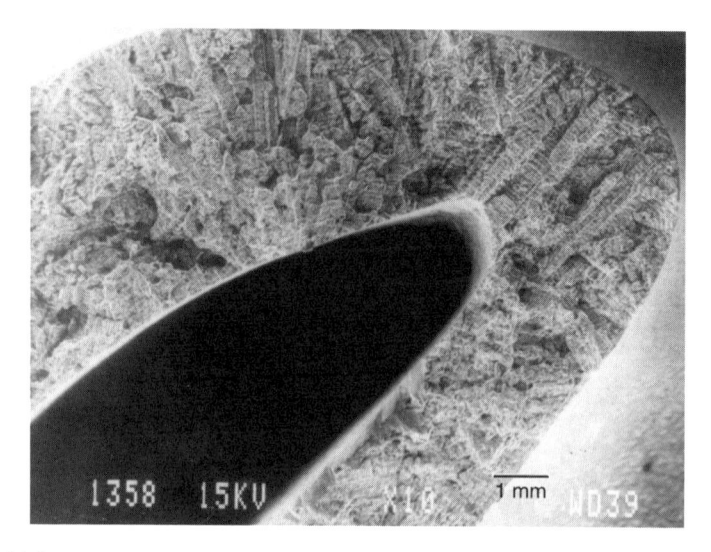

FIGURE C6.3
Fracture surface at and near the leading edge showing fracture along interdendritic boundaries.

FIGURE C6.4
Microstructure at the root region showing typical cast dendritic structure with carbides and γ' precipitates in γ matrix.

TABLE C6.1
Chemical Composition of the Turbine Blade Material

Element	Wt%	Element	Wt%
C	0.12	Mo	4.20
Cr	10.10	W	4.82
Ti	3.10	Co	4.31
Al	5.7	Ni	Bal

Discussion

The fractographic examination revealed that the fracture had originated at the first root groove below the trailing edge and propagated toward the leading edge. The presence of cavities and dark patches below the fracture origin suggested fretting damage, which in turn could have initiated cracking. The crack propagated under cyclic loading conditions toward the leading edge, as evidenced by the presence of beach marks and fatigue striations. The fatigue crack extended into the aerofoil, leading to a reduction in the cross section and final failure near the leading edge in interdendritic mode under overload conditions at elevated temperature. It is probable that the root region of the blade might not have been subjected to the microshot peening treatment stipulated in the production technology to improve the fatigue life. The fretting damage could have occurred due to loosening of the blade, causing relative motion between the blade and disc at the fir-tree joint.

Conclusion

The failure of the HPTR blade was caused by fatigue, originating at a region probably damaged by fretting wear at the first groove of the root of the blade below the trailing edge.

6.2 Failure of the Reheater Tube of a 220-MW Coal-Fired Boiler

Key words: Cr-Mo Steel, Craters and lips, Steam erosion, Decarburization

Introduction

Frequent failures of reheater tubes in the reheater circuit of a 220-MW boiler were observed [41]. All failures were observed close to a tube that was punctured and replaced about 6 months earlier. The boiler was in operation for about 15 years. The operating temperature and pressure of the system were 540°C and 96 kg/m^2 respectively. The reheater tubes were made of Cr-Mo steel conforming to T-22 quality of the SA 213 standard.

Experimental Results

The reheater tube layout showing the failed tube is shown in Figure C6.5. The failed tube was close to an earlier punctured and repaired tube. A hole 20 mm in diameter was observed in the failed tube (Figure C6.6). The wall thickness of the tube was significantly reduced around the hole. The outer

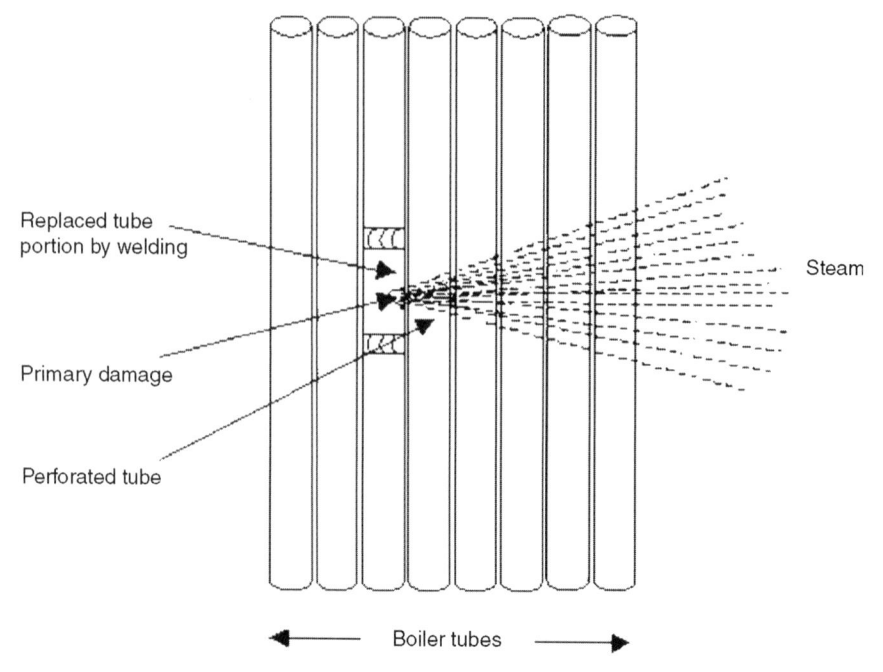

FIGURE C6.5
Schematic of the boiler tube layout showing the failed tube adjacent to an earlier repaired tube.

surface around the hole was rough and bright, while the rest of the outer surface and inner surface were covered with oxide. The material removal leading to wall thickness reduction had essentially occurred from the outer surface, as evidenced by the fact that the inner diameter of the tube had not undergone any change.

The average hardness measured at a 2.5-kg load close to and away from the hole was 109 HV and 141 HV, respectively. The chemical composition (wt%) of the failed tube is given in Table C6.2.

Samples from the damaged and undamaged regions of the tube were prepared for metallographic examination. The tube sample taken from a region away from the hole revealed recrystallized grains of ferrite and fine spheroidized pearlite (Figure C6.7a) and a thick adherent oxide layer on the inner surface (Figure C6.7b).

The sample near the hole revealed extensive decarburization and equiaxed grains of ferrite with a progressive increase in the amount of pearlite to the original level away from the hole (Figure C6.8). The outer surface of the boiler tube around the hole showed craters and lips with evidence of severe plastic deformation to a depth of about 100 μm (Figure C6.9).

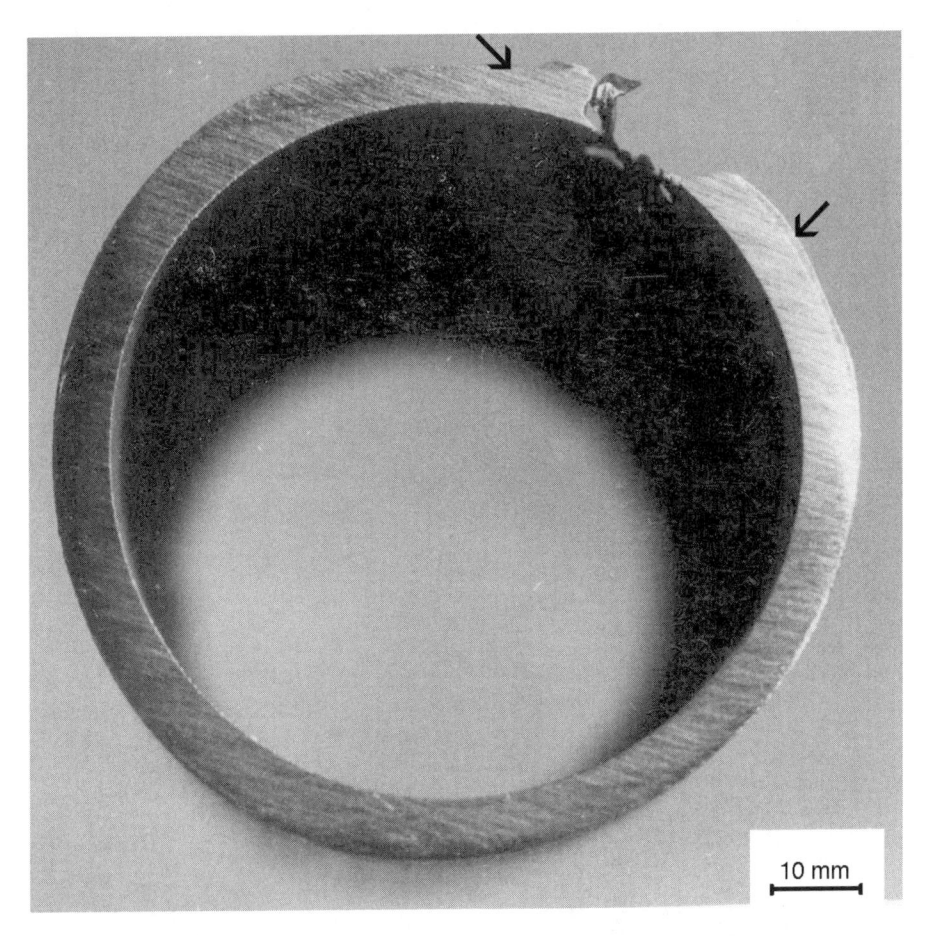

FIGURE C6.6
Cross section of the tube showing wall thickness reduction around the hole (arrows).

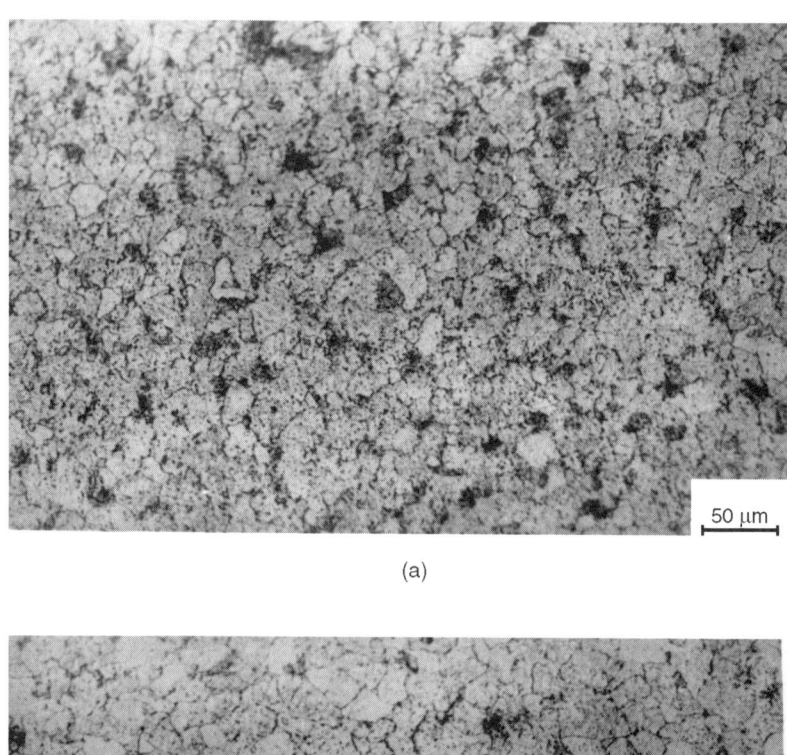

(a)

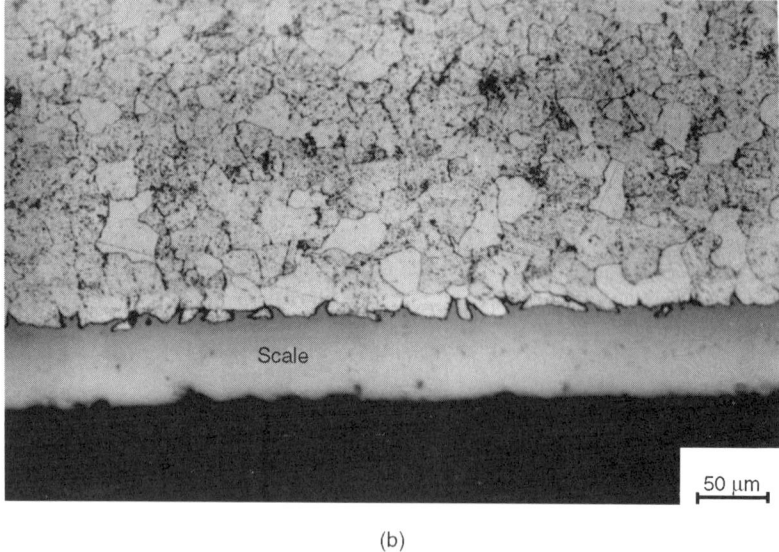

(b)

FIGURE C6.7
(a) Micrograph of the tube showing equiaxed ferrite grains and colonies of spheroidized pearlite.
(b) Micrograph showing adherent oxide scale on the inner surface of the tube.

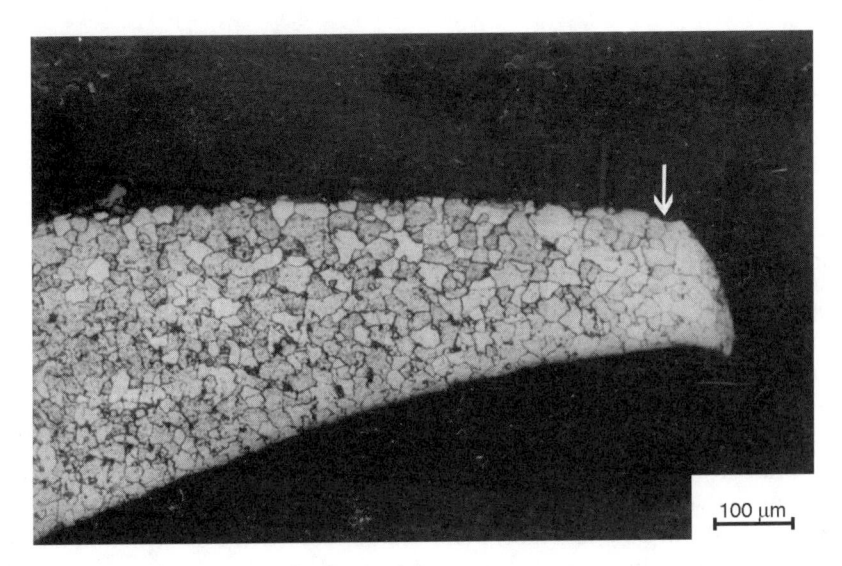

FIGURE C6.8
Micrograph showing total decarburization close to the perforation (arrow).

FIGURE C6.9
Micrograph showing craters and lips on the outer surface close to the hole.

TABLE C6.2

Chemical Composition of the Failed Tube

	C	Si	Mn	Cr	Mo	S	P	Fe
Wt%	0.09	0.28	0.43	2.00	0.81	0.005	0.016	Bal

Discussion

The composition of the reheater tube conformed to the specified standard. Visual and microscopic examination suggested that significant reduction in wall thickness had taken place from the outer surface near the perforation. The decarburization of the steel at the perforation was the result of the oxidizing character of steam at the high boiler hearth temperature. The spheroidization of carbides was due to prolonged exposure to temperatures below AC_1. Even though the steam temperature was about 500°C, the thick oxide layer at the inner surface, being a poor conductor of heat, acted as an insulator, raising the tube wall temperature above the steam temperature. The lips and craters at the outer surface with evidence of plastic deformation suggested erosion damage. The tube erosion had obviously occurred due to impingement of high-velocity steam laden with particulate matter leaking from the tube that had failed initially and subsequently been replaced. The reduction in wall thickness of the adjacent tubes to different degrees by this event went unnoticed, leading to a series of failures of the tubes on further usage.

The perforation of the tube thus occurred due to wall thickness reduction and its consequential inability to withstand the operating pressure at enhanced temperature because of the insulating effect of the oxide layer at the inner surface.

Conclusion

The reheater tube failed due to a reduction in wall thickness as a result of erosion on the outer surface of the tube.

References

1. R.E. Peterson, *Stress Concentration Factors*, John Wiley & Sons, New York, 1974.
2. T.P. Gabb et al., *Surface Enhancement of Metallic Materials*, Adv. Mater. & Proc., 160, 1, 2002, p. 69-72.
3. K.P. Balan, A.Venugopal Reddy, Vydehi Joshi, and M. Sreenivas, Technical Report: DMRL/MIG/97, 1997.
4. Trilok Singh and A. Venugopal Reddy, Technical Report: DMRL/TR/8746, 1987.
5. A. Venugopal Reddy, D.P. Lahiri and W. Krishnaswamy, Technical Report: DMRL/MIG/82, 1982.
6. A. Venugopal Reddy, K.P. Balan, P.G. Tekade, D. Pradeesh Kumar and N.B. Jagtap, Technical Report: DMRL/MIG/14/94, 1994.
7. K.P. Balan and A. Venugopal Reddy, Technical Report: DMRL/MIG/2/92, 1992.
8. A. Venugopal Reddy, D.P. Lahiri and G.S. Sharma, *Failure of Draw Hook of a Railway Coach*, Prakt. Met., 21, 1984, p. 148-152.
9. K.P. Balan and A. Venugopal Reddy, Technical Report: DMRL/MIG/7/92, 1992.
10. A. Venugopal Reddy and D.P. Lahiri, *Failure of Balancing Gear Rod of a Gun Carriage*, Trans. IIM, 38, 6, 1985, p. 531-532.
11. A. Venugopal Reddy and G. Krishna Murthy, Technical Report: DMRL/MIG/99, 1999.
12. H.H. Kolla and A. Venugopal Reddy, Technical Report: DMRL/MIG/3/99, 1999.
13. K.P. Balan, A. Venugopal Reddy and D. Pradeesh Kumar, Technical Report: DMRL/SFAG/9/98, 1998.
14. A. Venugopal Reddy and P.G. Tekade, Technical Report: DMRL/MIG/14/93, 1993.
15. K.P. Balan, Vydehi Joshi, B.V. Rao and A. Venugopal Reddy, *Failure of High Pressure Turbine Blades of an Aeroengine*, Prakt. Met., 30, 1993, p. 413-420.
16. A. Venugopal Reddy and K.P. Balan, Technical Report: DMRL/MIG/5/92, 1992.
17. A. Venugopal Reddy and N.B. Jagtap, Technical Report: DMRL/MIG/5/91, 1991.
18. A. Venugopal Reddy and D.P. Lahiri, *Failure of Landing Gear of an Aircraft*, Trans. IIM, 35, 1982, p. 99-102.
19. A. Venugopal Reddy and D.P. Lahiri, *Spalling of Work Roll of Cold Rolling Mill*, Prakt. Met., 21, 1984, p. 316-320.
20. A. Venugopal Reddy, Technical Report: DMRL/SFAG/8/97, 1997.
21. A. Venugopal Reddy, Technical Report: DMRL/MIG/3/97, 1997.
22. A. Venugopal Reddy, *Delayed Cracking of Maraging Steel Billets*, Prakt. Met., 21, 1984, p. 536-540.
23. K.P. Balan and A. Venugopal Reddy, Technical Report: DMRL-TR-98227, 1998.
24. P.G. Tekade and A. Venugopal Reddy, Technical Report: DMRL/MIG/9/95, 1995.
25. A. Venugopal Reddy, *Failure of High Pressure Turbine Blades of an Aeroengine*, Prakt. Met., 19, 1982, p. 659-663.

26. K.P. Balan and A. Venugopal Reddy, Technical Report: DMRL/SFAG/7/98, 1998.
27. K.P. Balan and A. Venugopal Reddy, Technical Report: DMRL/SFAG/20/98, 1998.
28. A. Venugopal Reddy and D.P. Lahiri, *Failure of an Aeroengine Centre Support Bearing*, Prakt. Met., 21, 1984, p. 485-488.
29. K.P. Balan, A. Venugopal Reddy and D.P. Lahiri, *Failure of Drive Shaft*, Prakt. Met., 23, 1986, p. 244-250.
30. A. Venugopal Reddy and P.G. Tekade: Technical Report: DMRL/MIG/1899, 1999.
31. H.H. Kolla and A. Venugopal Reddy, Technical Report: DMRL/SFAG/23/99, 1999.
32. A. Venugopal Reddy and D.P. Lahiri, *Failure of Undercarriage Cylinder of an Aircraft*, Trans. IIM, 38, 6, 1985, p. 528-530.
33. A. Venugopal Reddy, D.P. Lahiri and G.S. Sharma, *Failure of Flame Tube Retainer Bolts*, Prakt. Met., 20, 1983, p. 310-314.
34. K.P. Balan and A. Venugopal Reddy, Technical Report: DMRL/SFAG/22/97, 1997.
35. A. Venugopal Reddy and K.P. Balan, Technical Report: DMRL/SFAG/8/99, 1999.
36. A. Venugopal Reddy, K.P. Balan and D. Pradeesh Kumar, Technical Report: DMRL/SFAG/21/97, 1997.
37. D.P. Lahiri and A. Venugopal Reddy, *Exfoliation Corrosion in Aircraft Structural Member*, Trans. IIM, 35, 5, 1982, p. 456-460.
38. A. Venugopal Reddy, Technical Report: DMRL/MIG/24/93, 1993.
39. A. Venugopal Reddy and D.P. Lahiri, Technical Report: DMRL/MIG /3/83, 1983.
40. A. Venugopal Reddy, Technical Report: DMRL/SFAG/21/96, 1996.
41. A. Venugopal Reddy and D.P. Lahiri, Technical Report: DMRL/MIG /77, 1977.

Further Reading

Corrosion

1. Corrosion - Understanding the Basics: (Ed) J. R. Davis, *ASM International*, 2000.
2. Encyclopedia of Corrosion Technology: P.A. Schweitzer, *Marcel Dekker Inc.*, New York, 1998.
3. Emerging Trends in Corrosion Control, Evaluation, Monitoring Solutions: (Ed) A.S. Khanna et al., *Academia Books International*, 2000.
4. An Introduction to Metallic Corrosion: Ulick R. Evans, *Edward Arnold*, 1981.
5. Corrosion and Corrosion Control: Herbert H. Uhlig and R. Winston Revie, *John Wiley & Sons*, 1985.
6. Metallic Corrosion, Principles and Control: (Ed) A. S. Khanna et al., *Wiley Eastern*, 1994.
7. Corrosion and Corrosion Protection Handbook: (Ed) Philip A. Schweitzer, *Marcel Dekker Inc.*, 1983.
8. Uhligh's Corrosion Handbook: R. Winston Revie, *John Wiley & Sons*, 2nd Edition, 2000.

9. Stress Corrosion Cracking: (Ed) H.J. Russel, *ASM International*, 1993.
10. Stress Corrosion Cracking: (Ed) GM Ugiansky and J H Payer, STP 665, ASTM, Philadelphia, 1979.
11. Stress Corrosion Cracking: Materials, Performance and Evaluation (Ed) R.H. Jones, ASM International, 1992.
12. Corrosion of Stainless Steels: A.J. Sidriks, *John Wiley & Sons*, 1979.
13. Corrosion Source Book: (Ed) Seymour K. Coburn, *ASM International*, Metals Park, and *National Association of Corrosion Engineers*, Texas, 1984.
14. Corrosion: ASM Handbook, Vol. 13, *ASM International*, 1987.

Creep and Fatigue

1. Process of Creep and Fatigue of Metals: A. J. Kennedy, *John Wiley & Sons*, New York, 1963.
2. Effects of Notches on Low Cycle Fatigue: B.M. Wundt, STP 490, *ASTM*, Philadelphia, 1972.
3. Creep and Fatigue of High Temperature Alloys: (Ed) J. Bressers, *Applied Science*, 1978.
4. Creep of Engineering Materials and Structures: (Ed) G. Bernasconi and G. Piatti, *Applied Science*, 1978.
5. Mechanisms of Creep Fracture: H. E. Evans, *Elsevier*, Oxford, 1984.
6. Creep and Fracture of Engineering Materials: (Ed) B. Wilshire and D.R. J. Owen, Proc. Conf., SWANSEA, March 1981, *Pineridge Press*, UK.
7. Creep-Fatigue-Environmental Interactions: (Ed) R.M. Pelloux and N.S. Stoloff, Conf. Proc., *The Metallurgical Society of AIME*, 1980.
8. Creep and Fatigue in High Temperature Alloys: (Ed) J. Bressers, *Applied Science*, 1981.
9. Fatigue Design: E. Zahavi, *CRC Press*, New York, 1996.
10. Metal Fatigue in Engineering: H.O. Fuchs and R.I. Stephens, *John Wiley & Sons*, New York, 1980.
11. Fatigue of Structures and Materials: Jaap Schijve, *Kluwer Academic*, 2001.
12. Thermal Fatigue of Metals: A. J. Weronski and T. Hejwowski, *Marcel Dekker*, 1991.
13. Fatigue and Fracture: ASM Handbook, Vol. 19, *ASM International*, 1996.
14. Designing against Fatigue of Metals: R. W. Heywood, *Reinhold*, New York, 1962.

Failure Analysis

1. Metal Failures, Mechanisms, Analysis, and Prevention: A.J. Mc Evily, *John Wiley & Sons*, New York, 2002.
2. Failure Analysis, Case Histories and Methodology: Friedrich Karl Neuman, *ASM International*, 1983.
3. Failure Analysis: F. R. Hutchings, P. M. Unterweiser, *ASM International*, 1981.
4. Metallurgy of Failure Analysis: A. K. Das, *Tata Mc Graw-Hill*, New Delhi 1996.
5. Handbook of Case Histories in Failure Analysis: Robert C. Uhl et al., *ASM International*, 1992.
6. Fatigue Failure of Metals: S. Kocanda, Sijthoff and Noordhoff, Warsaw, 1978.
7. Understanding How Components Fail: D.J. Wulpi, *ASM International*, Metals Park, 1985

8. An Atlas of Metal Damage: Lothar Engel and Hermann Klingele, *Wolfe Publishing Ltd.*, 1981.
9. Failure Analysis, Techniques and Applications: (Ed) J. Ivan Dickson, Proc. of the First International Conf. on Failure Analysis, July 1991, Montreal, *ASM International*, 1992.
10. Ductile Fracture of Metals:, P. F. Thomson, *Pergamon*, 1990.
11. Fractography: (Ed) R. Koterazawa, Current Japanese Materials Research, Vol. 6, *Elsevier*, 1990.
12. Strength and Fracture of Engineering Solids: D.K. Felbeck and A.G. Atkins, *Prentice-Hall*, 1984.
13. Failure of Materials in Mechanical Design – Analysis, Prediction, Prevention: J.A. Collins, *Wiley-Interscience*, New York, 1993.
14. Fracture of Structural Materials: A.S. Tetelman and A.J. McEvily, *John Wiley & Sons*, New York, 1967.
15. Fractography: ASM Handbook, Vol. 12, *ASM International*, 1987.
16. Failure Analysis and Prevention: ASM Handbook, Vol. 11, *ASM International*, 1986.
17. Principles of Heat Treatment, G. Krauss, ASM International, 1980.
18. Microstructural Characterization of Materials, D.J. Brandon and W.D. Kaplan, *John Wiley & Sons*, New York, 1999.
19. Application of Fracture Mechanics for Selection of Metallic Structural Materials: (Ed) J. E. Campbell et al., *ASM International*, 1982.
20. Metallography in Failure Analyst: (Ed) J.L. Mc Call and P.M. French, *ASM International*, 1977.
21. Deformation and Fracture Mechanics of Engineering Materials: R.W. Hertzberg, *John Wiley & Sons*, 1976.
22. Mechanical Metallurgy: G.E. Dieter, *Mc Graw-Hill*, New York, 1986.
23. Principles of Mechanical Metallurgy: Le May, *Elsevier*, Oxford, 1981.
24. Deformation Mechanism Maps: H.J. Frost and M.F. Ashby, *Pergamon*, Oxford, 1982.

Residual Stress and Stress Concentration

1. Residual Stresses for Designers and Metallurgists: (Ed) C.J. Vande Walle, ASM International, 1981.
2. Residual Stresses and Fatigue of Metals: J.O. Almen and P.H. Slack, *Mc Graw – Hill*, 1963.
3. Stress Concentration Factors: R.E. Peterson, *John Wiley & Sons*, 1974.
4. Stress Intensity Factors Handbook: (Ed) Y. Murakami et al., *Society of Materials Science*, Koyoto, Japan, 1987.

Wear

1. Rolling Contact Phenomena: (Ed) J.B. Bidwel, *Elsevier*, 1962.
2. Friction and Wear of Materials: (Ed) Rabinowicz, *John Wiley & Sons*, 1965.
3. Friction, Lubrication and Wear Technology: ASM Handbook, Vol. 18, ASM International, 1992.
4. Erosion, Wear and Interfaces with Corrosion: STD 567, *ASTM*, 1974.

References and Further Reading

5. Source Book on Wear Control Technology: (Ed) D.A. Rigney, ASM International, 1981.
6. Wear of Metals: A.D. Sarkar, *Pergamon*, 1976.
7. Friction, Wear and Lubrication: K.C. Ludema, *CRC Press*, 1996.
8. Materials to Resist Wear: A.R. Lansdown and A.L. Price, *Pergamon*, 1986.
9. Fundamentals of Friction and Wear of Materials: (Ed) D. A. Rigney, *ASM International*, 1981.
10. Source Book on Wear Control Technology: (Ed) D.A. Rigney and W.A. Glaeser, *ASM International*, Metals Park, 1978.

Index

A

Auto Frettage, 57–58
Auger Electron Spectroscopy, 16
Austenitisation, 91

B

Beach Marks
    Bevel Gear, 226
    Boiler Tube, 323
    Compressor Blade, 165, 171, 179
    Gear Box Pinion, 220
    General, 33
    Turbine Blade, 186, 193, 329
    Work Roll, 214
Brittle Fracture
    Cleavage, 28, 32, 141, 148
    General, 28
    Quasi Cleavage, 31, 144, 203, 233, 239, 279, 285
Burning, 78

C

Carbide Banding, 84
Casting defects, 74
Chemical Inhomogeneity, 75
Chevron Pattern, 144
Corrosion
    Branching Cracks, 123, 285
    Biological, 37
    Debris, 128, 179, 186, 203, 285, 295, 303
    Exfoliation, 315
    Fretting, 37
    Galvanic, 37, 323
    General, 36
    Hot, 23, 191, 311
    Intergranular, 37, 208
    Pitting, 37, 38, 123, 320
    Stress, 37
Creep
    Crystal Structure Effect, 22
    Grain Size Effect, 22
    General, 21
    Round Type Cavitation, 22, 253
    Strengthening Effect, 22
    Wedge Type Cavitation, 22

D

Decarburisation, 89
Delamination, 77, 316
Defects
    Operational, 106
    Maintenance, 111
Design
    Deficiencies, 49–57
    Residual Stresses, 54
    Stages, 49
    Stress Concentration Factor, 52
    Stress Raiser, 51
Design Defects
    Machining Grooves, 52, 171
    Residual Stresses, 54
    Stress Raisers, 51
Dimples
    Elongated, 123, 152
    Equiaxed, 128, 141, 148, 197, 209, 189
    General, 28

E

Electron Hole Number, 73
Electron Probe Micro Analysis, 16
Embrittlement
    Hydrogen, 25, 27, 101, 233
    Liquid Metal, 23
    Liquid Oxide, 23
    Oxygen, 245
    Sensitisation, 25
    Sulphide, 24
    Temper, 24, 95, 238
    Tempered Martensite, 24, 95, 135
Energy Dispersive Spectroscopy, 16

F

Faceted Fracture, 128, 245
Failure Analysis Team, 7
Failure Analysis Techniques
    Chemical Analysis, 17
    Energy Dispersive Spectroscopy, 16
    Fractography, 16
    NDT, 16
    Structural Examination, 17
    Visual, 8
Failure Analyst, 5
Fatigue Failure
    Bevel Gear, 225
    Casting Defect Initiated, 192
    Corrosion Initiated, 178, 184, 203, 208
    First Stage Compressor Blade, 165
    Fuel Pipe, 196
    Gear Box Pinion, 220
    General, 31
    HP Turbine Blade, 184
    Impeller, 203
    Inclusion Initiated, 152, 171, 214, 225
    Landing Gear, 208
    Machining Grooves Initiated, 52, 197
    Microstructural Effects, 32
    Operational factors, 33
    Striations, 33, 155, 165, 169, 171, 179, 186, 193, 197, 203, 209, 226, 329
    Surface Effects, 31
    Third Stage Compressor Blade, 178
    Wear Initiated, 329
    Work Roll, 214
Finish Operational Defects
    Grinding Cracks, 65, 101, 215
    Hydrogen Pickup, 103
    Machining Grooves, 101, 171
    Sand Embedment, 100, 165
    Spalling, 215
FMAAM, 7
Forging Flash, 80
Fretting Corrosion, 38
Fretting Wear, 40, 329
Frictional Overheat Failure
    Center Main Bearing, 279
    Center Support Bearing, 270
    Drive Shaft, 274
FTA, 7

G

Grain Boundary crack
    Round-Type, 22
    Wedge-Type, 21, 253

Grain Boundary Oxidation, 79, 89
Grinding Burns, 101
Grinding Cracks, 65, 101, 215

H

Heat Cracks, 23
Heat Treatment
    Damages, 90
    Defects, 88
    Deficiencies, 64–65, 140
    Variables, 87
Heat Treatment Defects
    Decarburisation, 89
    Grain Boundary Oxidation, 88
    Incipient Melting, 89
    Quench cracks, 94, 157
    Retained Austenite, 92
    Segregation, 239
    Wrong Treatment, 140, 144, 147, 157, 238
Hydrogen Embrittlement
    13-8PH Steel, 27
    AISI 1018, 27
    AISI 1074, 27
    AISI 4130, 27
    AISI 8740, 27
    Fracture Modes, 27
    General, 25, 101, 103
    H11 Tool Steel, 27
    Intergranular Fracture, 27
    Maraging Steel, 27, 233
    Quasi Cleavage, 31

I

Incipient Melting, 90, 263
Induction Hardening Defects, 59, 157
Ingotism, 77, 79, 220
Intergranular Corrosion, 25
Intergranular Fracture
    Cam Shaft, 159
    Causative Factors, 20
    Center Main Bearing, 279
    Creep, 21, 253
    Drive Shaft, 275
    Flame Tube Retainer Bolt, 295
    General, 19
    Gun Barrel, 239
    Sensitization, 25
    Stress Rupture, 21, 258
    Sulphide Embrittlement, 24
    Temper Embrittlement, 24, 96, 98, 238

Index

Tempered Martensite Embrittlement, 24, 96, 135
Thermal Shock, 23
Track Shoe, 135
Turbine Blade, 253
Under Carriage Cylinder, 289
Internal Stresses, 94

J

Jet Engine Flame out
    Lean Fuel – Rich Air, 10
    Rich Fuel – Lean Air, 9

L

Liquid Embrittlement
    Drive Shaft Failure, 275
    Introduction, 23

M

Machining Defects, 101
Manufacturing Defects
    Adiabatic Shear Bands, 78
    Blow Holes, 74–75
    Freckles, 75–76
    Inclusions, 78, 82, 144, 152
    Ingotism, 77, 79
    Overheating, 78, 81–82
    Segregation, 73
    Shrinkage Cavities, 74–75
    Slag Pockets, 73
Maintenance Defects, 114
Material Selection
    Chemical Composition, 61–69
    General, 60
    Mechanical Properties, 61–69
    Microstructural Characteristics, 61–69
    Non-Conformity, 62–65, 128
    Physical Properties, 61–69
    Surface Properties, 61–69
Metal Joining Defects
    Brittle HAZ, 62
    Weld Bead, 245
Microstructural Banding
    Al-Cu-Mg Alloy, 316
    Carbide Banding, 84–85, 279
    Delta Ferrite, 84
    Ferrite-Pearlite, 83–84

General, 83
Ni-Cr-Mo Steel, 239

O

Operational Defects
    Corrosion, 105
    Debris Deposition, 110
    Erosion, 110
    Overheating, 107
    Score Marks, 110
    Wear, 106
Overheating
    Center Support Bearing, 270
    HP Turbine Blade, 253
    Introduction, 78
    Nozzle Guide Vane, 263
    NGV Mounting Bolt, 258
    Turbine Stator Ring, 107

P

Pitting Corrosion
    Balancing Weights, 38
    Emitting Electrode, 123
    Oxygen Cylinder Failure, 320

Q

Quench cracks, 95

R

Residual Stresses
    Auto Frettage, 58
    Burnishing, 59
    General, 54
    Laser Shock Peening, 59
    Measurement, 59
    Mechanical Factors, 57
    Metallurgical Causes, 57
    Roll Failure, 59–60
    Shot Peening, 59
Retained Austenite, 65, 92

S

Segregation
    General, 73

Phosphorus, 98, 139
Trace Elements, 239
Sensitisation, 25
Shear Lips, 140
Slag pocket, 77, 157, 171, 233
Spalling, 214
Stress Concentration factor
    Bomblet Body Failure, 54
    Compressor Disc Failure, 52
    Introduction, 50
Stress Corrosion Cracking
    AISI 303, 26
    AISI 316, 26
    AISI 321, 26
    AISI 4340, 26
    Al Alloys, 26
    Al-Brass, 26
    Blow off Vanes, 285
    Brass Impeller, 128
    Cu Alloys, 26
    Ferritic Stainless Steel, 26
    Flame Tube Retainer Bolt, 302
    Fourth Stage Stator Casing Bolt, 306
    Fracture Modes, 26
    General, 25
    IN 738, 26
    Intergranular, 25
    Side Strut, 303
    Emitting Electrode, 123
    Ti Alloys, 26, 285
    Transgranular, 34
    Under Carriage Cylinder, 289
Stress Raisers
    Geometrical, 51
    Microstructural, 51
    Stress Concentration Factor, 52
Stress Rupture, 22

T

Temper Embrittlement
    General, 97
    Gun Barrel, 238
Tempered Martensite Embrittlement, 94, 135
Transgranular Fracture
    Balancing Gear Rod, 151
    Causative Factors, 29
    Draw Hook, 134
    General, 28
    Gigle Saw, 152
    Maraging Steel, 215
    Stress Corrosion Cracking, 123

W

Wavelength Dispersive Spectroscopy, 16
Wear
    Abrasive, 41–42
    Adhesive, 41–42
    Combating Methods, 46
    Erosion, 41–43, 333
    Fretting, 41–43
    General, 40
    HP Turbine Blade Failure, 329
    Reheater Tube, 333
Weld Failure
    Anchor Bolt, 62
    Nose Fairing, 245

Y

Yawning Cracks, 124